T0298029

CAMBRIDGE LIBRARY COLLECTION

Books of enduring scholarly value

Botany and Horticulture

Until the nineteenth century, the investigation of natural phenomena, plants and animals was considered either the preserve of elite scholars or a pastime for the leisured upper classes. As increasing academic rigour and systematisation was brought to the study of 'natural history', its subdisciplines were adopted into university curricula, and learned societies (such as the Royal Horticultural Society, founded in 1804) were established to support research in these areas. A related development was strong enthusiasm for exotic garden plants, which resulted in plant collecting expeditions to every corner of the globe, sometimes with tragic consequences. This series includes accounts of some of those expeditions, detailed reference works on the flora of different regions, and practical advice for amateur and professional gardeners.

The Municipal Parks, Gardens, and Open Spaces of London

This survey, and fascinating history, of the public green spaces of London was published in 1898. Its author, John J. Sexby, the Chief Officer of Parks of the London County Council, is described as a lieutenant-colonel and a professional associate of the Surveyors' Institution, from which it can be deduced that he probably worked as a surveyor in the army. His skills as a horticulturalist and garden designer cannot be doubted, and he left his mark on many of the municipal parks and gardens about which he writes with such enthusiasm. Sexby focuses on the municipal parks (those maintained by local authorities) rather than the nationally managed parks in central London. He describes large open spaces such as Hampstead Heath as well as small, disused churchyards like that of St Dunstan's in Stepney, providing details of their former owners and use as well as their present condition.

Cambridge University Press has long been a pioneer in the reissuing of out-of-print titles from its own backlist, producing digital reprints of books that are still sought after by scholars and students but could not be reprinted economically using traditional technology. The Cambridge Library Collection extends this activity to a wider range of books which are still of importance to researchers and professionals, either for the source material they contain, or as landmarks in the history of their academic discipline.

Drawing from the world-renowned collections in the Cambridge University Library and other partner libraries, and guided by the advice of experts in each subject area, Cambridge University Press is using state-of-the-art scanning machines in its own Printing House to capture the content of each book selected for inclusion. The files are processed to give a consistently clear, crisp image, and the books finished to the high quality standard for which the Press is recognised around the world. The latest print-on-demand technology ensures that the books will remain available indefinitely, and that orders for single or multiple copies can quickly be supplied.

The Cambridge Library Collection brings back to life books of enduring scholarly value (including out-of-copyright works originally issued by other publishers) across a wide range of disciplines in the humanities and social sciences and in science and technology.

The Municipal Parks, Gardens, and Open Spaces of London

Their History and Associations

J.J. SEXBY

CAMBRIDGE
UNIVERSITY PRESS

CAMBRIDGE
UNIVERSITY PRESS

University Printing House, Cambridge, CB2 8BS, United Kingdom

Cambridge University Press is part of the University of Cambridge.

It furthers the University's mission by disseminating knowledge in the pursuit of
education, learning and research at the highest international levels of excellence.

www.cambridge.org
Information on this title: www.cambridge.org/9781108076135

This edition first published 1898
This digitally printed version 2015

ISBN 978-1-108-07613-5 Paperback

THE MUNICIPAL PARKS, GARDENS,

AND

OPEN SPACES OF LONDON.

I must gratefully acknowledge the kind assistance I have received from many friends in the preparation of this history : Mrs. Beck for permission to make extracts from her late husband's private records with regard to the acquisition of Clissold Park; Mr. John Burns, M.P., for information regarding the agitation for the right of public speaking on Clapham Common ; Mr. Arthur Cates for revising the chapters on Battersea, Kennington, and Victoria Parks ; Mr. George Chambers, the courteous honorary librarian of the Tyssen Library, Town Hall, Hackney, for placing at my disposal the unique collection of drawings, manuscripts, etc., relating to the Hackney district ; and to Mr. W. Minet, the donor of Myatt's Fields, for the particulars of the history of that place. My thanks are also due to the following for the loan of blocks and drawings illustrating this work : The proprietors of London for the illustrations of parks which have already appeared in that paper ; Mr. McDougall, J.P., L.C.C., for many of the photographs ; Mr. Martin, librarian of Hammersmith Public Library, for the illustration of the Red House, Battersea Park ; Messrs. Oetzmann and Co., Hampstead Road, for views of Hampstead and other places ; Mr. W. Sugg for photographs of views on Clapham Common ; Mr. W. T. Vincent, author of ' Records of the Woolwich District,' for the loan of illustrations from that work ; and the trustees of Whitefield's Tabernacle for an old view of their ancient place of worship.

J. J. S.

View of Bostal Woods.

THE

MUNICIPAL PARKS,

GARDENS, AND OPEN SPACES OF LONDON:

Their History and Associations.

BY

LIEUT.-COL. J. J. SEXBY, V.D.,

PROFESSIONAL ASSOCIATE OF THE SURVEYORS' INSTITUTION.

ILLUSTRATED BY NUMEROUS SKETCHES, PHOTOGRAPHS,
AND FACSIMILES.

LONDON:
ELLIOT STOCK, 62, PATERNOSTER ROW, E.C.
1898.

CONTENTS.

LIST OF ILLUSTRATIONS.

INTRODUCTION.

THE meaning of the title and scope of this book calls for some explanation, especially when it will be found that no mention is made of such places as Hyde Park, Regent's Park, or St. James's Park, which have a right to be considered the most important parks in London. In the first place, the history of these parks has been more than once written, and in the second place it must be pointed out that they are national rather than municipal 'lungs,' because they are kept up at the expense of the nation at large, and not by any one municipality. London's municipal parks and open spaces are those which are maintained by the London County Council at the expense of the Metropolitan ratepayers. In addition to the national and municipal places of recreation, there are a number of disused burial-grounds and other small grounds maintained by the various local vestries, whilst the number is completed by those under the care of Conservators, private bodies and individuals, the most important of which are Wimbledon Common and Putney Heath, together making a magnificent open space, 1,412 acres in extent. The Corporation of the City of London also possesses some 6,500 acres of parks and open spaces which are available for the use of Londoners, but are maintained out of 'city cash' and the funds derivable from metage on grain. The largest of these is Epping Forest; but, like the majority of the City parks, this is outside the County of London.

PARKS.

The London County Council, as is well known, succeeded
the Metropolitan Board of Works as the municipal authority
for London, and the history of the parks and open spaces
described in this volume must commence with the formation
of this latter body. The 144th section of the Metropolis
Management Act, 1855, authorized the Board to apply to
Parliament whenever it was of opinion that further powers
were required for the purpose of any work for the improvement of the Metropolis or the public benefit of the inhabitants. One of the first steps which the newly-constituted
authority proposed to take was to provide public parks in
districts where such places of recreation did not already
exist, but it was felt doubtful how far the authority to apply
to Parliament given by the above statute extended.   To
remove all question so far as parks were concerned, a clause
was inserted in an amending Act of 1856, in which it was
laid down that the powers given to the Board in their original
Act did extend to applications to Parliament for the purpose
of providing parks, pleasure-grounds, and open spaces. This
question having been settled, the Board applied to Parliament, and obtained power to purchase and lay out what is
known as Finsbury Park. Some seven years later, power
was obtained to provide another park in the South - East
District, known as Southwark Park. Both these places were
opened to the public in 1869, and they were the nucleus of
the municipal parks of London, which have now increased
to so extensive an area. At the time of the formation of
these parks they were, to a great extent, surrounded by open
ground, chiefly used for market-gardens, and the schemes for
their acquisition were by many people voted as extravagant
and unnecessary. But the wisdom of this policy has been
more than justified in the lapse of time, for these parks are
now in the midst of a large population, and are invaluable as
places of recreation.

The next addition to London's municipal parks was made

in 1887, when four places, which up to this time had been maintained by Her Majesty's Office of Works at the expense of the nation, were transferred to the late Board. These were Victoria, Battersea, and Kennington Parks, and the gardens surrounding the Bethnal Green Museum. Three more parks were acquired by the Board before passing out of office: Ravenscourt Park at Hammersmith, Clissold Park at Stoke Newington, and Dulwich Park, the land of which was a gift from the Dulwich College Governors.

The London County Council in their first year of office, 1889, were presented with two parks: Myatt's Fields and Waterlow Park; and since this time they have gone on adding to the number, by purchase or otherwise, as will be detailed later.

OPEN SPACES.

It has been a great advantage to London to have on its outskirts a number of commons and open spaces available for public resort. The commons have a peculiar charm in their freedom and their natural beauty as opposed to the restrictions and the artificialness of a made park. They are, moreover, part of the history of the country, for they are almost the only relics of the feudal system, and take us back to the time when England was tilled in common, and private ownership of land in the modern sense was unknown. Previously to the year 1866 the inhabitants of London had no rights in connection with these places, since the nature of agricultural holdings had gradually changed, and although they were, like other common land in England, open to the public by custom, the only legal rights were those of the Lords of the Manors and of the copyholders and commoners in each case. It is only of comparatively recent years, owing to the enormous increase of the Metropolis, that they have acquired value as building lands, and have consequently proved a source of temptation to the Lords of the Manor to enclose them. But the first general movement in the way of enclosure seems to date back to the fifteenth century, at the close of the Wars of the Roses, in which so many of the feudal aristocracy lost their lives. Previous to this, the Act of 20 Henry III., cap. 4, had been passed, commonly called the Statute of Merton, which enabled the lord of a manor to enclose common lands without either the assent of the commoners or the sanction of Parliament. In after years this proved the most disastrous law ever passed as regards common land, and it has been the cause of many a fine open space being lost to London. After the Wars of the Roses, the feudal system gradually began to undergo a change, as the necessity for maintaining a large number of armed dependents became less apparent, and the opening of the Continent to trade encouraged the tending of sheep for their wool. By degrees we find the common land being enclosed for pasture, but not without considerable protest on the part of many leading statesmen. So important a question did enclosure become that a Royal Commission was held at the instance of Protector Somerset 'for the redress of enclosures'; but nothing came of it, owing to the powerful influence of the nobles, who terrorized witnesses from giving evidence. These common lands which were thus being converted from public to private ownership were not what are now known as commons. The commons of the present day were the waste lands, perhaps not suitable for cultivation, in many cases covered with brushwood and undergrowth, which furnished fuel for the copyholders and commoners of the manor. In the case of those which were suitable, the commons were used for grazing; but in course of time, owing to the increase of population, they were not able to provide food for the cattle of all the manorial tenants, and, as a consequence, those around London began to lose their value for agricultural purposes. At the same time their value as building land, and the development of our network of railways, led to a far more serious enclosure movement, by which many of the smaller wastes were taken for building, railway, or other purposes. This resulted in a double loss to London, for not only were the lands not available for pasture, but the general health of the Metropolis was bound to suffer if all the breathing-places were built over. The enclosure movement has left its mark on several of the larger London commons, particularly Wandsworth and Tooting, which are intersected by railways in various directions, and, instead of presenting an unbroken extent of ground, are divided into small and almost separate areas.

This process would probably have continued, had it not been for the action of the late Board, aided by a number of public-spirited persons who saw that London, as well as other parts of the country, was in danger of losing its open spaces, which were being encroached upon year by year. It was apparent that further legislation was necessary, seeing that the lord of a manor could combine with his tenants and then sell or dispose of any part of the manorial wastes. The Board resolutely opposed any alienation of this kind within the limits of its jurisdiction, and as a consequence of its action, backed up by the efforts of private individuals, Parliament appointed in 1865 a Select Committee to inquire into the best means of preserving for the public use the forests, commons, and open spaces in and around London. After a lengthy inquiry, the Committee recommended that the Statute of Merton should be repealed, that no enclosure should take place under the provisions of the Enclosure Acts within the Metropolitan area, and that a body of trustees should be appointed for the preservation of open spaces within the area. In the following year the Government introduced a Bill which, after a good deal of discussion and alteration, became law under the title of the Metropolitan Commons Act, 1866, and which, whilst not quite following the lines suggested by the Select Committee, prescribed a mode of procedure under which the commons in the neighbourhood of London could be permanently secured for the public. This Act appointed the late Board the local authority for all commons situate wholly or in part within the Metropolitan area, and by its powers, supplemented by subsequent Acts, all the commons and open spaces on the outskirts of

London have been preserved for public use. The circumstances connected with them differed in almost every case. For instance, in the case of Blackheath, one of the first commons acquired, the Earl of Dartmouth, Lord of the Manor in which the greater part of the common is situated, generously refrained from making any claim with respect to his manorial rights, whilst as regards Hampstead Heath and the Hackney Commons, immense sums have had to be paid before they could be secured from encroachment.

THE MUNICIPAL PARKS, GARDENS, AND OPEN SPACES OF LONDON.

CHAPTER I.

BATTERSEA PARK.

THIS, the largest municipal park in the south of London, is 198 acres in extent, and occupies the site of Battersea Fields. What was forty years ago one of the dreariest and darkest spots in transpontine London, has now become a veritable oasis in the desert. If these lands had not been rescued from the hands of the builder, industrial dwellings, and the third-rate terraces, with their attendant general stores, which abound in Battersea, would have crept down to the water's edge. Battersea is not looked upon with much favour by its more aristocratic neighbours across the water, and this ill-favour for a very long while seemed to attach itself to the park without the slightest foundation, as its varied attractions make it one of particular interest.

Before we describe its present condition, we may just go back to the past and see what formerly took its place. In the sixteenth century Battersea Fields was to all intents and purposes that portion of the River Thames lying between low and high water mark, and at every recurring tide the land was under water. Somewhere about the year 1560 a

rough embankment was made to keep the water out, and the
land thus reclaimed became the property of the lord of the

The Carriage Drive, Battersea Park.

manor, subject to some rights of common exercised by the
inhabitants of Battersea at certain periods of the year.   It

is said that this land was gained for the parish of Battersea by the act of charitably burying a man who had been drowned there, whom the adjoining parish had refused to bury. Battersea certainly reaped a rich reward for this kindness, for this act was held in a subsequent lawsuit as sufficient to prove a right of ownership. The land thus recovered was naturally of a swampy nature, and was divided into a number of plots, called marshes or shots. A lane led from Nine Elms to the Red House, about which we shall have something more to say.

The lands forming the park were part of the common fields of the manor of Battersea, the history of which can be traced back to the time when William the Conqueror had his never-failing Doomsday Book compiled. At this period the manor was in the possession of Earl Harold, to whom it had probably descended from the powerful Earl Godwin. The Battle of Hastings, which ended the power of the Saxon Kings, was followed by the confiscation of their estates. The manor did not, however, pass to any of the Norman adventurer's followers, as he retained it for his private enjoyment till, attracted by the beauty of Windsor, he exchanged the Manor of Battersea for that of the now royal manor, which was then in the possession of the monks of St. Peter, Westminster. One hide of the Battersea land was not included in this exchange, and this was the property of the Abbot of Chertsey, who somehow had managed to acquire many a broad acre in the south of London. The manor before the Conquest was of great extent and of great value, including as it did, in all probability, part of Wandsworth, Lambeth, Camberwell, Peckham, Streatham, Penge, Tooting, and perhaps also Clapham. After the Conquest, the manor dwindled down to its present size, and the lands which have disappeared from the Court Rolls are probably those lying between the present parish and the outlying district of Penge, which is still considered part of the manor. We can see the difference by comparing the quantity of land held by Earl Harold, which was taxed for 72 hides and valued at £80, with that recorded in the survey, which was taxed for 18 hides only.

The monks remained in undisturbed possession of the manor for 450 years. On two separate occasions the grant of the manor was confirmed to them—once when Henry I. usurped the throne, and later on when Stephen imitated his example. The Church at this time being quite as powerful as the State, it was necessary for any whose titles to the

The Cascade, Battersea Park.

throne were not quite clear to make the clergy their friends. This will doubtless account for the fact of these confirmations. But the power of the Church declined, till the great blow fell which deprived the monasteries of their lands under the rule of Henry VIII. Westminster did not suffer so heavily as its sister convents, but the Manor of Battersea was taken from it and vested in the Crown, in whose hands it remained till the reign of Charles I. It was assigned, with other manors, for the maintenance of Prince

Henry in 1610, and in 1627 Charles I. granted it to Sir
Oliver St. John, afterwards Viscount Grandison. Upon the
death of this nobleman, in 1630, it passed into the possession
of his great-nephew, William Villiers, who died of a wound
received at the siege of Bristol, 1644. Sir John St. John,
nephew of the first Lord Grandison, inherited Battersea;
from him it passed in a regular descent to Sir Walter St.
John, and then to Henry, Viscount St. John, who had all his
estates in Battersea confiscated, owing to a murder which he
had committed, and they were only redeemed by paying the
King £16,000. His son, Viscount Bolingbroke, then succeeded
to the manor, followed by the latter's nephew, Frederick
Viscount Bolingbroke. By an Act of Parliament obtained
in 1762 he was enabled to sell his estates, whereupon, in
1763, the trustees of Earl Spencer purchased it, and it has
remained in this family ever since.*

The origin of the word 'Battersea' is involved in much
obscurity. Each antiquary has a different derivation for it;
and where doctors disagree, it is not the place for laymen to
intrude. We will let these authorities speak for themselves.
Spelman, in his Glossary, says it means a member of a manor
disjoined from the main body, a villa or hamlet. This would
be appropriate in the case of Battersea, as we have just
mentioned how the manor was dismembered before the
Conquest. But as no other historian has adopted this view,
it seems unlikely that it is the true one. Lambarde gives
another guess rather wide of the mark. 'Battersey,' says
he, 'quasi Botersey; because it was near the waterside, and
was the removing house of the Archbishop of York.' Unfortunately for this ingenious derivation, we need only point
out that the Archbishops of York did not possess any
property here till the reign of Edward IV. Others contend
that the true spelling is 'Battlesea,' and derive the name from
some battle which is supposed to have taken place near here.
A fourth solution is that the present word is a corruption of
the name by which the district was known at the time of the

* Lysons, 'Environs of London,' 1811, vol. i., part l., pp. 20, 21.

Doomsday survey, viz., Patricesy, *i.e.*, the 'sea or water of St. Peter or St. Patrick.' Opinion is divided as to which saint it is dedicated to. Those who favour St. Peter quote as a similar example the name of Petersham, which is known to have received its appellation from St. Peter's, Chertsey.* This is mentioned in the Doomsday Book as Patricesham, so that by analogy Patricesy would mean 'St. Peter's water.' We must just mention that Aubrey, on the other hand, makes it the 'water of St. Patrick,' arguing that the Norman chronicler made a mistake in the word, owing to the very unsettled state of spelling at that period. Seeing that we have not yet reached perfection on this point, Aubrey may be right; but as England has now been placed, by kind permission of the Pope, under the protecting wing of St. Peter, it might bring joy into the hearts of the inhabitants of Battersea to know that in these early times their neighbourhood was specially dear to him.

The old marshes had a picturesqueness of their own. A contemporary writer describes how, late in an autumn afternoon in Battersea Fields, he watched a Flemish broom-seller, seated with her brooms in her lap, with a background not unlike a view in the Low Countries. Behind her was a windmill, near the 'Red House,' with some dwarfish buildings among the willows on the bank of the Thames, thrown up to keep the river from overflowing the marsh flat.† Such a view as this could, of course, only be obtained on the outskirts of the marsh, the greater portion being bare, flat and uninteresting.

One of the earliest events connected with Battersea Fields is the attempted assassination of Charles II. by Colonel Blood. He hid in the reeds which fringed the shore, intending to shoot the King whilst bathing, as was his custom, in the Thames over against Chelsea; but 'his arm was checked by an awe of majesty.' So Blood, at least, had the impudence to relate when on his trial for his audacious

* Lysons, 'Environs of London,' 1811, vol. i., p. 19.
† Hone, 'Everyday Book, p. 810.

attempt to steal the regalia from the Tower.* He gained entrance to the fortress in the garb of a clergyman, and had actually got the crown concealed under his cassock. He put such a bold front on when tried before the King that he was pardoned.

Picturesque Corner, Battersea Park.

Almost the only other historical event recorded of Battersea Fields is a duel that took place between the Duke of Wellington and the young Marquis of Winchelsea on March 21, 1829. The lonely character of the Fields made them particularly suitable for the settlement of these affairs of honour. As we read the account of this ludicrous affair, we are reminded of the childish duels of the present day in France and the Fatherland, when a scratch suffices to satisfy the wounded honour of a passionate Gaul. This duel arose from a

* B. E. Martin, 'Old Chelsea,' pp. 162, 165.

political quarrel, brought about by the course taken by the
Duke during the passing of the Catholic Emancipation Bill.
The Marquis of Winchelsea, who was one of the leaders of
the anti-Catholic party, had, of course, taken a strong stand
in opposition to the Bill, and not content with opposing
the Duke in the House, he thought fit to publish various
imputations against the personal character of Wellington,
charging him with premeditated treachery to the Protestant
party, and treason against the Constitution. As he would
not retract these libels, the matter had to be settled in the
fashionable way. The hero of Waterloo had the first shot,
with which he pierced the hat of his opponent, who thereupon fired into the air, and then tendered an apology.*
Many attempts have been made to discover the exact spot
—in fact, a movement was once on foot to erect a permanent
memorial here; but we are glad to say that it was decided
not to waste public money in doing anything to perpetuate
the follies of great men. Some say that the spot is marked
by Wellington Street, near Battersea Bridge; but this is
mere conjecture.

Such distinguished visitors gave way, however, to coster-
mongers and roughs, who settled their differences here after
the example set them by the nobility. Adjoining the Fields
was the famous tavern known as the ' Red House,' so called
because it was built of red bricks. In its prime, the Red
House and its grounds formed a second Vauxhall Gardens,
and attracted quite a number of aristocrats to Battersea.
The gardens were laid out in small arbours decorated with
Flemish and other paintings, and fancifully-formed flower-
beds. In the centre of the garden was a fish-pond. The
walks were prettily disposed, and at the end of the principal
one was a painting, the perspective of which rendered the
walk in appearance much longer than it really was. Beyond
the east end of the house was situated a range of boxes or
alcoves—seven in number—which at night were illuminated
with oil-lamps. Each of these alcoves had a table in the

* ' Imperial Dictionary of Universal Biography,' vol. iii., p. 898.

centre, and seats for twelve. Some of the dishes provided
here became regular institutions. The 'Flounder Breakfast'
at ten o'clock used to attract several of the Guards from
Whitehall Stairs ; and certain noblemen dignified with their
presence and patronage the annual 'Sucking-pig Dinner,'
which generally took place in the month of August. But
the Red House was also a famous place for sports of all
kinds.* Part of its grounds (now included in the park)
were devoted to pigeon-shooting, and attracted the cream
of society till the more fashionable Hurlingham took its
place. Colburn's 'Kalendar of Amusements,' published in
1840, has the following : 'Pigeon-shooting is carried on to

The Red House, Battersea. (From an old woodcut.)

a great extent in the neighbourhood of London ; but the
Red House at Battersea appears to take the lead in the
quantity and quality of this sport, inasmuch as the crack
shots about London assemble there to determine matches
of importance, and it not unfrequently occurs that not a
single bird escapes the shooter.' In addition to pigeons,
sparrows and starlings were also shot at, pigeons being sold
at 15s. per dozen, starlings at 4s., and sparrows at 2s.

Being situated on the river's bank, nearly opposite the
gardens of Chelsea Hospital, the Red House was chosen as
the winning-post of many a race on the Thames. In the

* H. S. Simmonds, 'All about Battersea,' 1882, p. 77.

'Good Fellow's Calender' of 1826 we read that on August 18, in the previous year, ' Mr. Kean, the performer ' (not a very flattering designation), gave a prize wherry, which was ' rowed for by seven pairs of oars. The first heat was from Westminster Bridge round a boat moored near Lawn Cottage, and down to the Red House.' The other heats, too, all ended here, and the Calender adds that, ' although Westminster Bridge was crowded with spectators, the Red House was the place where all the prime-of-life lads assembled.' In front of the house, by the riverside, was a tall flagstaff standing on a small space which was embanked and enclosed with railings. This space formed a kind of jetty, divided in the centre by a flight of steps from the river, and it was also approached by steps at both ends, so as to accommodate the numerous visitors by water. So far we have described the Red House at its best, but its latter end was a sad contrast to these palmy days. At one time ' the ripe corn waved to and fro in the broad, low-lying meadows of Battersea '; now they became the scene of everything that was low. Horse-racing, donkey-riding, fortune-telling, gambling, cock-shying, swings, roundabouts, boxing, and all the accompaniments of a seventeenth-rate fair, were the constant order of the day here on Sundays. A former City missionary in Battersea* thus describes the place : ' That which made this part of Battersea Fields so notorious was the gaming, sporting, and pleasure-grounds at the Red House and Balloon public-houses, and Sunday fairs, held throughout the summer months. These have been the resort of hundreds and thousands, from royalty and nobility down to the poorest pauper and the meanest beggar. And surely if ever there was a place out of hell that surpassed Sodom and Gomorrah in ungodliness and abomination, this was it. Here the worst men and the vilest of the human race seemed to try to outvie each other in wicked deeds. I have gone to this sad spot on the afternoon and evening of the Lord's day, when there have been from 60 to 120 horses and donkeys

* Mr. Thomas Kirk.

racing, foot-racing, walking matches, flying boats, flying horses, roundabouts, theatres, comic actors, shameless dancers, conjurers, fortune-tellers, gamblers of every description, drinking-booths, stalls, hawkers and vendors of all kinds of articles. It would take a more graphic pen than mine to describe the mingled shouts and noises and the unmentionable doings of this pandemonium on earth. I once asked the pierman how many people were landed on Sunday from that pier. He told me that, according to the weather, he had landed from 10,000 to 15,000 people. This influx was besides that by the various land roads, by which hundreds and thousands used to come until the numbers have been computed at 40,000 or 50,000.'* This writer is evidently not afraid to call a spade a spade, and his description of the scenes on the Fields is not a particularly flattering one.

Things came to such a pitch that it became necessary for Government to interfere, as the Red House was only one of the many beershops on the Fields. The others, afterwards taken by the Government for the formation of the park, were the Albert Tavern, the British Flag, and Tivoli Gardens on the river front, and the Balloon Tea-Gardens and another beershop on the marshland. It had been suggested by Mr. Thomas Cubitt in 1843 to Her Majesty's Commission for improving the Metropolis, that the laying out of Battersea Fields as pleasure-grounds would be a very advisable step. If ever a place had room for improvement, this did, and many other influential gentlemen pressed the matter upon the Commissioners, including the Rev. Mr. Eden, then Vicar of Battersea, afterwards Bishop of Sodor and Man. Fortunately for Battersea, the demand for open spaces in the outskirts of the Metropolis had taken firm hold of public attention, and in 1846 an Act was passed to enable 'the Commissioners of Her Majesty's Woods to form a Royal Park in Battersea Fields, in the Parish of Saint Mary, Battersea, in the County of Surrey.' They were authorized for this purpose to expend a sum not exceeding £200,000 in

* *London City Mission Magazine*, September, 1870.

the purchase of lands, laying out and planting the same, and
forming an embankment along the Thames. A further Act
was, however, required to provide for payment for Lammas
and other commonable rights, as it was doubtful who were
the right parties to receive the compensation for their
extinguishment. Consequently the Battersea Park Act of
1853 settled the matter by enacting that the Battersea Park
Commissioners were to pay £1,500 to the churchwardens of
St. Mary, Battersea, to be applied to such purposes as a

The Avenue, Battersea Park.

specially-convened vestry might direct, and that thereupon
all rights of common were to be extinguished. The land
thus taken comprised about 320 acres, of which 198 were
devoted to Battersea Park, the remainder being let for
building sites. It was originally intended to lease also
certain of the frontages of the present area of the park;
but, owing to the opposition, this idea was subsequently
abandoned. Of the amount of £246,517 paid for the land,
£10,000 went for the purchase of the Red House, with its

shooting-grounds and adjacent premises, while the laying out, extending over a period of six years, cost £66,373, so that the expenditure involved in the scheme, without regarding the recoupment, was about £312,000. In addition to building an embankment, it was necessary to raise the whole surface of the Fields, and many hundred thousand cubic yards of earth were required for this purpose, the greater portion of which came from the Victoria Docks Extension works.*

The park was laid out under the direction and from the designs of Sir James Pennethorne, Architect of the Office of Works. Perhaps the two principal features in the design are the avenue and the subtropical garden. This avenue, of English elms, whose branches meet in a leafy arch, forms the chief promenade of the park, and at the end the charming vista is completed with a sight of a Gothic fountain tastefully executed in wrought iron. The subtropical garden, designed by Mr. John Gibson, for many years the Park Superintendent, some 4 acres in extent, was opened in 1864, and forms the chief botanical feature of the park. Situated at the head of the ornamental water, and surrounded by sloping banks, it is designedly sheltered on every side from the keen winds, and on the coldest day it is comparatively warm. An attempt is made here to try and present to Londoners some of the hardiest of tropical plants. Without going into botanical details, we may mention that in the summer palms, tree-ferns, gigantic grasses, and other specimens of tropical vegetation which have braved the winter frosts in the shelter of the palm-house, are planted out from year to year. The rockwork forms another attraction of the park. It is hardly necessary to add that this is entirely artificial, although the imitation of Nature is very close. The rocks represent a mountain-side, as if it had been rent asunder by some volcanic eruption, and the water meanders between the rugged walls into the lake below. An Alpine garden has been laid out with very good effect, the intention being to present the varying vegeta-

* Simmonds, 'All about Battersea,' pp. 80-82.

tion of a snow-clad peak, the snow-plant taking the place of snow, and various other specimens garnish the sides of the miniature mountain, thus representing the plant-life of the different zones.

The Subtropical Gardens, Battersea Park.

The ornamental waters comprise a large lake of fifteen acres, and a smaller lake of one acre, called the Ladies' Pond. The former of these is kept to an average depth of

2½ feet by means of water taken from the Thames. In summer it is covered with pleasure-boats, the shallow depth providing for safe and enjoyable amusement. A portion is reserved for the many kinds of water-fowl, and water-lilies flourish in luxuriance here. In the winter frosts this large area is a perfect paradise for skaters, who crowd here in thousands, well aware that an immersion would only mean an unpleasant cold bath, and that they are quite safe from the consequences that would await them in the event of a similar mishap on other sheets of water of this size. Other forms of recreation are well provided for. There are large cricket and football grounds of several acres, and the local matches here are as keenly watched as any county contest at the Oval. Lawn-tennis has a ground to itself near the engine-house, well shaded with trees, and another near the pier. A band-stand, on which bands play two or three times a week, provides pleasure for those who are of a musical turn of mind. A series of horse-rides encircle the park, which attracts a goodly number of equestrians. There is a gymnasium for adults, whilst the little ones are not forgotten, for two children's playgrounds have been formed where they can swing and skip to their hearts' content. Lastly, a quoit-ground has been laid out, and also a bowling-green, whilst the wants of the inner man are provided for by three refresh ment houses of reasonable tariff. It is hardly, then, to be wondered at, with so many attractions, Battersea Park is a popular one, or that the inhabitants of the surrounding district are proud of their 'Garden of Eden.' The park is now under municipal control, having been transferred in 1887 from the Government to the late Metropolitan Board of Works and their successors, the London County Council.

No history of Battersea Park would be complete without some reference to the part it played in connection with the development of the cycling craze. For some extraordinary reason it sprang into sudden favour—especially with the ladies—as a place where beginners might master the rudiments of the bicycling art. It has now passed through this

View of the Lake, Battersea Park.

stage, and the number who come here to learn are comparatively few. Now the roads which formerly were the

nursery of cycling have developed into a fashionable promenade, and although privileges have been extended to cyclists in Hyde Park, the cream of society still come to Battersea for their morning ride. This is only right, as Battersea Park was the place where the first experiment of

Cycling in Battersea Park.

making cycling a fashionable pastime was carried out. But the age of experiments is over, since cycling has become part of the national life.

It was a much-cherished wish of the late Prince Consort that the exhibition building of 1851 (held in Hyde Park, and

2

afterwards transferred to Sydenham as the Crystal Palace) should be erected in Battersea Park. For this purpose an elaborate plan was prepared, showing the main building as a huge palace of glass, situated in the centre of a raised gravelled promenade. From this steps led down to the rest of the grounds, where the chief feature was a spacious oblong lake, crossed by a bridge, containing two islands planted with trees, and running the whole length of the building. The exhibition if held here would have had special facilities of access, as, in addition to trains, trams, and omnibuses, steamboats could have brought visitors almost to the door. Perhaps, however, the change of site to Hyde Park was for the better, as Battersea is hardly in sufficient favour with the classes to have secured their patronage, and the support of the masses, as was proved afterwards in the case of the Albert Palace, was not enough to ensure a financial success. It was mainly owing to this idea of the late Prince Consort's that the Albert Palace was brought here. Although not built in the park, it was erected on part of the land which was acquired by Government for its formation, the materials being brought from a former Dublin exhibition. The whole of the design was never carried out, the building consisting of a central transept, containing the Connaught Hall, and one wing, whilst the erection of the other wing was left to the time when the success of the undertaking rendered more space imperative. Unfortunately for the promoters of the scheme, these prosperous times never came. It dragged on a weary existence for three years—from 1885 to 1888—but was never a financial success. The Connaught Hall contained one of the finest organs in the world, known as the Holmes organ, after its former proprietor. A peculiar feature of this was its echo organ, situated at the opposite end of the hall to the main portion, though it was played from the same keyboard. As its name implies, it was used for producing echo effects. The Palace had a fine collection of paintings illustrative of the winning of the Victoria Cross, which has been secured by the Crystal

Palace. When the ill-fated Albert Palace was closed, the building was allowed to go to rack and ruin; birds made their nests in the pipes of the organ, and eventually the materials were sold to pay the arrears of rent. The handsome building now erected on this site is the Battersea Polytechnic. This institution is the last of three towards the erection and endowment of which the Charity Commissioners contributed £150,000, the others having been erected at New Cross and Borough Road.

At the time when the park was first laid out, the only means of access from the other side of the river was the old Battersea Bridge. The Victoria Suspension Bridge was then in course of erection, whilst the Albert, also a suspension bridge, was not built till 1873. For several centuries before the building of the old wooden bridge, the other side was reached by means of a ferry. This ferry, whose history can be traced back to the fifteenth century, has passed through many hands. Previous to 1603, when the ferry was in full working order, it was the property of the Crown, for in that year James I. sold it for £40, to his ' dear relations, Thomas, Earl of Lincoln, John Eldred and Robt. Henley, Esqs.' At this time the Earl of Lincoln owned Sir Thomas More's house at Chelsea, and fifteen years afterwards the Earl sold the ferry to William Blake. The next owner was Bartholomew Nutt; and then it seems to have become about 1700 part of the manorial estates of Battersea, and after passing through the hands of some of the St. John family, was sold together with the manor to Earl Spencer. In 1766 he obtained an Act of Parliament authorizing him to replace the ferry with a bridge, and so unite the two parishes of Battersea and Chelsea, the boundaries of which meet in the middle of the river.* According to the *Gentleman's Magazine*, the original design for the bridge was of stone, and it was to have been built at Earl Spencer's own cost. In view, however, of the great outlay, estimated at £83,000, a company was formed, consisting chiefly of

* Simmonds, ' All about Battersea,' p. 67.

adjacent landowners, who carried out the project. It was eventually built of wood, at a cost of £20,000, and opened for traffic in 1772. Although at first it was not a profitable undertaking, it soon amply repaid the proprietors by means of the tolls levied. In 1873 it was purchased by the Albert Bridge Company, whose interests were in turn secured under an Act by the late Metropolitan Board of Works, who freed it from toll on the Queen's birthday, 1879.

The former wooden bridge formed a picturesque addition to old views of Battersea. Turner was especially fond of

Old Battersea Bridge.

painting it, and it is said that in his last illness he crept out on the roof of his house and took one long farewell gaze at th old bridge and the broad river he had so often trans-ferred to canvas. But the picturesque has to give way to the useful in these up-to-date times, and the old bridge, which was often a serious hindrance to navigation, was doomed to destruction. Its place has now been taken by the new Battersea Bridge, opened in July 1890 by Lord Rosebery. The work took four years to complete, and the cost was about £143,000. It is a composite structure, of

stone and iron, and has five spans in place of the nineteen
of its predecessor.

Among the owners of property whose interests were purchased at the time of the formation of the park, we find
mention of the Archbishop of York, who owned what is
described in the schedule of the Act as 'wharf, dock, kiln,
and rough land.' York Road, one of the principal thoroughfares of Battersea, still reminds us of the former connection
of the Archbishops of York with Battersea. They had a
residence here, York House, which stood near the water-
side, on the spot now occupied by Price's candle factory.
It was supposed to have been built for Lawrence Booth,
Bishop of Durham, about the year 1475, and when he was
translated to be Archbishop of York he took this house as
his town residence, so that he might be near the Court when
wanted. The house was standing at the end of the last
century, although the Archbishops had long ceased to live
here, and had let it to tenants. One of the holders of the see,
Archbishop Holgate, was committed to the Tower by Queen
Mary, in 1553, and this house was rifled of its valuables by
those who were sent to arrest him, including gold coin,
plate, a particularly fine mitre, and the seal and signet of the
diocese.*　He was afterwards deprived of the Archbishopric
of York, and never restored to it.

Visitors to Battersea Park a few years ago must have
noticed a number of broken stone pillars lying prostrate by
the river-side. These were the stones which formed the
colonnade or peristyle of old Burlington House in Picca-
dilly. This magnificent pile of buildings, which forms the
headquarters of most of the learned societies and the home
of the Royal Academy, was built by Denham for Lord
Burlington about 1664. It was rebuilt by the third Earl
of Burlington, the architect, who gave it a new front and
added this colonnade. Horace Walpole, in his reminiscences, says of this : 'As we have few examples of architecture more antique and imposing than that colonnade, I

* Simmonds, ' All about Battersea,' p. 58.

cannot help mentioning the effect it had on myself. I had
not only never seen it, but had never heard of it—at least,

Breakfast by the Lake, Battersea Park.

with any attention—when, soon after my return from Italy,
I was invited to a ball at Burlington House. As I passed

under the gate at night it could not strike me. At daybreak, looking out of the window to see the sun rise, I was surprised by the vision of the colonnade that fronted me.' Another eminent authority, Sir William Chambers, architect of Somerset House, called it one of the finest pieces of architecture in Europe. Upon the death of Lord Burlington the mansion passed into the hands of the Dukes of Devonshire. In 1854 it was purchased by the Government, and, although several proposals were made to pull it down, it was kept intact till 1866, when arrangements were made for the preservation of old Burlington House, while the Royal Academy and the University of London obtained a lease of the grounds which formed the gardens. It was found necessary to remove the colonnade, however, and with the hope that it might be re-erected, the stones were numbered and removed to Battersea Park. Unprotected as they were, they naturally suffered much through the rough usage of crowds of holiday-makers, so that it would have been almost impossible to have re-erected them in their original state. The scheme proposed was to form them into a ruin, somewhat similar to those in the Parc Monceaux at Paris, and it was hoped that Government would have helped towards the cost of re-erecting them ; but as they did not see their way to contributing, the project was abandoned, and all that remained of this masterpiece was used for building purposes.

From this river front we have a fine view of Chelsea Hospital with its grounds, Chelsea Church, and Cheyne Row, all of them teeming with historical associations, but the consideration of these must be deferred to another chapter.

CHAPTER II.

BLACKHEATH.

THIS heath is one of the oldest and one of the largest of London's open spaces, having an area of 267 acres. The scene of many a State reception, of many an angry outbreak on the part of a down-trodden people, and the resort of highwaymen, it is now reserved for a less eventful career. During the summer months it is a favourite resort of holiday-makers, although it suffers much from its proximity to its more aristocratic neighbour Greenwich Park, from which it is only divided by a wall. From the highest parts of the heath, especially from the isolated portion know as the Point, extensive views of the counties of Kent and Surrey can be obtained. In the distance the banqueting-hall of the once famous Eltham Palace may be discerned, looking like a huge barn against the sky, but the majority of the views are decidedly inferior to those of Greenwich Park.

It forms an extensive elevated plateau, fairly level except for the extensive excavations for gravel, which Nature has transformed into grassy dells. There are several fine clumps of trees dotted about on the heath, which greatly relieve the otherwise bare appearance. Its name is variously derived from its bleak site, or from the blackness of its soil.*

The acquisition of this desirable recreation-ground was brought about by the Metropolitan Commons Supplemental Act, 1871. The freehold of Blackheath is vested in the

* Lysons, 'Environs of London,' 1811, vol. i., part ii., p. 542.

lords of the manor, who, however, have given free use of the heath, and it is preserved as an open space for ever. If these active steps had not been taken, it is very probable that its area would have been seriously diminished. As in the case of so many other Metropolitan commons, large encroachments have been made at various times, and in addition to these, the surface of the heath has been much disfigured owing to the Crown having let the right to remove an unlimited quantity of gravel for a sum of £56 a year.

The Dover Road crossing the heath is supposed to have been the Roman Watling Street. Along this, as well as in Greenwich Park, were several tumuli or barrows. In January, 1784, fifty of these were opened by the Rev. J. Douglas, with the permission of the surveyor of the Crown lands, and some interesting relics were discovered, although he had been forestalled in his search some seventy years before by one of the park-keepers, Hearne, who had no doubt removed all available valuables. The majority of the barrows were small and conical, with a circular trench at the base, and were settled by the archæologists as having been of Roman or Early British origin. In some of them were found traces of human hair (although the skeletons had disappeared), iron spear-heads, some beads of dark blue-green colour, and some patches of woollen cloth. On another occasion some labourers were digging in the kitchen-garden at Dartmouth House, and discovered several Roman urns and other remains, about one or two feet below the gravel. The larger ones contained charred fragments of bone. These were afterwards exhibited before the Society of Antiquaries by the Earl of Dartmouth in 1803, and some of them are now in the British Museum.*

Another curious discovery was made in 1780, at the Point. This was a cavern cut out of the solid chalk, which consisted of several large rooms, connected by narrow, arched passages, extending some 160 feet underground from

* Hasted's ' History of Kent,' by Streatfield and Larking, edited by H. H. Drake, 1886, p. 83.

the entrance. Some of them had circular domes supported
by columns. In the farthest of the rooms was a well
27 feet deep. The bottom of the cavern, which is of fine
dry sand, was formerly reached by a narrow shaft, but a
flight of steps was afterwards formed.* This is probably
the same as the ' chalkpytte' underneath Blackheath mentioned in a lease of Shene Priory when it was let with all
the sand there, for a term of seven years, at an annual rent
of 13s. 4d.† Some fifty years ago this cavern was open to
the public at a charge of 4d. and 6d. each, the tour of
inspection being undertaken by torchlight. One of the
residents who visited the cavern about this time relates that
a ball was given there, but owing to the want of ventilation
the lights went out, and a panic very nearly ensued. The
entrance was then filled up, it is believed by the local
authorities, but the fact does not seem to be recorded in
their minutes.

Blackheath lies in no less than four separate manors.
Part of the heath in Greenwich parish is within the Manor
of Greenwich; another part, in the parish of Lewisham, is
in the Earl of Dartmouth's Manor of Lewisham. A third
part, in Greenwich parish, is in the Manor of West Combe,
and the remainder, which is the portion called ' The Point,'
at the top of Maidenstone Hill (also in Greenwich), forms
a part of the Manor of Old Court.

The Manor of *East Greenwich* (so called to distinguish
it from West Greenwich, or Deptford) is a royal one. It
was considered as an appendage to the Manor of Lewisham,
and was given with that to the Abbey of St. Peter at Ghent
by the niece of King Alfred, who was herself buried in the
church. The grant was confirmed in 964 by King Edgar.
It is not mentioned in Doomsday Book, which would be
easily accounted for if it was but subsidiary to another
manor, and so it must be included in the list of the

* Richardson, ' Greenwich,' pp. 81, 82.

† Hasted's ' History of Kent,' by Streatfield and Larking, edited by
H. H. Drake, p. 84.

Abbot of Ghent's possessions under the general title of Lewisham. Its subsequent history is the same as that of Lewisham. On the dissolution of the alien priories, in 1414, it was taken by the Crown, and in the following year was granted to the Carthusian Priory of Shene, whose property it remained till the twenty-third year of Henry VIII.'s reign, when it reverted to the Crown by exchange. On the death of Charles I. it was seized by the State, but once

Princess of Wales Pond, Blackheath.

more came back to the Crown at the Restoration, and it has remained a royal manor ever since.*

The Manor of *West Coombe* is called Coombe West in the rolls of the Manor of Dartford or Richmond's, in Kent, of which manor it was held by a quit-rent of 9s. 2d. At some very remote time it belonged to the Dean and Chapter of Westminster. The Kentish historian Hasted says it belonged to the family of Badelesmere, but fell to the Crown, 15 Edward II., by the attainder and execution of Bartholomew, Lord Badelesmere, and continued part of the

* Lysons, 'Environs of London,' 1811, vol. i., part ii., p. 497.

royal possessions till Richard II. granted it to Sir Robert Belknap, the judge, on whose attainder in 10 Richard II. it again reverted to the Crown. We next find the manor as the property of Gregory Ballard, and afterwards of John Lambarde, draper and Alderman of London, father of William Lambarde, author of ' The Perambulation of Kent,' the earliest county history known. He resided at the Manor House, West Coombe. The next owner appears to have been Mr. Theophilus Biddulph, created a Baronet in 1664. On his death, in 1718, the manor was sold for £12,000 to Sir Gregory Page, of Wricklemarsh, in the adjoining parish of Charlton, who left it to his nephew Sir Gregory Turner, in whose family it has since remained.* The portion of the heath within this manor is that to the east of Greenwich Park wall, on both sides of the road leading to Charlton.

The Manor of *Old Court* came to the Crown in 23 Henry VIII. by exchange with the Prior of Shene. The King granted it in 1536 free of rent, with the tithes of hay and corn, to Sir Richard Long for life. On his death a similar grant was made by Edward VI. to Sir Thomas Speke in 1547 and three years later the reversion in fee was granted to the Earl of Warwick, who, however, exchanged this manor with the King for the Castle and Manor of Tunbridge. Upon the death of Speke the manor was conferred upon Lord Darcy of Chiche, and subsequent owners were Sir Henry Jerningham, 1554; Sir George Howard, 1572; Sir Christopher Hatton, 1580; Lord Buck-hurst, afterwards Earl of Dorset, 1594; and then Viscount Cranbourne, afterwards Earl of Salisbury. After this long list of noble owners we find it next settled upon Anne of Denmark, Queen of James I., in 1613. Upon the death of the Queen, in 1619, Old Court was settled in trust on Prince Charles, who in 1629 granted Greenwich Park and House, this manor, and other lands, to his Queen Henrietta. During the Commonwealth it was sold to Robert Tichborne,

---

* Hasted's ' Kent,' by Streatfield and Larking, edited by H. H. Drake, 1886, pp. 50-53.

but reverted to the Queen Dowager at the Restoration. After some minor fluctuations between the Crown and grantees, the manor was sold with its demesnes and the lands called Queen's Lands in 1699 to Sir John Morden for £1,276 10s. After having spent an additional £9,000 in the purchase of another interest, he left it by will to the trustees of Morden College.* There has been some dispute as to whether the claim made by the trustees of the college to Maidenstone Hill, as part of the Manor of Old Court, was a legitimate one. Proceedings were taken against them in 1751 by the

Vanbrugh Park, Blackheath.

Crown to restrain them from granting leases to dig for sand and chalk under the hill, and for erecting houses round it. A compromise was effected in 1771, by the trustees admitting the right of the Crown to the waste (*i.e.*, the Point), and receiving a grant of fifty years of Maidenstone Hill, and the surrounding houses. In 1823 the trustees purchased the Crown's interest in the houses for £5,053 5s. 5d., the plain on the top being reserved for public use, or a church, or other public building, so that if the scheme for the manage-

* Hasted's ' Kent,' annotated by Streatfield and Larking, pp. 44-46.

ment of Blackheath had not been formulated, this part must have been preserved from private building. The remaining portion of the heath in the parish of Lewisham is within the Manor of *Lewisham*.*

Situated at so short a distance from London, on the main road from Dover and Canterbury, Blackheath has from the earliest times furnished a splendid site for military gatherings, and those gorgeous state pageants in which our forefathers delighted. The earliest of these of which any record remains was the encampment of the Danish army in the reign of Ethelred, when their fleet was moored at Greenwich.† The part of the heath where they entrenched themselves was probably the high ground at East and West Coombe.‡ At these places distinct traces of entrenchments have been found from time to time, some of which may date back to these early ages, whilst the remainder must be attributed to the various bodies of insurgents who have encamped here.

It was at Blackheath that Richard II., with the daughter of the King of France, whom he was about to take as his second wife, were met by the Lord Mayor and Aldermen, duly attired in their scarlet robes. They accompanied the King to Newington, where he dismissed them, as he and his bride were to 'rest at Kennyngtoun,' where the royal palace was.

Another incident in connection with Blackheath during this reign was of a less pleasant character. The body of insurgents who resented the imposition of the poll-tax of 3 groats on all persons above fifteen assembled on Black-heath in June 1381. The Kentish contingent, headed by Wat Tyler, the blacksmith of Dartford, united their forces with the men of Essex led by Jack Straw, and the combined body estimated at 100,000 marched upon London. They afterwards separated into three parties, one of them being stationed at the Tower, a second proceeding to the

* For descent of this manor, see p. 158.
† Lysons, ' Environs of London,' vol. i., part ii., pp. 496, 497.
‡ Coombe, from Anglo-Saxon *coomb*, a camp.

Temple, which they burnt to the ground together with its library and documents, whilst the third burnt the monastery of St. John of Jerusalem, at Clerkenwell. Both the leaders afterwards suffered for the prominent part they had taken, Wat Tyler being stabbed by the Lord Mayor in Smithfield, and Jack Straw together with many others beheaded.

In the next reign the Emperor of Constantinople, Manuel Palæologus, came over to England to solicit the aid of Henry IV. against Bajazet, Emperor of the Turks. He was met here by the King in 1400, who conducted him to London with great pomp and ceremony.*

On November 3, 1415, Henry V. was met here on his return from the glorious victory of Agincourt by the Mayor, Aldermen, and Sheriffs, accompanied by great numbers of the citizens of London, who came to Blackheath to welcome their hero. The aristocratic citizens were mounted and clothed in scarlet robes, while the meaner sort numbering some 20,000 attended on foot all ' with the devices of their craft.' The victor had had one long triumphal procession all the way from Dover, and was doubtless rather wearied of the sweets of triumph by the time he reached Blackheath. He bore the honours thus thrust upon him with exemplary modesty, and nipped all the preparations of the Mayor in the bud. The meeting, according to Holinshed, seems to have been rather a failure, for he tells us that ' the King, like a grave and sober personage, and as one remembering from Whom all victories are sent, seemed little to regard such vaine pompe and shewes as were in triumphant sort devised for his welcoming home from so prosperous a journie ; insomuch that he would not suffer his helmet to be carried before him, whereby might have appeared to the people the blowes and dints that were to be seene in the same ; neither would he suffer any ditties to be made and sung by minstrels of his glorious victorie, for that he would have the praise and thanks altogether given to God.'†

* Thorne, ' Environs of London,' vol. i., p. 46.
† Holinshed, vol. iii., p. 556.

These worthy citizens were particularly fond of these
magnificent receptions, and their ardour was evidently not
damped by the coolness of Henry, for in the following May
we find them again here, this time to meet the Emperor
Sigismund, who had come over to mediate a peace between
France and England.   He was particularly well looked
after, for at Dover he was received by Humphry, Duke of
Gloucester ; at Rochester by John, Duke of Bedford ; and
at Dartford by Thomas, Duke of Clarence—the King's three
brothers, who, with many other lords, conducted him to the
King at Lambeth.*

'Good Duke Humphry' attended by 500 men wearing his
livery is again here on May 18, 1428, together with the
Mayor and Aldermen, to meet Margaret of Anjou before her
coronation.   He conducted her to his palace, which she
afterwards obtained for herself.

In the following reign Blackheath was much in requisition,
but the meetings were of a more turbulent nature.   The new
King, Henry VI., returned to London after his coronation in
Paris, and received a royal reception at the hands of the
citizens of London on the heath in February, 1431.   The
various dresses must have made the gathering very picturesque.   The Mayor was attired in crimson velvet, with
a girdle of gold about his waist ; the Aldermen were in
scarlet robes, and the citizens had white gowns with scarlet
hoods, all of them wearing the badge of the particular company to which they belonged.†

The next gathering was of a very different nature.   Jack
Cade, representing himself to be a kinsman of the Duke of
York, laid before the royal council the complaint of the commons of Kent, called the 'Blackheath Petition.'   His insurrection soon became formidable, and he headed about 20,000
men, who encamped on the ' plaine of Blackheath, between
Eltham and Greenwiche.'   Their objects were 'to punish
evil ministers, and procure a redress of grievances.'   They

* Holinshed, vol. iii., p. 556.
† Thorne, 'Environs of London,' vol. i., p. 46.

defeated and slew the King's leader Sir Humphry Stafford, at Sevenoaks in June, 1450, who, according to Shakespeare, calls them 'rebellious hinds, the filth and scum of Kent, mark'd for the gallows.' Much allowance must be made for poetic license in this description, as the insurgents included many men of high standing. After their first success they entered London in triumph, beheaded Lord Saye, the Lord Treasurer, amongst others, defaced the records of the law, burnt down the office of arms, and destroyed the rolls, registers, and books of armoury. Their reason for these destructive acts appears to have been to destroy all title-deeds and evidences, and so to place everybody on a glorious equality. Soon after this the insurgents lost ground, and when a general pardon was proclaimed Cade was deserted by his followers and fled. He was discovered, but as he refused to surrender he was slain by the Sheriff of Kent. After his death many of the rebels came once more to Blackheath, 'naked save their shirts,' and, with halters on their necks, knelt to the King to receive their doom of life or death.* It is pleasant to record that they were pardoned.

Once more we find Henry VI. at Blackheath. The following year, in 1452, his cousin the Duke of York, father of Edward IV., who had openly claimed the Crown, drew up his forces in the neighbourhood of Dartford, while Henry encamped upon Blackheath. On this occasion the Duke was induced to enter the royal tent unarmed, and was seized and carried prisoner to London.

In 1471 the bastard Falconbridge, who had taken up arms in the cause of Henry VI., encamped here with his army;† and three years later the new King, Edward IV., was met here by the Mayor and citizens, when returning from France, where he had been with an army of 30,000, to conclude a treaty of peace with Louis.‡

The Wars of the Roses now being finished, a fresh

* Stow, 'Annals.' † Holinshed, vol. iii., p. 690.
‡ *Ibid.*, p. 701.

rebellion arose, and we find another insurgent crowd on Blackheath in 1497. The Cornishmen resented the taxes levied to pay the Scottish war expenses. Their leaders were Thomas Flammock, a lawyer, and Michael Joseph, a farrier, and to the number of 6,000 they marched towards London. At Wells the chief command was given to Lord Audley. Henry VII. gave them battle on Blackheath, where many of them were slain, and the remainder were forced to surrender. Lord Audley was executed on Tower Hill. He had been clad in a suit of paper displaying his coat-of-arms reversed. The two other leaders were hanged at Tyburn.* The Kentish historian Lambarde, who lived at West Coombe, and was, of course, familiar with the locality, says 'there remaineth yet to be seen upon the heath the place of the smith's tent, called commonly his forge, and the grave-hills of such as were buried after the overthrow.'† This smith's forge is an earthen mound marked with fir-trees, close to the end of Chesterfield Walk. This spot has also been called Whitefield's Mount, from the fact that it had been used as an open-air pulpit by that prince of preachers. To the west of it may be seen ridges, which may be the remains of the encampments referred to by Lambarde. It has also been put to other uses, as a butt for artillery practice, for our old friend Evelyn mentions, under date March 16, 1687, in his diary : 'I saw a trial of those devilish, murdering, mischief-doing engines called bombs, shot out of a mortar-piece on Blackheath.'

But to return to our state receptions, we must now pass on to the reign of Henry VIII. In the early part of this reign there are two of these to record, both of religious dignitaries. One of these was to meet a solemn Embassy, consisting of Lord Bonevet, Admiral of France, the Bishop of Paris, and others, with a train of 1,200. The honours on this occasion were entrusted to the Earl of Surrey, Lord Admiral of England. 'The young gallants of France had

---

* Stow, 'Annals,' 4to., p. 802.
† 'Perambulation of Kent,' 1596, p. 392.

coats guarded with one colour, cut in ten or twelve parts, very richly to behold, and so all the Englishmen accoupled themselves with the Frenchmen lovingly together, and so rode to London.'* An equally brilliant pageant was seen when the Papal Legate, Cardinal Campegius, was met here by 'the Duke of Norfolk with a great number of prelates, knights, and gentlemen all richly apparelled. And in the way he was brought into a rich tent of cloth of gold, where he shifted himself into a robe of a cardinal edged with ermines, and so took his moyle (mule), riding toward London.'†

The much-married Henry himself was here in 1540, and a second edition of the Field of the Cloth of Gold took place on the occasion of his meeting his fourth wife, Anne of Cleves. Henry pretended that this was the first time he had set eyes upon her, but he had already inspected her privately at Rochester, when he made up his mind to put her away speedily; but for all that her reception on Blackheath was conducted with all propriety and decorum. On the eastern side of the heath 'was pitched a rich cloth of gold, and divers other tents and pavilions, in the which were made fires and perfumes for her, and such ladies as should receive her grace.' Our chronicler gives such a full and minute description of the ceremony that one would almost have thought that he was present. Henry's dress and appearance in general will serve as a good guide 'to those about to marry.' The account runs as follows:

'The King's highness was mounted on a goodly courser, trapped in rich cloth of gold . . . all over embroidered with gold of damask, pearled on every side of the embroidery; the buckles and pendants were all of fine gold. His person was apparelled in a coat of purple velvet . . . all over embroidered with flat gold of damask, with small lace mixed between of the same gold . . . about which garment was a rich guard very curiously embroidered. The sleeves and breast were cut, lined with cloth of gold, and tyed together

---

* Hall's 'Chronicle,' p. 594, reprint.   † *Ibid.*, p. 592.

with great buttons of diamonds, rubies, and orient pearl, his sword and sword girdle adorned with stones and especial emerodes . . . but his bonnet was so rich with jewels that few men could value them.'*

After the meeting, the royal pair went in procession to Greenwich Palace, there to commence that happy married life which lasted but seven months. We must quote one more passage:

' O what a sight was this, to see so goodly a prince and so noble a king to ride, with so fair a lady, of so goodly a stature and so womanly a countenance, and in especial of so good qualities; I think no creature could see them but his heart rejoiced.'

This becomes interesting in the light of subsequent events, for although all the huge crowd of spectators may have rejoiced, it is certain that the two principals never did. Henry vented his wrath upon Cromwell, who had brought about the match, and he paid the penalty of ill-success with his life. Anne was compensated for the loss of the doubtful joys of married life by the gift of many a manor in Kent and Sussex.

The last state reception we have to record in connection with Blackheath eclipses all the others in splendour and magnificence. This was at the Restoration, when the re-action against Puritanism had triumphed, and all London made holiday one fine day in May, 1660, to greet their exiled King. Those familiar with Sir Walter Scott's ' Woodstock ' will remember how well the scene is described there. Charles, who had slept the night at Rochester, rode on to the heath escorted by his brothers, the Dukes of York and Gloucester, and there saw drawn up to meet him that same army which he had good cause to remember. Lord Macaulay gives us a vivid picture of their attitude in his ' History of England ' :

' Everywhere flags were flying, bells and music sounding, wine and ale flowing in rivers to the health of him whose return was the return of peace, of law, and of freedom. But

* Hall, pp. 833-836, reprint.

in the midst of the general joy, one spot presented a dark and threatening aspect. On Blackheath the army was drawn up to welcome the Sovereign. He smiled, bowed, and extended his hand graciously to the lips of the Colonels and Majors; but all his courtesy was vain. The countenances of the soldiers were sad and lowering, and had they given way to their feelings, the festive pageant, of which they reluctantly made a part, would have had a mournful and bloody end.'* From this, however, he was mercifully preserved, and continued his triumphal procession to London.

In addition to these state receptions, Blackheath has furnished a suitable ground for reviews and military parades. The good Queen Bess, when at Greenwich, came to Blackheath and reviewed the city militia, completely armed, to the number of 4,000 or 5,000.

On May 1, 1645, Colonel Blunt, to please the Kentish people, who were fond of old customs, particularly May games, drew out two regiments of foot, and exercised them on Blackheath, representing a mock fight between the Cavaliers and the Roundheads. The old writer adds that 'the people were as much pleased as if they had gone a-maying.'†

Evelyn mentions several of these encampments. On June 10, 1673, he records in his diary :

'We went after dinner to see the formal and formidable camp on Blackheath, raised to invade Holland, or, as others suspected, for another designe.'

They encamped here again on their return (July, 1685). He also tells of another camp of about 4,000 men formed here when London was agitated by the rumour that the English fleet had sought refuge in the Thames from the French fleet under De Tourville in 1690.

In 1798, owing to the war scare caused by the rebellion in Ireland, and the success of the French arms, the Government encouraged the formation of volunteer corps. Three

* 'History of England,' 1858, vol. i., p. 156.
† Quoted in Lysons, vol. i., part ii., p. 544.

of these were raised in Greenwich, one of which was called the Blackheath Cavalry. This body consisted of about fifty troopers, residents of the neighbourhood. It was strengthened by the addition of the Woolwich troop of about the same number in 1802, but was disbanded in 1809.*

It will be seen, therefore, that Blackheath has played an important part in the annals of England. It is only natural that the social position to which these meetings and the proximity of Greenwich Palace raised our common should attract to the neighbourhood many noblemen of note, whose residences clustered round the spot.

One of the most famous of these is the residence of the Ranger of Greenwich Park, facing the heath, situated in Chesterfield Walk, a shady pathway running along under the park wall from the top of Croom's Hill. The name of this delightful avenue is a reminiscence of a former occupant, the celebrated Philip, Earl of Chesterfield, principally remembered now for those extraordinary letters of advice written for the guidance of his son. This, together with the adjacent mansions, was built on part of the waste lands of the manor, which were allowed to be enclosed in consideration of the payment of 40 bushels of coal annually to the poor of Greenwich. The manor court in 1676 threatened to level the houses, or exact £50, but we find a lease of the ground granted in 1688 by the Queen's trustees for £3 a year.† In 1697 Colonel Stanley, afterwards Earl of Derby, resided here, and the Earl of Chesterfield in 1753 bought the assignment of part of this ground with a house standing thereon, which he improved and enlarged for his occasional residence. Although it was known to the general public as Chesterfield House, the owner himself called it in his letters ' Babiole,' and afterwards ' La Petite Chartreuse.' Its name was changed in 1807 to Brunswick House, when the Dowager Duchess of Brunswick, sister of George III., bought the

* Richardson, ' Greenwich,' pp. 22, 23.

† Hasted's ' History of Kent,' by Streatfield and Larking, edited by H. H. Drake, p. 82.

mansion. She came here so as to be near her daughter Caroline, Princess of Wales, who had been appointed Ranger of the park in the previous year, and occupied the adjoining mansion, Montague House. In 1815 the house was purchased by the Crown as an official residence for the Ranger, and it was subsequently occupied by Princess Sophia, who was appointed Ranger in 1816 till her death in 1844.* In more

Chesterfield Walk, Blackheath.

recent years it has been the residence of Ranger Lord Haddo, afterwards Earl of Aberdeen, the Duke of Connaught (whilst studying at Woolwich for the Engineers), the Countess of Mayo, and lastly of Lord Wolseley. The grounds of the mansion, some 15 acres in extent, were added by order of the Queen to Greenwich Park in 1897, and the house was

* Thorne, 'Environs of London,' vol. i., p. 49.

handed over to H.M. Office of Works. It is needless to say these gracious acts were much appreciated.

Immediately to the south of the Ranger's residence stood Montague House, so named after the Montague family, to whom the lease was assigned in 1714, from whom it descended to the Duchess of Buccleugh. The house, which was purchased by the Crown in 1815, and pulled down to enlarge the grounds of Ranger's Lodge, was an irregular brick building whitened over.* We have seen that the Princess Caroline lived here as tenant of the Duchess of Buccleugh. She had been married to George, Prince of Wales, afterwards George IV., in 1795, by whom she was treated from the very first with indifference, which developed afterwards into hatred, so that they were forced to separate. Her husband had no scruples in making the gravest charges against her life at Blackheath; but, after a rigid scrutiny on the part of a secret commission appointed by him, she was acquitted from all guilt. During her residence here, the Princess enlarged the grounds by enclosing a few acres of the park known as the Little Wilderness. Although Montague House has disappeared, it has given its name to Montague Corner, at the south-east end of Chesterfield Walk.†

Another house in Chesterfield Walk, also facing part of the heath, was once the residence of Major-General Wolfe. His son, the hero of Quebec, who is buried in Greenwich Church, occasionally lived here. It afterwards passed into the hands of Lord Lyttelton, from whom it was named Lyttelton House.‡ It is now called Macartney House.

Turning now to the eastern side of the park, Vanbrugh Fields, at the north-east corner of Blackheath, are so named after Sir John Vanbrugh, the architect of Blenheim Palace and the Mansion House, who took a lease of 12 acres of ground here in 1714, and built the grotesque castellated building called by him Vanbrugh Castle, but popularly

* 'Beauties of England and Wales.'
† Thorne, 'Environs of London,' vol. i., p. 49.
‡ Lysons, 'Environs of London,' 1811, vol. i., part ii., p. 537.

known as the Bastille, from its supposed resemblance to the
French prison, in which Vanbrugh had on one occasion been
confined, owing to his examining a fortification too closely,
and being thus mistaken for a spy. It is approached by an
embattled gateway overgrown with ivy, and the whole building, with its round tower and spire, has the appearance of a
fortification.

Vanbrugh Castle, Blackheath.

Close by is an equally curious building, also built by
Vanbrugh, called Vanbrugh House. This, too, had a nick-
name, Mince-pie House,* which it may have received as a
place of entertainment. It is built of brick with raised bands,
and has a round tower at each end, with a central porch.

* Lysons, ' Environs of London,' 1811, vol. i., part ii., p. 526.

The embattled archway to this, with a lodge on each side, stands at some distance from the house. The two houses south of this arch were built in 1719 for the Duchess of Bolton (Polly Peacham) and Sir James Thornhill. It is probable that the heath at some time reached to here, and that the present gateway was the entrance from it.

The greater part of the northern boundary of the heath is formed by the wall of Greenwich Park. Although any lengthy mention of this would be out of place in the present work, we must just briefly notice it in passing, inasmuch as it may be rightly termed part of the heath itself. Humphry, Duke of Gloucester, who had received the grant of the manor of Plesaunce (subsidiary to that of East Greenwich), had a license, in 1433, to enclose and empark 200 acres of land, and erect therein 'towers of stone and lime after the form and tenure of a schedule to this present bill annexed.' The Duke enclosed his park, but did not build his palace on the hill, but, preferring the view of 'the silver Thames,' chose the site now occupied by the west wing of Greenwich Hospital. The wall round the park was built by James I., but we owe the present state of the park to Charles II., who commissioned Le Notre to lay it out. Upon the hill where now stands the world-famed observatory he built a tower called Greenwich Castle. When the Duke died, the manor and palace reverted to the Crown, and successive royal owners beautified and enlarged the palace, till it was demolished about 1664.* Charles II. is also responsible for the observatory, which was commenced in 1675 and completed in 1676.

At the south-east corner of Blackheath, seen through a screen of sheltering elms, is Morden College, founded by Sir John Morden, and erected in 1694. He made his fortune in Aleppo, in spite of a clause in the Navigation Act of Charles II. prohibiting the indirect importation of African, Asian, or American products, under penalty of forfeiture of the ship. The following tradition is current in the college

---

* Thorne, 'Environs of London,' vol. i., p. 249.

to account for its foundation. Sir John Morden, having resided many years at Aleppo, decided to return and settle in England. Having shipped the whole of his merchandise on board three of his ships, he sent them on a trading voyage, after which they were to return to the Port of London. Years passed without tidings of them, till they were given up for lost, and Sir John, being reduced to extreme poverty, was employed as a traveller by a trades-man. While waiting in the hall of a gentleman's house he heard him exclaim, ' Here is an astonishing circumstance !' and read from a paragraph in a newspaper stating that three ships had just arrived, supposed to be lost, for they had not been heard of for ten years or more. Sir John rushed into the city, and found they were his own long-lost vessels, and in the joy of the moment he vowed to build an asylum for decayed merchants.* Whether this was so or not, we find the building erected and endowed, and occupied by twelve Turkey merchants during Sir John's lifetime, but after his death Lady Morden was obliged to reduce the number to four, owing to the estate not answering anticipations. At her death the college obtained the whole property, and the number of occupants was increased. The benefits of the institution are intended, in the first place, for merchants trading in the Levant, whose fortunes have been ruined by perils of the sea or other unavoidable accidents. It must not for a moment be supposed that Morden College is a kind of private workhouse. On the contrary, the inmates, numbering about forty, receive all the comforts of home. The college, which was designed by Sir Christopher Wren, and erected by his master-mason, Edward Strong, is a large quadrangular building of brick, with stone quoins and cornices. Over the entrance are statues of the founder and his wife. Inside the walls is a chapel, wainscoted with oak to a height of about 9 feet. The carvings of the oak altar-piece, door, and cornices are attributed to Grinling Gibbons.

* ' Memoir of Sir J. Morden, Bart.,' by H. W. Smith, Esq., treasurer of the college.

Each resident member of the college is allowed £100 annually, in addition to grants for washing and candle money. There are servants to clean the apartments, which consist of a sitting-room, bedroom, pantry, cupboard, and a cellar for each member. Every provision is made for the comfort and recreation of the inmates—library, billiard-

Morden College, Blackheath.

rooms, card-rooms, and well-laid-out pleasure-grounds for promenading. In addition to the resident inmates, there are nearly 100 out-pensioners, who receive annual sums varying from £20 to £80. The income of the college is steadily increasing, and amounts to considerably over £10,000, thanks to the munificence of the donor and of the other benefactors. The 'canal' which is shown in old engravings in front of the college has now disappeared. It was drained when the tunnel for the North Kent Railway was made under the grounds, the sand from which was used to form the undulating lawn of the college.* Altogether Morden College forms an ideal retreat for those who, having once been in prosperous circumstances, have now come down in the world.

The crescent known as the Paragon adjoining Blackheath, close to Morden College, occupies the site of the manor-house and grounds of Wricklemarsh. This is supposed to be identical with the estate called Witenemers in Doomsday

Wricklemarsh, the seat of Sir Gregory Page, 1730.

Book, held at that time by the son of Turald, of Rochester. The manor-house together with four tenements was sold for £1,950 in 1669 to Sir John Morden, who in his will devised ' his mansion-house, called Wricklemarsh, with all the orchards, gardens, walks, ponds, and appurtenances, and as many acres of land, next adjoining to the said house, as amounted to the yearly value of £100 at the least,' to his wife for life. Upon the death of Lady Morden the estate was sold under a decree in Chancery to Sir Gregory Page Turner, who pulled down the old house and erected in its place a magnificent mansion, which was then one of the finest seats

* Thorne, ' Environs of London,' vol. i., p. 50.

in England belonging to a private gentleman. It was built from the designs of John James, after Houghton, and completed in one year. It is described* as 'consisting of a basement, state and attic story. The wings contained the offices and stables, which were joined to the body of the house by a colonnade; the back front had an iron portico of four columns, but without a pediment. It stood in the midst of the park, with a large piece of water before it, on a beautiful rise, about a quarter of a mile distance from the heath, which from the pales of the park rises again up to the London road.' A newspaper cutting of 1783 records that 'the fine house built by Sir Gregory Page, and lately inhabited by Lord Townsend, was on Monday sold by auction, together with the enclosure where it stands, for £22,550.' The house cost Sir Gregory Page £90,000. Four years later the mansion was pulled down, and the present houses built on the site.†

At the corner of the heath, near Blackheath Hill, is the Green Man Hotel, which occupies the site of an ancient inn of the same name, and another place of entertainment, the Chocolate House. This latter is mentioned by the Duke of Richmond, Master-General of the Ordnance, in a private letter. The name of this house was long kept in memory by Chocolate Row.

Blackheath at the present day is most intimately associated with the game of football. Among its historical associations must not be forgotten its connection with the game of golf. There are people who say that the introduction of golf into England was the only good thing the Stuart Kings ever did for the country, whilst, on the other hand, others maintain that it is the worst thing they ever did, which is saying much. King James VI. of Scotland brought golf down South on his accession to the English throne, and played it on Blackheath. The Royal Blackheath Golf Club is the

* 'Beauties of England and Wales.

† Hasted's 'History of Kent,' by Streatfield and Larking, edited by H. H. Drake, p. 125.

oldest in the world, though the first real links on which the game was played in England were those of the Royal North Devon Club at Westward Ho.

Blackheath, in the same way as Kennington Common, was made one of the polling-places for members of Parliament for the Western Division of Kent under the Reform Bill of 1832.

Dickens has made us familiar with the associations of Blackheath with the old coaching days. Before we had been accustomed to rattle along at the rate of sixty miles an hour on our modern railroads, an advertisement like the following (quite oblivious to the requirements of grammar or punctuation) might often have been seen :

'A STAGE COACH

WILL SET OUT

for Dover every Wednesday and Friday from Christopher Shaws the Golden Cross at four in the morning to go over Westminster Bridge to Rochester to dinner to Canterbury at night and to Dover the next morning early; will take passengers for Rochester, Sittingbourne, Ospringe, and Canterbury—and returns on Tuesdays and Thursdays.'*

Travellers by this coach would have good cause to remember Blackheath, so notorious a resort for highwaymen in the last century. As the old coach lumbered up Shooter's Hill, the tremulous passengers would hide their watches and purses in their boots, and thankful indeed would they be if they escaped without the surrender of their valuables. These attacks by the knights of the road became so numerous that the inhabitants of the neighbourhood combined together in 1753, and subscribed a fund to suppress the lawlessness for which Blackheath was so notorious. Rewards were offered for the conviction of highwaymen and footpads caught within a prescribed radius, but now the extension of the Metropolitan Police Act has put an entire stop to these disorders. Other methods were adopted in past days

* *London Evening Post*, March 28, 1751.

to deter them from their predatory excursions. It was no infrequent sight to see a gibbet adorning Shooter's Hill, from which was hanging the body of some highwayman who had been so unfortunate as to have been caught red-handed. Friend Pepys (who had been on the merriest of all the journeys he had ever made) felt a slight shudder as he 'rode under the man that hangs upon Shooter's Hill, and a filthy sight it was to see how his flesh is shrunk to his bones.'* But the days of highwaymen on Blackheath are passed, although a gang of young ruffians did attempt a revival in 1877; but as they were all brought to justice, the midnight wayfarer, it is hoped, may now cross the heath in perfect security.

The fairs formerly held on Blackheath were of very ancient standing. They were held on that part of the heath which lies in the parish of Lewisham. George, Lord Dartmouth, obtained a grant from Charles II. to hold a fair twice a year on that part of the heath within the manor of which he was lord. Evelyn, who had a particularly intimate acquaintance with Blackheath, mentions his visit to the fair on May 1, 1683: 'Blackheath to the new fair, being the first procured by the Earl of Dartmouth. This was the first day, pretended for the sale of cattle, but I think, in truth, to enrich the new tavern (the Green Man) at the bowling green, erected by Snape, his Majesty's farrier, a man full of projects. There appeared nothing but an innumerable assembly of drinking people from London, pedlars, etc., and I suppose it is too neere London to be of any greate use to the country.' The following is an example of the attractions provided:

'GEO. II. R.

'This is to give notice to all gentlemen, ladies, and others, that there is to be seen from eight in the morning till nine at night, at the end of the great booth on Blackheath, a West of England woman 38 years of age, alive, with two heads, one above the other, having no hands,

---

* Diary, April, 11, 1661.

fingers, nor toes; yet can she dress or undress, knit, sew, read, sing (? a duet with her two mouths). She has had the honour to be seen by Sir Hans Sloane, and several of the Royal Society.

'N.B.—Gentlemen and ladies may see her at their own houses if they please. This great wonder never was shown in England before this, the 13th day of May, 1741. *Vivat Rex !'**

The fair continued to flourish as a 'hog and pleasure' fair being held regularly on May 12 and October 11, till it was suppressed by Government in 1872. Some idea of the condition of Blackheath in the old time of fairs may be gathered from a visit on a Bank Holiday, when the numbers who flock to it will bear favourable comparison with those that came for the state receptions, however much they may be behind them in the matter of dress. Swings, roundabouts, cockshies, and donkey-rides are then the order of the day, and it is on occasions like these that the full benefit of this roomy open space to the masses of London appears most strikingly.

* 'Merrie England in the Olden Time.'

CHAPTER III.

BOSTAL HEATH AND WOODS.

THESE open spaces are the most attractive of the Kentish commons. Indeed, we may go so far as to say that every other common of the Metropolis, with the possible exception of Epping Forest, must yield to them the palm of beauty. There are few places so close to the busy hum of London which have retained so sylvan a character, and although they are a favourite resort of those living near, they are a *terra incognita* to the general body of Londoners. As a place for a picnic they are ideal. Rising gradually to a considerable height, they are crowned by extensive stretches of pines and larches, whilst the view from the top is unsurpassed for many a mile round. Winding at one's feet is the Thames, rendered beautiful by that distance which lends enchantment to the view. Beyond lie the Essex marshes, whilst in the distance can be seen the forests of Epping and Hainault. Within the woods we are favoured with the softest possible carpet of pine-needles, shed by the larches and pines, which are the homes of the squirrel and many a feathered songster. Our reveries in this delightful spot will be broken from time to time by the rush of the timid bunny, who makes for the nearest burrow, frightened at our approach. At every opening we meet thick clumps of gorse and bracken, whilst in the sandier spots the purple heather and red sorrel greet us with ever-varying effect.

We may quote a writer in the *English Illustrated Magazine*, who says : ' The Kentish group of commons is redeemed by Bostal Heath from the charge of bareness and monotony, and may boast that it contributes to the circle of London commons one of the prettiest little bits to be found any- where. It would be difficult to find a more delightful example of the wild wooded common. Fortunately, it is too far from Woolwich, too hilly, and perhaps too small, to offer any temptations to drill-sergeants. It has therefore been

Main Walk to the Pines, Bostal Wood.

left in its natural condition, and most charming it is ; situate, like Plumstead, on the top of the sand-hills, its knolls are higher, and most delightful views of the marshes and river may be had from them. On the other hand, the little gorges which penetrate its sides are covered with wild verdure. Young birches wave their delicate leaves and reflect the light from their silvery stems. Purple heather mingles with bright green or yellowing bracken and dark furze ; young oaks give richness of foliage, and sandy scaurs add a touch

4—2

of orange. On the west the common is flanked by the
Scotch firs in the plantation of Sir Julian Goldsmid (Bostal
Woods), while one or two modern villas of bright red bricks
with gable roofs do no harm to the scene.'

All this and more can be said about Bostal Wood, which
is more thickly wooded than the heath. This portion is
also deeply scored with gorges, which, however, have the
advantage of being beautifully timbered with specimens of
oak, ash, birch, and chestnut, whilst the thick undergrowth
of holly gives it a verdant appearance even in the depth of
winter. The wood has several footpaths running through it,
which enable good views of its lovely valleys to be obtained.

Robert Bloomfield, the poet of Woolwich, sang of the

'Brown heaths that upward rise
And overlook the winding Thames.'

A poet might well revel in the beauties of Bostal, although
dejected with continued ill-health, as Bloomfield was.

The Woolwich group of commons seems to delight in the
possession of names the derivation of which baffles the
antiquary's skill. Bostal in this respect is similar to its
neighbour Plumstead. Bostal, or Borstal, is probably
derived from some word meaning 'woody,' which describes
its character, and is the nearest conjecture that can be
hit upon.*

Bostal Heath was one of the wastes of the Manor of
Plumstead, the property of Queen's College, Oxford, whose
rights were purchased under the Metropolitan Commons
Supplemental Act, 1877, for the sum of £5,000. The wood
was purchased under the London County Council General
Powers Act, 1891, the 62¼ acres costing £12,000, while
another 16 acres, the portion known as the Clam Field, were
added to the heath in 1894, at an additional cost of £3,350.
The Plumstead District Board contributed largely as the local
authority towards the several acquisitions.

A shade of romance has been thrown over Bostal Wood

* Vincent, ' Records of the Woolwich District,' vol. ii., p. 529.

through the tradition that the caves under its edge furnished a hiding-place for the notorious highwayman, Dick Turpin. He must have visited a good many places, if we are to believe all the traditions connected with his name. These great chalk-pits are of curious formation, and many examples exist in the locality. Some antiquaries call them 'dene holes.' A central shaft is sunk from 20 to 40 feet, and at the bottom the ground is excavated to form a chamber with a dome-like roof, from which branch off corridors terminating

The Pines, looking South, Bostal Wood.

in smaller chambers. These are always found close to the main road, and would so be extremely favourable for a highwayman, who in due time could pounce upon a passer-by and lighten him of some of his valuables. Must we bid this romance vanish by saying that these caves have in all probability been excavated for their chalk for the repair of the roads of the district? Their peculiar shape may be accounted for by the practice of 'picketing' under the surface, which miners generally follow till this day. Far more wonderful even than these caves are the galleries made for the chalk under the brickfields of Wickham Lane. Miles of these subterraneous passages have been worked, and many thousands of tons of chalk have been thus extracted.*

Bostal Wood may have decreased in area from its original size, for Hasted, the Kentish historian, shows it in his map of 1778 as extending across the valley to Wickham Lane, and up the valley to Lodge Lane and Wickham Church. This Wickham Lane must prove one of the most attractive in the locality to a thoughtful mind. We are evidently here on the site of an old Roman highway. The discovery of Roman remains within 100 yards would make this conjecture appear the more probable. To go back earlier than this, it may have been the bed of a stream into which the smaller rivulets which have carved out the combes emptied their waters. The chalk sections yield many fossils, and in the gravel-beds exist many remains of the great elk and wild oxen which peopled these woods in prehistoric times.†

Wickham Lane possesses a building known as the Old Manor-House, which would dispute with the house adjoining St. Nicholas' Church the honour of being the manor-house of Plumstead. At present it is a very decrepit structure, and serves for two poor cottages, but it may be the remnant or successor of some stately mansion. The name would not come to it by pure accident, so perhaps it is the site of what was once a building worthy of its present appellation. It is a very picturesque object, and very dear to artists, who may often be seen in the summer engaged in sketching it. It may, perhaps, be the manor-house attached to the Manor of Borstal, or Bostal, which was at the beginning of the sixteenth century the property of John Cutte. In 1504 the manor was purchased by the Abbot and Convent of Westminster, and after the convent was dissolved it was still in the possession of the Dean and Chapter. In consideration of their being discharged from the maintenance of certain students in the Universities of Oxford and Cambridge, they conveyed this manor, in 1545, to the King, who in the same

* Vincent, 'Records of the Woolwich District,' vol. ii., p. 532.　† *Ibid.*

year granted it to Joan Wilkinson. It is now the property of the Clothworkers' Company.*

This district contributed its quota to the 20,000 men of Kent who were led on by Jack Cade in his ill-fated insurrection. When the insurgents lost ground, a general pardon was proclaimed, and although Cade was put to death, many

Old Manor-House in Wickham Lane, 1886.

of his followers were reprieved. Among these seventy-four who thus received the royal clemency was ' John Crabbe, of Borstall.'†

From the high ground of the heath a good view can be obtained of all that remains of Lesness Abbey, situate on an adjoining hill. The district of Lesness or Lesnes (called

* Vincent, ' Records of the Woolwich District.'   † *Ibid.*, p. 16

Loisnes in Doomsday) was of considerable extent, and gave
its name to the hundred.   The name of Abbey Wood, by
which the district is now known, is, of course, a relic of the
time when the abbey was the chief centre of interest here.
This ancient institution was founded in 1178 by Richard de
Lucy, Lord Chief Justice of England, at the time when he
was Regent of the kingdom during the absence of Henry II.
De Lucy seems to have resided at West Wood (now known
as Abbey Wood), and was as distinguished a soldier as he
was statesman.*   Lesness Abbey took its title from the
Lessenesse, or ' little nose ' of land, now Crossness Point.

Lesness Abbey Ruins.   (Drawn by Dr. Stukely in 1750.)

Very little remains of this once flourishing abbey.   The
founder, acting in accordance with the spirit of the age in
which he lived, had built and endowed the monastery two
years before his death.   Not content with this, he is said
to have retired from active life, and become the Prior of his
own convent, thinking that the taking of the monastic vow
would aid his passage to heaven.   It is also remarkable to
note that he dedicated the church of the abbey to the Virgin
Mary and St. Thomas à Becket, though he had formerly
been excommunicated by him for ' being a favourer of his
sovereign, and a contriver of those heretical pravities, the
Constitutions of Clarendon.'   In 1801 some of the dilapidated
walls were incorporated in farm buildings, and in 1844 the

* Thorne, ' Environs of London,' vol. i., p. 1.

greater portion of the remainder were cleared away to make
way for the present farmhouse (Abbey Farm). The abbey
flourished for nearly 350 years, and was one of the first
to suffer when the monasteries were dissolved by Henry VIII.
The revenues were taken to endow Christ's College, Oxford,
founded by Cardinal Wolsey in 1525. When Wolsey fell
into disfavour, however, the revenues, amounting to about
£200 a year, were taken from the college, and granted to
William Brereton, a gentleman of the Privy Chamber. He
did not live long to enjoy them, for the fickle monarch trumped
up a false charge against him two years later, and he was

Bostal Woods from Plumstead Common.

executed. After passing through various hands, the estates
came into possession of Christ's Hospital, the present owners.

The fragments that remain are part of the original outer
walls, and an open path leads up to them. The walls of
the convent garden, the most perfect relic, still enclose a
vegetable-garden and orchard. The modern farmhouse on
the hillside facing the marsh, called Abbey Farm, is on the
site of the Abbey Grange. A pleasant country lane leads
up from Abbey Wood Railway-station to the high ground
of Lesness Heath and Bostal Heath, from which charming
views of the surrounding country can be obtained. On

Lesness Heath there are some particularly fine trees, one gigantic old yew being especially conspicuous. It was split into two parts during a strong wind in 1882, but was formerly of great girth.

To the north of the heath, on the slopes of the Bostal hills, is Suffolk Place Farm, so called from having once been the property of the Dukes of Suffolk, and possibly a ducal residence. In the Admiralty accounts mention is made that the Duke of Suffolk sent a contribution of timber from 'Plumstede in Kent' for the building of the *Great Harry*, and a State Paper of the reign of Henry VIII. informs us that the King's agent at Plumstead, Sir Edward Boughton, had a charge against 'my lord of Suffolk' in 1534 for his share in the cost of repairing the marsh wall.* The Duke of Suffolk referred to in these entries was the celebrated Charles Brandon, the favourite of Henry VIII. In childhood he had been the playmate of his future Sovereign and his sister, Princess Mary, whom he ultimately married. In 1313 he took part in a desperate conflict with a French squadron off Brest, and on his return was created Viscount Lisle. Shortly after he accompanied the King in the invasion of France, and was next rewarded by being made Duke of Suffolk. Meanwhile, Princess Mary had been married to the old French King, Louis XII., who witnessed from a coach the gallant exploits of Brandon at the tournaments. Louis died in less than three months, and his young and beautiful widow was privately married two months afterwards to her old playmate and first love, Brandon. Henry at first showed signs of disapproval; but he soon forgave them, and they were publicly married at Greenwich, the Duke receiving at the same time from the King a grant of the great estates which had formerly belonged to Edmund de la Pole, Earl of Suffolk, together with an immense dowry. The Duke readily gave his support to all the measures which led to the Reformation, and was rewarded with large grants of the forfeited monastic lands. This will probably account

* Vincent, 'Records of the Woolwich District,' vol. ii., p. 527.

for his possession of an estate at Plumstead. He sold Suffolk
Place in 1535 to Sir Martin Bowes, of Woolwich. In the
middle of the seventeenth century it was in the possession
of Sir Robert Jocelyn, by whom it was conveyed to the New
England Company, or, to give them their full title, the
Company for the Propagation of the Gospel at Boston in
New England.

On Bostal Heath.

The lane leading on to the heath from Wickham Lane
is known as Lodge Lane, and its name has been conjecturally
attributed to Goldie Leigh Lodge, a pleasant house in the
grounds east of the wood. The place is, however, described
in the book for the churchwardens of Plumstead, 1701, as
Logge's Hill, and was first called Loge Hill in 1736.
The house is named after one of the family of Sir John
Leigh, who was a large owner of property in the neighbour-

hood.* To the right of the lane is the immense field to which it was proposed to transfer the Epsom races, including the Derby and the Oaks.

PLUMSTEAD COMMON.

Plumstead Common, a fine open space of some 100 acres, is a wide-stretching, elevated plateau, broken in places by depressions in the surface. The greater portion of the common is flat and bare, but some parts are very beautiful, especially the steep fragment at the east end known as the Slade.

The origin of the name of Plumstead has baffled the skill of most antiquaries. Lysons does not attempt to give any explanation, so that we must content ourselves with conjectures. The name has not undergone any changes during the past centuries. In Doomsday Book it is written ' Plumstede'; in 1631 it occurs as ' Plumsted'; so that we are not able to get out of the difficulty by this means. Many authorities can give no better origin than the *stead*, or place, of plums. This is well known as a rich fruit-producing neighbourhood, and it is surmised therefore that the plums of its orchards are the source of the name of Plumstead. It would require a considerable stretch of imagination to suppose that the orchards of Plumstead were established at the time of the Doomsday survey. Of course an antiquary could not accept so simple a derivation, so we must mention some other solutions of the problem. A former curate of the parish suggests that the people who lived here collected from the wild-geese and herons of the marshes the *plumes* to ornament the Court beauties, and to furnish quills for the scribes of the monasteries, and so the district came to be known as *Plume-place* or *Plumestead*. One other ingenious conjecture is that the name is connected with *plump*, or clump of trees, and that the woods which crowned the hills or clothed their sides supplied the name.†

* Vincent, ' Records of the Woolwich District,' vol. ii., p. 531.
† *Ibid.*, pp. 486, 487.

The earliest record we have of this district dates back to 960, when the Manor of Plumstead was given by King Edgar to the Abbot and Convent of St. Augustine in Canterbury. The abbey was robbed of this possession by Godwin, Earl of Kent, who settled it upon his son Tostins, or Tostan. He was slain in a rebellion against his brother Harold, when Plumstead and the rest of his estates reverted to the Crown. William the Conqueror gave the manor to Odo, Bishop of Bayeux, who was persuaded, through the intercession of Archbishop Lanfranc, to restore a moiety of it to St. Augustine's, which grant of the Bishop's was confirmed by the Conqueror's charter. In 1074 he gave the other half to the monks, and the whole manor remained in their hands till Henry VIII., the first Defender of the Faith, deprived all the monasteries of their landed possessions. The King only retained the manor for a few months, for in 1539 he granted it to Sir Edward Boughton, who was agent not only for the King, but also for Wolsey and Thomas Cromwell at Plumstead. The estate continued in the Boughton family till 1685, when it was sold to John Michel of Richmond, Surrey, who by his will dated 1736 devised the Manor of Plumstead to the Provost and scholars of Queen's College, Oxford, for the maintenance of eight master fellows and four bachelor scholars, with allowances of £50 and £30 a year respectively, the surplus to be laid out in the purchase of livings.* In course of time Plumstead Common shared the fate of many other commons of the Metropolis, and many bits on its borders began to be nibbled away, and given in the early part of this century to poor widows to keep them from the workhouse. But after 1850 more serious enclosures began to be made, and the trustees of the college commenced selling large plots of the land in several places. Plumstead was at this time but a little village, and was very slow to take any action about these appropriations of its common land. At length in 1866 an action was brought by John Warrick, Mr. (afterwards Sir)

* Lysons, ' Environs of London,' 1811, vol. i., part ii., pp. 573, 574.

Julian Goldsmid, Messrs. Dawson and Jacobs, against the college, which was decided in favour of the plaintiffs in 1870. This judgment was confirmed when the college appealed, and they were ordered to remove all the fences they had erected on the common since 1866. This decision would probably have settled all disputes if a fresh claim to

Jacobs' Smithy, Plumstead Common.

the common had not been put in. This was by the War Office, who claimed the right of exercising troops on the common, by virtue of an immemorial user.* Although the military authorities had undoubtedly had this privilege, they had made very little use of it till 1870, when the outbreak of the Franco-German War roused England to defensive action.

* Preamble of Plumstead Common Act, 1878.

Woolwich Common was not large enough to accommodate all the troops for drilling, and so Plumstead began to be freely used for this purpose, with the result that its whole area became a barren desert. This once more roused the indignation of the inhabitants, and protests were sent to the War Office, and also to the Metropolitan Board of Works, urging them to formulate a scheme for the preservation of the common. It is well known that public bodies are not over hasty in their movements, and their progress did not satisfy the anxious inhabitants, who became so impatient that it needed very little for them to be roused into taking the law into their own hands. They found their champion in a John de Morgan, who by his inflammatory speeches incited the people into committing several acts of violence in the way of pulling down and destroying the fences and gates of several properties which were supposed to have formed part of the common at one time. These forcible measures could not pass by unnoticed, and the leaders were brought to trial, with the result that Morgan was sentenced to two months' imprisonment, which was, however, revoked after seventeen days, owing to the strenuous efforts of his friends.* In the meantime the long-desired object had been attained by the Plumstead Common Act, 1878, which placed the common under the jurisdiction of the late Metropolitan Board of Works, whose duties are now discharged by the London County Council, the victories of peace being in this case at any rate greater than those of war. Under the provisions of this Act the Provost and scholars of Queen's College, Oxford, received £10,000 for their manorial rights, £6,000 being found by the late Board, and the remainder by the War Office. Of the 100 acres thus preserved from encroachment for public use, the War Office have the power of drilling over 77. In 1884 an addition was made to the common by the purchase of a small plot rejoicing in the uninviting title of Sots' Hole. This was formerly a hollow place on the northern side of the highway, which had been

---

* Vincent, ' Records of the Woolwich District,' vol. ii., pp. 588, 590.

left when the road was raised. It was afterwards used as a dust-shoot until it became level ground. There were a couple of old cottages standing in it, one of which had been a tavern, and was said to have given to the place its ugly name.* A very peculiar circumstance happened here in 1858. A man had been collecting road-scrapings with a horse and cart on the common opposite these two cottages, which were then some 12 feet below the level of the roadway, and he attempted to back the cart into the channel. The cart, which was loaded, crossed the footpath (an inclined plane) and after breaking through the hedge, fell over and went through the front-wall of the old building, horse, cart, and contents alighting in the family parlour. Three children who were sitting by the fire escaped almost by a miracle, but it took two hours to extricate the horse.†

The late Board, in their Various Powers Act, 1885, obtained sanction to carry out some important alterations with regard to Plumstead Common. The first of these was an exchange with the London School Board, by which a portion of the common was given to them for the erection of schools in return for some of their land, which was added to the common. The second was the abandonment of the gravel-digging, which was claimed as a right by the Woolwich Local Board of Health. The rights of the Local Board were extinguished by the payment of £500. Another exchange was carried out under Parliamentary powers in 1891, which enabled the Provost and Fellows of Queen's College, Oxford, to develop a building estate adjoining Old Mill Road. Subsequent additions have been made by the acquisition of Jacobs' Sand-pit, and the purchase of a piece of land at the corner of Purrett Road from the British Land Company.

Plumstead Common and its surrounding neighbourhood are rich in historical associations. Excavations which have been made from time to time tend to prove that this was formerly a Roman settlement. The unearthing in 1887 of

* Vincent, ' Records of the Woolwich District,' vol. ii., p. 583.
† *Kentish Independent,* February 13, 1858.

a leaden coffin and skeleton and other remains at Wickham Lane, make it probable that this district was either the site of a Roman cemetery, or of a villa of some wealthy Roman. The leaden coffin, which was in an excellent state of preservation, had upon the lid a border of blue beadwork, and has been assigned by eminent authorities to some date between A.D. 200 and 400. The coffin is now in the Maidstone Museum, while the ancient bones were interred in the churchyard.*

An ancient mound or barrow in the centre of the eastern division of the common, beyond the Slade, is a relic of another class of burials. The nature of its loamy soil, quite unlike the sand and gravel of the common, points to the improbability of its being a natural hillock. The material was perhaps brought from a distance to build the earthen tumulus of some great chieftain. There are depressions across it north to south and east to west, which may either mark the lines of excavations, or show where the cross-passages of the mausoleum have fallen in. This barrow has been put to practical uses in times past as a butt for artillery practice. A map at the Royal Artillery Institution, drawn about 1740, represents a party of gentlemen cadets at mortar practice on Plumstead Common using the mound as a butt.†

The part between the common and the river Thames at some time or other was under water at every high-tide. It is a question of much doubt at what period the river was embanked. One theory would place it as early as 400 A.D., saying that the great embankment from London to the Medway was all constructed by the Romans. But as no mention of this embankment is made in Doomsday Book, it would certainly appear likely that it was made after 1086. Even after this the marshes were often flooded, and it is probable that the monks of Lesness Abbey more than once reclaimed the land from the river. Lambarde tells us that in 1279 'the Abbat and Covent of *Lyesnes* inclosed a great

---

* Vincent, 'Records of the Woolwich District,' vol. i., pp. 11-13.
† *Ibid.*, pp. 14, 15.

part of their marshe in Plumsted, and within 12 yeeres after they inned the rest also to their great benefite.'* Plumstead Church must have stood on the very beach. It is dedicated to St. Nicholas, the patron saint of fishermen, and it may have been from the high ground here that the early fishers took to their boats. The chalk spur which descends from the common, and serves as a solid foundation for the church, would give some colour to this conjecture. The projecting 'rock' on the common, too, is supposed by some to be a

View of Plumstead Marshes from the Common, 1851.

portion of a headland which existed when the river was not embanked. The fishermen's cottages were built along the strand, and the large field known as the Strand Field, which slopes from the highroad to the marshland, was probably the site of the fisher village. The adjacent High Street was formerly known as Strand Place. In this Strand Field old tiles and other remnants of buildings, together with many coins of early date, are often found.†

* 'Perambulation of Kent,' 1596, p. 440.
† Vincent, 'Records of the Woolwich District,' vol. ii., pp. 490-492.

A workhouse once stood on the common near the Slade, on the site of Winn's Cottages. It was only a small affair, and its successor situate at Cage Lane, where Agnes Place now stands, was partly built of the old materials. In connection with this workhouse, we may just quote an entry from the accounts of the churchwardens and overseers of Plumstead for 1701. It relates to the convenient practice of buying a husband for a workhouse inmate. It runs as follows :

	£	s.	d.
Wedding ring for Mary Tatterson	0	6	0
Marriage fees do. do.	0	11	0
Paid her husband ·	5	0	0
Do. for the wedding dinner	1	1	0
	£6	18	0

We hope he never had any occasion to regret this bargain, and that they lived happily ever afterwards.*

The old mill on the common, which added such picturesqueness to the view, was the scene of a startling occurrence, in 1827, on the occasion of a sham fight on the common. A number of persons were gathered on the staging round the mill to see the fight, when it suddenly gave way, and many of the spectators were seriously injured. The ever-advancing invasion of the suburbs by the workers of the Metropolis made a mill out of date, and it was allowed to fall into decay, till, about 1848, it was converted into a public-house.†

The Vicarage of St. Margaret's Church, situated at the north-west corner of the common, is one of the best remaining houses of Old Plumstead. The mansion is now known as Bramblebury House, although formerly called Bramblebriars. In the grounds is a cypress-tree, said to be 150 years old. This was planted by a relation of Captain Dickinson, superintendent of Ordnance shipping, who resided here in 1811. After his death his widow, Lady Dickinson, lived here for a

* Vincent, ' Records of the Woolwich District,' vol. ii., p. 524.
† *Ibid.*, p. 533.

number of years.   The whole of the district called Vicarage
Park was formerly included in its grounds.*

Adjacent to St. Nicholas' Church is Plumstead Manor-
House, known as Court Lodge or Manor-House; but this
house must once have been the vicarage of the parish.
Beneath the house are extensive vaults, now used as cellars,
containing carved arches and doorways of stone.   It is sur-

The Old Mill, Plumstead Common, in 1820.

mised that the church, or some monastic building, must have
covered the spot at some remote period, as these vaults could
never have been designed for mere cellars.

Several of the roads in the vicinity of the common have
names which are relics of agricultural days.   Timbercroft
Lane, Swingate Lane, and Plum Lane may be taken as
examples.   We have mentioned before, in connection with

* Vincent, 'Records of the Woolwich District,' vol. ii., p. 550.

the derivation of the word ' Plumstead,' that this was a rich fruit-producing neighbourhood. Cherries were first acclimatized in this part of England; and pippins, too, were first grown here, when they were brought over the sea in the sixteenth century.* A youth of Plumstead, aged ten, who is buried in the churchyard of St. Nicholas, fell a victim to his fondness for his native fruit. He was killed in a tree whilst taking cherries, by the owner of the orchard. The stone was restored in 1870 by public subscription, upon which is inscribed this peculiar epitaph :

> ' The hammer of death was given to me
> For eating the cherris off the tree.'†

In addition to these records of former industries, other names remain to remind us of bygone times. Cage Lane tells of the old cage, or lock-up, which stood at the lower end. The stocks were on the western side of the cage, and the parish well was in the rear. Plumstead not being sufficiently rural to retain its stocks, they have disappeared, together with the cage, and only the name survives. Burrage Road and Burrage Town, or West Plumstead, take us back to the early days of Edward III. and the Black Prince. The name should be ' Burghesh,' after a Norman nobleman, Lord Bartholomew de Burghesh, who had his seat at Plumstead. His son who inherited the property changed his name to Lord Burwash, and the family pedigree goes to show that the poet Chaucer was among his descendants. From Burwash the name was gradually softened to Burrage. The original name of Burghesh is retained as the second title of the Earl of Westmoreland.‡

Another name which has changed is Skittles Lane. This was in 1849 Kiddel's Lane, from Mr. Kiddel, who owned a small farm on it. Both forms have now disappeared, for at the request of the inhabitants it has been renamed Riverdale Road.

* Vincent, ' Records of the Woolwich District,' vol. ii., p. 486.
† Thorne, ' Environs of London,' vol. ii., p. 472.
‡ Vincent, ' Records of the Woolwich District,' vol. ii., pp. 543, 544.

Before we leave the common we may just call attention to the sandpits which formerly existed at the eastern end, which have now been filled up and covered with furze. These local sandpits furnish a peculiar loam, which is in much request for metal casting; but they were probably worked in ancient times for the valuable silver sand they contained.

Included in the purchase of Plumstead Common was a small open space of some 5 acres in extent, called Shoulder of Mutton Green. Its distinctive shape accounts for this peculiar appellation. As a matter of fact it is outside the county of London, although in the same manor as Plumstead Common, and is the village green of Wickham parish. This rural spot has seen nothing out of the common to distinguish it from any other quiet village green. At the time when Plumstead Common was threatened, an attempt was made to enclose this green, but the railings were promptly pulled up by the inhabitants. In years to come this may become a busy spot, but it is to be hoped that it will retain its rural quietness for many ages.

CHAPTER IV.

BROCKWELL PARK—DULWICH PARK.

BROCKWELL PARK.

THIS comparatively new park, 84 acres in extent, has already established itself in the esteem of South Londoners. No doubt its favourable situation, close to an important railway junction, has something to do with this, for although its nearest neighbour, Dulwich Park, may appear to some more attractive, it has not the advantage of ready access which Brockwell possesses, and so is not so well known. The beauty of Brockwell Park consists in its wildness, if such an expression can be used of a well-kept park. In other words, one does not come here to see gay flower-beds, stately palms, and all the other attendant advantages of 'laid-out' gardens, but to admire the beauties of Nature unadorned : long stretches of undulating lawn, dotted here and there with fine specimen forest trees. When it was bought for the people of London, it was already a park—not a park site. From different points in the grounds there are several ' little bits' that an artist might be delighted to paint just as he finds them. Not only is the park a beautiful one, but its surroundings are, at least for the present, equally charming. Ruskin, who resided for some time at 30, Herne Hill, has described the beauties of Croxted Lane in his ' Præterita.'

The main entrance to the park is almost opposite the gates of Herne Hill Station, which, by the way, is not on a hill at all, but in a depression between Herne Hill and Brockwell

Park. From the three main roads which skirt the park—
Norwood Road, Dulwich Road, and Trinity Road—the
ground gently slopes upward to a mansion which crowns
the hill. Leaving the mansion, there is a depression, and
then the park slopes up again to Tulse Hill. This is probably
the only spot within four miles of Bow Bells where 'the
building rooks still caw'; but to judge by the nests, there
are yet in this park two rookeries of the right sort. The best
views are obtained from the hill on which the mansion stands.
From the side towards London, the Victoria Tower of the
Houses of Parliament stands out conspicuously, and on a
fine day the view extends as far as the hills of Harrow,
Highgate, and Hampstead, while the intervening distance is
dotted with graceful church spires. On the other side is the
ridge of the Sydenham and Norwood hills, crowned with
the gigantic Crystal Palace, flashing back the sunlight from
its thousands of panes. The bare, lumpy, clay height of
Thurlow Hill in the mid-distance only sets off to greater
advantage the wooded slopes beyond. In its original form
the new park only wanted a little water to give it life and
variety, and this has been introduced in the form of a large
ornamental lake with two or three smaller pools artistically
arranged with rustic bridges and waterfalls between them.
The large lake is available for bathing at certain hours, and
is of such depth as to admit of a good header from a stage
projecting out from the bank. The necessary shelters have
been erected at one side, and the numerous seats around the
lake are extensively patronized when bathing is not in progress. Brockwell Park is a regular home of sports, large
areas being reserved for cricket, football, and lawn-tennis.
There is a bandstand, constructed of rustic materials to
harmonize with the surroundings, the upper part of which
forms a home for pigeons, who, we hope, are lovers of music.
The amusements of the children are also looked after, and a
gymnasium is provided for them, where they may romp and
swing to their hearts' content.

And now the last feature to be described in connection

with the park is one which is peculiar to itself—viz., the old garden. When the park was taken over, this was used as the kitchen-garden. It is walled in on all four sides, but the walls are covered with roses and fruit-trees, so that there is no bareness to offend the eye. Inside the walls, we find ourselves within a garden laid out in the old-fashioned formal

In the Old Garden, Brockwell Park.

geometrical style, and it is a quaint and pleasing specimen of the kind found at many stately old castles and halls. As this is the only park which boasts of such a garden, a few words on the formal style of gardening adopted here may not be out of place. The difference between the 'formal' and the 'natural' styles is practically this:

'The formal school insists upon design; the house and

the grounds should be designed together, and in relation to each other. No attempt should be made to conceal the design of the garden, there being no reason for doing so ; but the bounding lines, whether it is the garden wall or the lines of paths and parterres, should be shown frankly and unreservedly. . . . The landscape gardener, on the other hand, turns his back upon architecture at the earliest opportunity, and devotes his energies to making the garden suggest natural scenery, to giving a false impression as to its size by sedulously concealing all boundary lines, and to modifying the scenery beyond the garden itself by planting or cutting down trees, as may be necessary to what he calls his picture.'*

Whatever faults may be found with the ' formal' style of gardening, this small example in Brockwell Park is a great favourite with the general public. The paths, edged with box, are rigidly straight, leading to circular flower-beds, and one can well imagine that the whole area was designed with a ruler and pair of compasses. The planting has been confined to roses and old English garden plants, all flourishing in their natural wildness. In the centre of the garden a small fountain adds life to the scene, and in the corner a well which was here formerly has been capped with the old-fashioned rope and bucket winding-gear as appropriate to the rustic simplicity of the place. Round the walks at intervals rustic shelters are placed which have been quite covered with creepers, and on one of the walls is an old sundial, which only needs an inscription to be complete. The impression left on the mind by a visit to this little corner of the park will be quite as pleasing as the most elaborate carpet-bedding to which it forms so striking a contrast.

Brockwell Park is a lasting memorial to the energy and enterprise of South London. It was a cause of no little rivalry between certain sections of the community. Some few years ago Raleigh House, a historical mansion situated in the Brixton Road, was in the market together with its grounds, amounting in all to about 10 acres. Part of this

* Blomfield and Thomas, ' The Formal Garden in England,' pp. 10, 11.

property could not be built upon, owing to the provisions of the Rush Common Act. A movement was at once set on foot to purchase the grounds for a public park. A subscription-list was opened, contributions were promised by local

A Band Performance in Brockwell Park.

bodies, and the Raleigh Park Act, 1888, was passed, one of the clauses of which empowered the late Metropolitan Board of Works to contribute £1,000 per acre towards the cost. While these preparations were in full swing, attention was called to a more favourable site, namely Brockwell Park, some 78 acres of which were for sale at a cost of £1,500 per acre, as compared with £4,000 for Raleigh Park. After lengthy negotiations extending over some years, the details of which cannot be given here, the scheme for obtaining Raleigh Park was abandoned, the Act was repealed, and the beautiful grounds of Brockwell Park purchased at a cost of nearly £120,000, towards which the London County Council gave £61,000, the Charity Commissioners £25,000, Lambeth Vestry £20,000, Camberwell Vestry £6,000, Newington Vestry £5,000, the Ecclesiastical Commissioners £500, to which were added many private subscriptions.

It was a great event in South London when on Whit-Monday, June 6, 1892, the Earl of Rosebery, Chairman of the Council, formally opened the park to the public. The proceedings were, however, marred by one painful incident. Soon after the termination of Lord Rosebery's speech, Mr. Bristowe, M.P. for Norwood, who had taken so active a part in the acquisition of the park, was suddenly seized with a fit, and in spite of the efforts of many medical men who were present, he never rallied and expired about noon from heart disease. He was a man of such quiet and unassuming manners that he never made any reference to the work he did in connection with the park, but it is right to mention that when there was a great degree of uncertainty as to whether the park would be purchased at all, he himself guaranteed a sum of £60,000 in order to secure it. The first object visible on entering the park from Herne Hill is a memorial to him in the shape of a drinking-fountain. It consists of a bust, over life-size, surmounted on a high pedestal with ornate capital. On the front of the pedestal stands in full relief a life-size figure of Perseverance presenting a branch of laurel ; in the base under the statue is a bronze panel, gilt, of children at play. The entire memorial, which is the work of Messrs. Farmer and Brindley, stands about 16 feet high, and is constructed throughout of white mountain limestone. The inscription is as follows :

THOMAS LYNN BRISTOWE,

M.P. FOR NORWOOD

1885—1892.

READY TO EVERY GOOD WORK,
HE LED THE MOVEMENT FOR THE
ACQUISITION OF THESE BROAD
ACRES AS A PUBLIC PARK, WITH
GREAT TACT AND ENERGY, AND
DIED SUDDENLY IN THE VERY
MOMENT OF HIS UNSELFISH TRIUMPH
AT THE OPENING OF THE PARK ON
WHIT MONDAY, 1892.
HIS FRIENDS AND NEIGHBOURS,
OF EVERY SHADE OF POLITICAL
OPINION, DEDICATE THIS FOUNTAIN
TO HIS MEMORY
1893.

Some four years after Brockwell Park was opened to the public, a much-needed improvement was effected which much increased the value of the park to the residents of Brixton. This was the formation of a new entrance across a narrow neck of land between Arlingford Road and the bathing lake. After some lengthy negotiations, this strip about 3½ acres in extent was purchased together with two other small plots for the sum of £6,000 and £200 costs. The western boundary of the land had some very fine old trees, and there were also two small ponds which were afterwards enlarged and formed into ornamental lakes connected by a rocky channel, and embellished with cascades and marginal planting. These waters on the new ground are connected with the bathing lake and the other pools by a miniature waterfall, and the whole forms a chain of lakes at varying levels, with a very pleasing effect. The water leaves the last of the lakes by a small stream which winds throughout the remainder of the land. The new entrance, which was laid out at a cost of about £1,800, was opened on March 14, 1896.

Brockwell Park must rest for its attraction on the beauty of its landscape, and not upon any wealth of historical asso-

ciations.   Many years ago the river Effra flowed past what
is now the boundary of the park.   The house of the proprietor of the rustic timber-works facing the park is built
over this stream.   Its course can be traced by means of old
maps.   Rising in the high ground of Norwood, it ran down
Croxted Lane, receiving an affluent from the east by the
Half-Moon Inn.   From the Half-Moon, skirting Brockwell

The Cascade, Brockwell Park.

Park, it ran along Water Lane, down to the Brixton Road,
and eventually made its way into the Thames at Vauxhall
Creek.*

If it is true, as some say, that Queen Elizabeth sailed up
the Effra to a point beyond Brockwell Park, she must necessarily have gazed upon its verdant slopes.   The Effra, having

* *Notes and Queries*, No. 198, p. 282.

diminished to a brook, is now used as a sewer, and has therefore practically disappeared.

Some 300 years ago there stood in Brockwell Park, close to what is now Norwood Road, an old manor-house, in which dwelt Count Lilly. His portrait hangs to this day in Brockwell House. From this time to the beginning of the present century we must draw a veil. Then we find the mansion and surrounding lands in the possession of a Mr. Ogbourne. He sold the property in 1809 to John Blades, who was Sheriff of the City of London in 1812. It was this owner who pulled down the old house and caused the present mansion to be erected from the designs of Mr. D. Reddell Roper, of Great Stamford Street.* The new site is infinitely superior to the old one, commanding as it does such a variety of views from its elevated position. Upon the death of Mr. Blades in 1829, the park came into the hands of his son-in-law Mr. Joshua Blackburn, from whom it descended to the late owner, Mr. Joshua John Blades Blackburn. The mansion has not much pretension to architectural beauty, but is one of those plain and solid structures which are characteristic of the time it was erected. Part of it is now used as a residence for some of the park officials, whilst a large room on the ground-floor serves for a refreshment-house. The walls have recently been embellished by fine mural paintings representing typical English scenes. These were the work and gift of Mr. J. St. Loe Strachey and his brother.

There is a story current which has been put forward by some as an explanation of the name which the park bears. It is said that the property was once called 'Badger's Well,' from being the home of the badger in old time, and that as 'brock' is the old Saxon word for 'badger,' Brockwell was probably substituted for brevity afterwards. In the absence of any other derivation we must accept this as the probable origin of the name.

These few details, then, represent the whole of the history of Brockwell Park. A place of such charm and public interest deserves to have more romance connected with it.

* Brayley, 'History of Surrey,' vol. iii., p. 379.

DULWICH PARK.

Dulwich Park is pleasantly situated in the shallow valley which stretches from Dulwich village towards Lordship Lane Station. The principal entrance is facing the old college chapel, hard by the famous picture-gallery, and taking our stand here we obtain an excellent view across the park to the surrounding hills, dotted with graceful church spires.

College Gate and Superintendent's Lodge, Dulwich Park.

The Crystal Palace is within sight, only a mile distant, while closer at hand is Dulwich College School, hidden from view by the belts of giant trees which give the neighbourhood so rural an appearance. Another good view of the park is obtained from the high ground of Lordship Lane, from which point of advantage the ground is seen spread out as a map at one's feet. Dulwich is certainly a delightful suburb; all the year round the thrush charms us with her song, and the fine old houses with their grounds cluster round the park with a repetition of its beauties on a smaller scale. It is particularly free from the loafing population, which lolls upon the grass in St. James's Park thick as wind-falls in an orchard. Though open to all, it is specially frequented by a superior class of visitors from the immediate neighbourhood; but now that the park is becoming better known, the circle has been extended, and on Sundays especially, Dulwich is blocked with vehicles and omnibuses plying from various parts.

The park was a free gift, so far as the land is concerned, on the part of the governors of Dulwich College. This gift required the confirmation of Parliament, which assent was obtained in May, 1885. Among other provisions of the Acts governing the transfer, it is enacted that no music or public meetings are to be allowed in the park; although the former restriction is of doubtful advantage, the result is certainly attained of keeping out the noisy element.

The park comprises 72 acres, and has been valued at £1,000 an acre, so that it is apparent that the gift was no inconsiderable one. The land, which consisted of a series of meadows covered with old oaks of good size, conveniently undulating in surface, adapted itself very easily to laying-out. Near the centre of the park is a lake covering 3 acres, which, although not deep enough for boating or swimming, quite suffices for the dignity of several swans which ride proudly on its surface, and the broods of ducks which make it their home. The water quits the lake by means of a miniature waterfall, below which a tiny rivulet winds for some distance through a well-kept lawn to its exit from the park. A rustic bridge is thrown across this, whilst the carriage-road traverses it on a stone bridge, which bears on its parapet the carved arms of the late Metropolitan Board of Works. Large areas are set apart for cricket and tennis, which are crammed to their fullest extent on Saturday afternoons. The American garden with its rhododendrons, azaleas and roses, presents gorgeous masses of bloom in season; but the feature of the park from a horticultural point of view is the rockwork

6

planted with showy Alpine and rock plants, which are
acknowledged to be second only to those of Kew. The
banks of bloom which they furnish in the early spring are
certainly worth a visit. The most extensive gardening of
this kind is on either side of the carriage-road, passing
through the Snake's Lane and Court Lane entrances. The
broad strips on each side of the roads have been planted in
a semi-wild style. Appearing alternately with the rockwork
are little dells, bright with daisies, primroses, polyanthus,

View of Lake, Dulwich Park.

crocuses and snowdrops in the spring, whilst stumps of
trees covered with ivies and creepers add a picturesque
wildness to the scene. Such are some of the beauties of a
park where everything is bright and gay. The buildings
which are very complete comprise two entrance lodges, a
large refreshment-house, lavatories, rustic shelters, a cricket
pavilion, and an aviary well stocked with British and other
birds. Altogether some £40,000 has been spent in laying
out and planting the park, providing fences and lodges, and
the thousand and one things which complete the public

enjoyment. Although the negotiations for the transfer of the land were carried on by the late Metropolitan Board of Works, who also executed the greater portion of the work of laying out, the duty of opening fell upon Lord Rosebery, the first Chairman of the London County Council, the ceremony taking place in June, 1890.

Indirectly, Dulwich Park forms another memorial to the generosity of Edward Alleyn, the friend and comrade of Shakespeare, who in 1606 purchased the Manor of Dulwich, and devoted it to the foundation of a college for the maintenance of twelve poor men and women, and the education and support of as many children, with a master, a warden, and four fellows. This was the beginning of what he called ' God's gift,' which has developed now to such gigantic proportions. The rich estates are administered by the Dulwich College Estate governors. Alleyn was born in 1566, and became an actor of no mean attainments. He took a prominent part in several of Shakespeare's plays the first time they were acted, and rose to possess considerable shares in the leading theatres of his time. He died in November, 1626, and was buried in the chapel of the college.

The Manor of Dulwich was part of the possessions of the rich monks of Bermondsey, to whom it was given by Henry I. in the year 1127. No mention of the manor is made in Doomsday Book, from which we may infer that at that time it was an insignificant village. Indeed, so late as the reign of Charles II. the number of persons who were assessed to the hearth-tax did not reach forty. The monks of Bermondsey retained Dulwich in their hands till their monastery was suppressed and their lands confiscated in 1537-8. Henry VIII. did not hold the manor for very long, for he granted it in 1545, together with the manor-house known as ' The Hall Place,' to Thomas Calton, to be held at the annual rent of £1 13s. 9d. The grant also included the advowson of the vicarage of Camberwell, which had been granted to Bermondsey Abbey by Robert, Earl of Gloucester, son of Henry I. Thomas Calton's grandson, Sir Francis Calton, sold the

manor to Edward Alleyn for the sum of £5,000, in addition
to an amount of 800 marks (£533 6s. 8d.) for the patronage.*
In referring to the transaction, Alleyn says he paid for the
manor 'one thousand pounds more than any other man
would have given for it.' As we have seen, the lordship
of the manor is now vested in the Dulwich College Estate
governors.

The derivation of the name Dulwich is involved in some
obscurity. In various deeds the word appears in the following forms: Dylways, Dilwisshe, Dilewistre, Dullag. The last
part *wick* occurs both in Anglo-Saxon and Norman names.
In the former it means a station on land, a house or village,
whilst with the Normans it was a station for ships, a creek
or bay. The way out of the difficulty, and a most reasonable
one, too, is to suppose that Dul was the name of a river, and
that the *wick* was the station or village situate on its banks.†
Other writers connect the word with Delawyk. In the reign
of Henry III., Henry de la Wyk, called also Henry de Dile-
wisse, accounted for two knights' fees in Camberwell.
Delawyk is more like Dulwich in appearance than pro-
nunciation; but Allport says 'the transition from Delawyk
to Dulwich appears to be so easy and natural as at once to
settle this etymology.'‡

The greater part of Dulwich was in ancient times an
immense wood intersected with devious paths. The present
road leading from Dulwich village to the Crystal Palace is
marked on Rocque's map (1745) as ending abruptly in a field
just before the entrance to the wood. What little remains of
the wood is practically closed to everyone, but its memory is
still preserved in the names of Dulwich Wood Park, Kings-
wood Road, and Crescent Wood Road. The Court in the
time of Charles I. paid frequent visits to Dulwich and to its
Woods, which afforded excellent sport. A royal warrant was
issued to one of the yeomen huntsmen-in-ordinary, Anthony
Holland, to command the inhabitants of Dulwich ' that they

---

* Blanch, 'Camerwell,' pp. 377, 378.   † *Ibid.*, p. 377.
‡ Allport, 'Camerwell,' p. 46.

forbeare to hunt, chace, molest, or hurt the king's stagges
with greyhounds, hounds, gunnes, or any other means what-
soever '; and he was also empowered ' to take from any person
or persons offending therein their dogges, hounds, gunnes,
crossbowes, or other engynes.'*

Another open space at Dulwich which has disappeared is
Dulwich Common. This formerly stretched along the College
Road, commencing close by the present principal entrance to
the park, and nearly touching Dulwich Wood. Dulwich
College School and grounds occupy most of its site. The
vestry minutes of December 27, 1804, contain the following
interesting record relating to the common :

' The committee reported that they had made diligent
search and inquiry, and from good information find that it
has been private property more than 300 years, and therefore
the committee are of opinion that the parish have no right
whatever to Dulwich Common.'†

By an Act passed in the following year, the college authorities
were empowered to enclose the common which, it is said, consisted of 130 acres. And so now that both Dulwich Wood
and Dulwich Common have gone, the importance of securing
Dulwich Park as a recreation-ground is all the more apparent.

The principal entrance to the park faces the chapel of
Dulwich College, which also serves as the parish church
of Dulwich. The old college, of which the chapel forms
part, cannot lay much claim to architectural beauty. It is
attributed by some to the celebrated Inigo Jones, but it is
very unlikely that he would turn out so poor a design, especially
when we find that the tower fell down in 1638. Moreover,
the original specification is still preserved giving particulars
of payments made to the real architect John Benson, of
Westminster. It forms three sides of a quadrangle : the
entrance and gates (upon which are the founder's arms and
motto, ' God's Gift ') closing in the fourth side. There is an
amusing story related by Aubrey, that Alleyn was frightened
into making this charitable and generous bequest by an
apparition of his satanic majesty among six theatrical demons

* Quoted in Blanch's ' Camerwell,' pp. 375, 376.   † *Ibid.*, p. 379.

in a certain piece he was playing.   In his terror he hastily
made a vow, which he redeemed by founding the College of
God's Gift.   An old writer* declares that 'no hospital is tyed
with better or stricter laws, that it may not sagg (swerve)
from the intention of the founder.'   The 123 orders deal
with the most minute details, even specifying 'that none of

Court Lane, Dulwich, in Winter.

the fellowes, poore brethren, or sisters, shall keepe any Doggs,
poultry, or any other noisome cattel, within the said college,
besides a cat.'   Some of these lengthy lists of ordinances are
rather interesting when read in the light of to-day.†   For
instance, the dietary of the boys, superintended by a 'surveyor

* Fuller.

† These are printed in full in an appendix to Blanch's 'Camerwell.
from which these extracts are taken.

of diett,' stipulates that they are to have 'a cup of beere' at breakfast, 'beere without stint' at dinner, with various added luxuries on high-days and holidays. At the same time it must not be supposed that the founder wished to encourage intemperance. It is provided that the pensioners 'shall not frequente any tavernes, or ale-houses, and if any of them be drunk and convicted thereof . . . then he or she so offending shall forfeyt for the 1st, 2nd, or 3rd offence, three daies pension for each of those times, for the fourth offence shall

Dulwich College.

be set in the stocks, in the outer court of the said college, by the space of one houre and also loose three daies pension.' By an Act of Parliament passed in 1857, the foundation was completely reconstituted, and the revenue is now divided into four parts, three of which are devoted to educational and the fourth to charitable purposes purely. One of the outcomes of this change was the foundation of the splendid pile of buildings popularly known as Dulwich College, costing about £100,000, the greater portion of which was received as compensation money from the railways. They form three distinct blocks, the architecture being of the Northern Italian style of the thirteenth century, after the designs of Mr. Charles Barry. The first stone was laid in June, 1866, and the completed building was opened by the Prince of Wales in June, 1870. Judging from the high place taken by Dulwich College as an educational establishment, the diversion of the funds from the original intentions of the founder will cause but little regret. The original college buildings have passed through some troublous times. During the Civil Wars, the fellows took up arms for the King, and as a consequence their fellowships were sequestered, and only a schoolmaster and usher were appointed. These two presented a petition to the ruling powers for a double allowance for diet, on the plea that they occupied the place of four fellows. Their petition was refused at first, but afterwards granted, as being in accordance with the terms of the trust. In 1647, a company of soldiers under Captain Atkinson, forming part of General Fairfax's army, which was then at Putney and Fulham, was quartered in the college. During their stay here they committed terrible havoc. It is said that they destroyed the organ, and took up the leaden coffins of the chapel to be melted into bullets.* A neighbouring mansion Belair contains some very curious specimens of the pollard oak, which tradition says were so cut by these soldiers whilst quartered in the neighbourhood. The college received 19s. 8d., which was very small recompense for the damage sustained.

Leaving now the old college, we pass round the park, and are so shut in by the walls of trees as to quite forget the outside world. Dulwich Park was not unknown to history before, for its meadows, so rural and so quiet, seem to have been especially suitable for the settlement of *affaires d'honneur*. In 'Captain Blake,' published by Bentley, 1838, is the following:

'"Now I prefer for the Surrey side, and there is not a prettier shooting - ground in Britain than the Dulwich

* Allport's ' Camerwell.'

meadows. I think I could mark off as sweet a sod there as ever a gentleman was stretched upon."

' " You are truly considerate, Colonel. But where shall our rendezvous be ?"

' " Oh, the Greyhound. Capital house that ! Civil people, excellent wine, and if a man's nicked, the greatest attention." '

The Greyhound still stands, although about to be pulled down, and must be at least 150 years old. It was famous as the meeting-place of the Dulwich Club, which has entertained at its table many distinguished men, including Dr. Glennie, Campbell the poet, Dr. Babington, and others; and Dickens, Thackeray, Mark Lemon among literary celebrities were frequent visitors to the house. The presence of Dickens will account for the fact that Mr. Pickwick is described as finding a quiet retreat at Dulwich in his old age, where he had ' a large garden situated in one of the most pleasant spots near London.' We find him ' visiting frequently the Dulwich picture-gallery, and enjoying walks about the pleasant neighbourhood.'

The two other entrances to the park are from Lordship Lane and Court Lane. The former takes its name from the lordship of Friern Manor, of which it is the boundary; Court Lane is so called after Dulwich Court, now turned into the Court Farm. Another house in this lane was formerly occupied by a school, and was afterwards taken as a summer residence by the Turkish Ambassador. Mr. Batt's school was much patronized by the nobility, several of whom received their education here. Dulwich then in times past has, as we shall afterwards show, taken no small place in educating the youth of England, and the present Dulwich College School will be more than able to sustain the past reputation.

Dulwich Court is mentioned in many ancient documents. At the commencement of the seventeenth century it was in the possession of the Calton family. ' Dulwich Corte Hall Place ' and three other messuages in Dulwich were mortgaged by Sir Francis Calton to Robert Lee, Lord Mayor of London on December 17, 1602, for £660. Alleyn paid off the mortgage in 1605, and acquired full possession of the property shortly afterwards. The house is shown on a plan as late as 1808, but whether it has been merged in Court Farm, or whether, as its position on the map would seem to indicate, it was a house nearer to Dulwich village, does not seem to be known. In this latter case the old house must have been demolished.*

It is a much-disputed point as to whether Alleyn ever lived at Dulwich Court. Manning and Bray assert that he resided ' either at Hall Place or what is now Dulwich Court.' Tradition has always averred that the manor-house was Alleyn's residence, but there is a lease still extant from Francis Calton to John Bone, of Camberwell, of Hall Place at a rent of £20 for twenty-one years, dated May 12, 1597, and Alleyn let the same house to William Lawton.†

At the corner of the triangle formed by the eastern boundary of the park, Lordship Lane and Dulwich Common Road, is situated Bew's Corner, which has played an important part in the history of Dulwich. On this site was formerly a tavern of some note, called ' The Green Man,' the green man in question probably being Robin Hood, who is always represented with his merry men dressed in suits of Lincoln green. It is figured on Rocque's map of 1745, and on May 19, 1752, we find that it was resolved ' that the vestry (Camberwell) be adjourned to Mr. Cox's, at the Green Man at Dulwich, in order to make out the rate-books.' Perhaps it is possible under these conditions to make the subject of rates an interesting one. A few years prior to this the Green Man had become sufficiently noted to find a place in a popular ballad in conjunction with Vauxhall and other well-known places of entertainment :

> ' That Vauxhall and Ruckhalt, and Ranelagh too,
> And Hoxton and Sadlers, both old and new,
> My Lord Cobham's Head, and the Dulwich Green Man,
> May make as much pleasure as ever they can.' ‡

---

* A. M. Galer, ' Norwood and Dulwich : Past and Present,' 1890, p. 56.
† *Ibid.*, p. 58.   ‡ ' Musick in Good Time : a new Ballad,' 1745.

One very amusing story is told in connection with this tavern in the ' Percy Anecdotes' which is quite worth repeating. A well-known literary man had received an invitation from a friend to dine with him on the following Sunday, the house being described as opposite the Green Man at Dulwich. Our literary friend trusted to his memory for the address, which unfortunately failed him when Sunday came round. At last a happy thought struck him. ' I have it,' he exclaimed excitedly, ' it's opposite the Dull-man at Greenwich !' and so to Greenwich he went in post haste. All effort to find the Dull-man proved fruitless, and at last he was asked if he didn't mean the Green Man at Dulwich, when the truth became apparent to him that he had lost his way. He lost his dinner too, and we much regret to say his temper also.*

Within the grounds of the Green Man was situate one of the famous Dulwich wells, of which the following account is given :

' In the autumn of 1739 Mr. Cox, master of the Green Man, about a mile south of the village of Dulwich, having occasion to sink a well for his family, dug down 60 feet without finding water. Discouraged at this, he covered it up, and so left it. In the following spring, however, he opened it again, when, the Botanical Professor in the University of Cambridge being present, it was found to contain about 25 feet of water of a sulphureous taste and smell. Upon analysis it was found to be beneficial medicinally, the waters being chalybeate.'†

Dulwich waters, however, were sold in the streets of London some fifty years before this discovery. The Guildhall Library contains a quaint little volume giving an account of an outrage committed at ' Dulledg wells ' in 1678, of which the title-page runs as follows : ' Strange and lamentable news from Dulledg wells ; or the cruel and barbarous father. A true relation. How a person which used to cry Dulledg

---

* Quoted in Blanch's ' Camerwell,' p. 367.
† Manning and Bray's ' Surrey.'

water about the streets of London, killed his own son on Tuesday, the second of this instant July, in a most inhumane manner, for which he was the next day committed and now remains a prisoner, in order to a Tryal.   London: printed for D. M., 1678.'

The proximity of the wells brought considerable custom to the ' Green Man,' and it flourished to such an extent that a current publication informs us that the proprietor 'has lately built a handsome room on one end of his bowling-green for breakfasts, dancing, and entertainment—a part of the fashionable luxury of the present age, which every village for ten miles round London has something of.'*   But the popularity of the waters waned, and the tavern for this cause became less flourishing till we find it at length disappear altogether, and its site occupied by another famous institution, viz., Dr. Glennie's Academy.   Byron was at one time a pupil at this school.   Dr. Glennie, writing to Tom Moore, speaks thus of his illustrious pupil: ' He is anxious to excel in all athletic exercises, notwithstanding his lameness—an ambition which I have found to prevail in general in young persons labouring under similar defects of nature.'   It is said that Byron and his school-fellows went in for a system of mimic brigandage, and the boyish demands to ' Stand and deliver' caused great alarm to the timid passers-by. Although this was mere mischief on the part of the boys, highway robbery was carried on here by genuine footpads in real earnest, so that Sydenham Hill at the beginning of this century had a reputation scarcely less inferior to that of Hounslow Heath.   The murder of an old recluse, who had lived for thirty years in a secluded cave in Dulwich Wood, roused Byron and his companions to a pitch of excitement, and they now were excessively anxious to suppress all the highwaymen and footpads of the district, so much so that in their zeal they nearly murdered an unfortunate apprentice who was suspected of belonging to that class.   It is to be feared that Byron's two years of school-life here were wasted.

* Quoted in Blanch's ' Camberwell,' p. 386.

He had the misfortune of being spoilt by his mother, Lady Byron, who frequently kept him away from school to go into society, and as a consequence his education was bound to be neglected. Any remonstrance with his mother on the part of his tutor ended in her breaking into a fit of temper, and some of these outbursts reached the ears of the schoolboys, one of whom remarked to Byron: 'Your mother is a fool,' to which he answered characteristically: 'I know it, but you shan't say so.' Though she was mainly responsible for it, his mother was dissatisfied with the progress made by the budding poet at Dr. Glennie's, and he was sent to Harrow. Master and pupil do not appear to have met again in after-life, though Dr. Glennie watched his subsequent career with peculiar interest, and he was often pleasantly chaffed in society because he had not made a better man of him. Young Byron would, of course, have known every spot in Dulwich, and it is interesting to think that he was familiar with the meadows that now form Dulwich Park. We must only mention the names of some other pupils of the school who became famous in after-life: General le Marchant, Sir Donald M'Leod, and Captain Barclay, the celebrated pedestrian, are among the chief.*

Dr. Glennie's Academy in turn disappeared, and a man employed at the college, of the name of Bew, opened a small public-house here, making use of some of the out-buildings of the school for the purpose, and turning the grounds into a tea-garden. It is from this proprietor that we obtain the name of Bew's Corner. This rural ale-house has gone too, and on its site is the Grove Tavern, the cricket-ground of which runs up to the park.

We must not forget to mention that Lord Chancellor Thurlow is said to have once lived at this spot. Priscilla Wakefield, in her 'Perambulations' (1809), writing about Dulwich, says: 'The house which has the sign of the Green Man was for some time the residence of Lord Thurlow. A fine avenue through the wood faces this

* Blanch, ' Camerwell,' pp. 389, 390.

house, and leads to a charming prospect.' It appears that Lord Thurlow had commissioned Henry Holland, the architect of Carlton House and of old Drury Lane Theatre, to build him a mansion called Knight's Hill. As the cost exceeded the stipulated price, Lord Thurlow took a dislike to the mansion and never resided in it. Lord Chancellor Eldon in telling the story says: ' He was first cheated by his architect, and then he cheated himself; for the house cost more than he expected, so he never would go into it. Very foolish, but so it was.' He contented himself with a smaller house called Knight's Hill Farm.*

* Thorne, ' Environs of London,' vol. ii., p. 454

CHAPTER V.

CLAPHAM COMMON.

THIS deservedly popular open space is 220 acres in extent, and is one of the most frequented of all the commons. Its use is by no means confined to the inhabitants of the immediate vicinity, for owing to the many convenient methods of access visitors from all parts of South London flock to it, and on Saturday afternoons especially it is teeming with London toilers. Every variety of sport is allowed ; cricket, football, and lawn-tennis are, of course, the chief of these, but they by no means exhaust the list. The ponds afford special facilities for model yachting and fishing, whilst at certain hours in the evening bathing is allowed. It is quite a sight to watch the thousands of youngsters around the Mount pond on a summer evening waiting for the signal which tells them they are free to cool themselves in the water. In a few moments their scanty clothing is off, and the pond is a mass of nude wriggling forms, splashing and paddling to their hearts' content. There is a horse-ride and a bandstand to make the attractions complete. But apart from these, the common itself is full of natural beauties. Although fairly level, it can boast of great variety. Its trees are well matured, and embrace most of the different kinds of English growth. Parts of it are still covered with furze, and several clumps of tall elms are worthy of notice.

It is something more than a common, and it has been

suggested more than once that it should be transformed into a park. If this were ever carried out, a fitting name would be Baldwin Park, as it was mainly due to the exertions of the late Christopher Baldwin that it was forwarded to its present state. Through his influence as a magistrate, in 1722 it was planted, drained, and the roads were improved, the expenses being defrayed by private munificence. As a consequence, the value of property in the neighbourhood increased so much that Mr. Baldwin disposed of some of his land at a greatly advanced price, 14 acres being sold for £5,000, *i.e.*, at the rate of £357 an acre, since which land has been gradually improving in value. It seemed, however, in later years a difficult matter to raise the necessary funds to maintain the common properly. In the great religious days of Clapham these fell off so much that in 1835-36 leases of the manorial rights were applied for and granted for twelve years at an annual rent of £65. During the next two years to December 1838, things mended, and the liberal donations of £896 were given, and in 1839-40 further sums, to the extent of £248 6s. were given, and in 1844 £138 more was obtained. Afterwards about £150 a year was collected, by which the common was kept in some order till the time of its final transfer to the late Metropolitan Board of Works in 1877. Extensive improvements have been carried out under their supervision, including a regular system of subsoil drainage, the filling up of the old ditches, and the planting of trees. The ponds have been cleaned out, and post-and-rail fencing erected. Altogether the common of this century is a great improvement upon Clapham Common of the last. It was then little better than a swamp, and the roads over it were almost impassable.

Much doubt exists as to the derivation of the name of Clapham. In Doomsday Book it is called *Clopeham*, and in one of the oldest documents in which the name occurs it is spelt *Clappenham*. These forms of the word make it appear rather doubtful that it can be derived from the name of a Danish lord Osgod Clappa, which is the explanation usually put forward. It was at the marriage feast of this lord's daughter that King Hardicanute died.*

Clapham Common is in two manors. The western half formed part of the waste lands of the manor of Battersea and Wandsworth, and was at one time called Battersea East Common. The eastern portion is in the parish and manor of Clapham. This manor in Saxon times was valued at £10, and was held of the Confessor by Turbanus. Geffrey de Mandeville owned it at the time of the Doomsday survey, which mentions, however, a report that he held it unjustly to the prejudice of one Asgar. For all this, he and his heirs continued in possession of it for some time and even after its alienation it was still held of the honour of Mandeville. In the time of Stephen, it was the property of Faramas de Bolonia, whose daughter and heiress married Ingram de Fiennes, who was slain in the Holy Land in 1190. A charter of King Richard's is extant which restored to the widow all the privileges of the manor as enjoyed by her husband and father. William de Fiennes died seised of this manor, 30 Edward I. It appears to have been granted soon afterwards to Thomas Romayne, though the Fiennes family reserved to themselves the right as mesne lords. Juliana, the widow of Thomas Romayne, died in the reign of Edward II., and left two daughters, between whom her property was divided. Clapham manor fell to the share of Margaret, who married William de Weston, and was the property of her descendants in the reign of Henry VI. The next owner we find is Richard Gower, who was lord of the manor in the reign of Edward IV., and he sold it to Sir George Ireland, Alderman of London. It afterwards belonged to Sir Thomas Cockayne, who alienated it to Philip Okeover, and Richard Crompton, who probably purchased it in trust for Bartholomew Clerk, who died seised of it in the reign of Elizabeth. Henry Atkins, physician to James I., bought the manor for the sum of £6,000, which money is said by a family tradition to have been the produce of presents bestowed

* Thorne, 'Environs of London,' 1876, part i., p. 110.

on him by the King after his return from Scotland, whither he had been sent to attend Charles I., then an infant, who lay dangerously ill of a fever. The manor remained in this illustrious family for many years, when on the death of Lady Rivers, sister of Sir Richard Atkins, the last Baronet of this family, the property was transferred to the Bowyer family, the present owners.*

The manorial rights over the whole of the common were purchased under a scheme confirmed by the Metropolitan Commons Supplemental Act, 1877, for the sum of £18,000— £10,000 for the part in the Battersea and Wandsworth manor, and £8,000 for that in Clapham.

The common was the cause of a very lengthy lawsuit between the two parishes of Clapham and Battersea. The inhabitants of Clapham resented the claim of their rivals to half the common which was said to be in the parish of Battersea. From words they came to deeds, and in 1716 the latter party enclosed what they called their portion of the common with a ditch and bank, in order to keep out the cattle of the Clapham folk. Upon this their antagonists filled up the ditch and levelled the bank, for which an action of trespass was brought against them by Lord Viscount St. John, then Lord of the Manor of Battersea. The cause was heard at the Lent Assizes for the county of Surrey, held at Kingston in 1718, when, after a long trial, Lord St. John was non-suited because his principal witnesses who were inhabitants of Battersea were disqualified from giving evidence.†

When we compare the present state of Clapham Common with that of its neighbour Wandsworth, it will be seen how fortunate the former has been as regards the matter of encroachments and enclosures. This can be accounted for partly by the fact of the dual control exercised by the inhabitants of the two manors, and partly because the common has been surrounded for nearly a century by good

* Lysons, 'Environs of London,' 1810, vol. i., pp. 117, 118.
† *Ibid.*, p. 116.

houses, the residences of wealthy City bankers and merchants. Many of these roomy old mansions have disappeared, and their grounds have been converted into profitable building estates. But for all this, there have been encroachments, as a glance at the map will show. It is difficult now to distinguish between the illegal pilferings from the common-land, and the enclosures made in proper form with the consent of the lord and copyholders of the manor.

One of the earliest of these was made by Mr. Henton Brown, of the firm of Brown and Tritton, bankers of Lombard Street. He fenced round the Mount Pond, upon which he placed a pleasure-boat, and in 1748 the parish gave him leave to substitute a close pale fence for the open one, and so convert this water into a private lake. The consideration for this concession was a payment of 5s. to the poor of the parish, which does not say much for the value placed upon the common then. Fortunately a subsequent vestry refused to ratify this agreement, and so saved the common from a permanent disfigurement. Mr. Brown's example was soon followed by a Mr. Fawkes, who had dug a trench 166 feet long upon the common. This particularly exasperated the inhabitants, and led to a committee being formed in 1768 to ' maintain the just rights and privileges of the parishioners entitled to commonage within the Manor of Clapham.' They seem to have been rather lax in their jurisdiction, for it was found necessary in 1790 to call another meeting to consider the various enclosures, encroachments and nuisances that had been made on the common and waste lands of the parish ; and it was ordered that in future the surveyors of the highway should notify the church-wardens of any attempted encroachment on the common. Still the enclosures went on—sometimes those benefited were called upon to pay a nominal sum of 1s. per year to the parish, and in other cases this was dispensed with or the encroachments were not noticed. Consequently more stringent measures had to be taken, and in 1796 it was resolved that in future no part of the common was to be

enclosed without a petition being presented to the vestry,
and a plan produced, specifying exactly the dimensions.
This had to be signed by the Lord of the Manor, or his
agent, and after these formalities had been gone through,

The Rookery, Clapham Common.

the petition was read, and referred to a subsequent vestry.
At the same time a standing committee was appointed to
raise subscriptions for the improvement and maintenance of
the common, and also to appropriate the moneys received

from the granting of the common and waste lands. But in spite of all these regulations, the enclosures went on, and in the following year no less than four were made of the land near the windmill. The serious consequences of these encroachments may be seen in the terrace of twelve staring red-brick houses which surround the site of the windmill pond. These houses have been built upon the grounds of what was formerly one residence—Windmill Place. Now that the common is subject to strict supervision, it is safe from any future curtailing.

We have before referred to the beauty and variety of the trees on Clapham Common. In addition to their natural beauty there are two which have a claim to be considered historical. On the south side an old poplar has been railed in which goes by the name of Spurgeon's Tree. The story connected with it is that on Sunday morning, July 3, 1859, it was struck by lightning during a violent storm. A man named Hutton, aged twenty-six, butler to W. Herbert, Esq., of Cavendish House, was very imprudently taking shelter underneath it, and was killed. He left a widow and four children, and on the following Sunday the Rev. C. H. Spurgeon, who lived for many years in the beautiful Nightingale Lane which connects Clapham and Wandsworth Commons, preached a sermon on their behalf from the very appropriate text, ' Be ye also ready, for in such an hour as ye think not the Son of man cometh ' (Matt. xxiv. 44).

The other tree, which stands in the pathway near the church, is variously called Captain Cook's or the Seat Tree. The latter name is correct enough, because a seat had been fixed round the base of the trunk, but the name of Captain Cook is connected with it simply by tradition. It is probable that this and the other fine trees standing by the highroad on the north side of the church were planted by the eldest son of the great navigator, Captain James Cook, R.N. The tradition that links the name of Captain Cook with this tree, asserts that he lived at Clarence House, on the common, now the Clapham Middle School for Girls, and that he called the curious balcony at the back of the house his quarter-deck. The tree was blown down in a violent gale in February, 1893. As Captain James Cook died about the commencement of this century, the tree must be nearly 100 years old. The stump has been covered with a zinc capping to prevent its decay.

Before a special site was set apart for public meetings, this tree was a rendezvous for those who wished to air their social, political or religious opinions. In the years 1877 and 1878 these meetings first began to assume importance, not so much on account of the particular views set forth, as because of the organized opposition the speakers met with on the part of medical students and other rowdy members of society. The consequence of this was that an attempt was made to prohibit public meetings altogether, a step which was energetically resisted by Mr. John Burns, M.P., then a young man about twenty. While attempting to address a meeting under this tree, he was forcibly apprehended by the police, detained in the local station, but acquitted on his appearance before the magistrate. It was mainly owing to the insistence upon the right to speak on the common, and the arrest of Mr. Burns in connection therewith, that the present spot for public meetings was set apart. John Burns is never tired of telling how that on the occasion of his arrest, he first met his future wife, and this will account for his great interest in the common where he learned the manly sports of swimming, skating, cricket and football of which he is no mean exponent.

From the trees we may turn to another important feature of the common—the ponds. Close by the church is the Cock Pond, taking its name from the adjoining inn. This was excavated in order to find material for raising the ground on which the church now stands.

The Mount Pond, which is a conspicuous object all over the common, was originally a gravel-pit, excavated to furnish gravel for the main road from London to Tooting. This is the pond which Mr. Henton Brown nearly succeeded in converting into private poverty. He spent a great deal of money in planting the mound, which he connected with the bank by a bridge, but after his failure the works were allowed to fall into decay. The pond affords good sport for the disciples of Izaak Walton, and is also available for bathing in the summer and skating in winter.

The Long Pond, which runs parallel to the above-mentioned highway, is chiefly devoted to model yacht sailing, and is generally covered with a large number of miniature sailing-boats. This was formerly called the Boat-house Pond, because the Lord of the Manor enclosed it with a quick

The Mount Pond, Clapham Common.

hedge, planted a shrubbery round it, and built a summer-house and boat-house on its banks.*

Behind the red-brick houses near the Windmill Inn was the Windmill Pond, now filled in, upon the site of which the lavatories and tool-shed have been built. This was completely hidden by trees, and was a very picturesque feature of this part of the common. It afforded much amusement to the youngsters who came here to fish for sticklebacks, and it is much to be regretted that it was found necessary to fill it up.

* J. W. Grover, 'Old Clapham,' p. 17.

Close by is another pond of some charm, called the Eagle House Pond from the fine mansion which it faced. This has a small island in the centre, and is also used for fishing. Eagle House has been pulled down and its grounds have been turned into a building estate. These three ponds and others which have been filled up were probably excavated to provide gravel for the formation of the highway which runs close to them.

An old description of the village of Clapham* mentions that it was supplied with water from a fine well situated on the north side of the common. The original reservoir was of ancient construction, and was repaired by the parish in 1717. About 1789 a new well was opened on the common, which was paid for out of the voluntary subscriptions of the inhabitants. This new well continued to satisfy the wants of Clapham till 1825, when the increase of population made further supplies necessary. In addition to this, the well became a nuisance, because of the numbers of men who came here with their carts waiting for their turn, and so blocked the traffic. Subsequently another well was formed some 100 yards further up, which produced a daily supply of 150 butts. This in its turn has had to give way to more modern means of supply, and the old well has been covered in.

One of the few buildings actually on the common is the parish church dedicated to the Holy Trinity. It is comparatively new for such an ancient place as Clapham, being little over 100 years old. Before its erection the parish church was St. Paul's, close by the Wandsworth Road. About 1768 this fell into a bad state of repair, and a Mr. Couse, an architect, was employed to shore it up 'to quiet the minds of the inhabitants.' This must evidently have been carried out satisfactorily, for in a few years' time he was commissioned to design the new parish church, and his instructions seem to have been to build a 'new strong church.' Both of these conditions were fulfilled, but that is

* 'Clapham, with its Common and Environs,' published by D. Batten, 1859.

all that can be said, for the most ardent of his admirers could scarcely call his work beautiful. In justification of the architect it must be said that at this time church architecture had reached its lowest depth. But what the church lacks in beauty is made up in comfort and solidity. Much of its plainness is hidden by the trees which surround it, and altogether it adds picturesqueness to this part of the common. Inside it is as plain as outside, the only redeeming feature being the massive carved pillars which are quite in keeping with the air of solid strength which is the chief characteristic of the church. The site was staked out on the common in 1774, railed in, and the building of the church commenced in accordance with an Act of Parliament obtained for the purpose. The church accommodates about 1,400 persons, and cost £11,000. Its old associations endeared it to Macaulay who writes from Clapham, February 1849: 'To church this morning. I love the church for the sake of old times; I love even that absurd painted window, with the dove, the lamb, the urn, the two cornucopias, and the profusion of sun-flowers, passion-flowers and peonies.' This work of art no longer adorns the church.

There are singularly few memorials or tombs inside; most of the great Claphamites were buried in the ancient church on the site of St. Paul's, and not a few have their last resting-place in Westminster Abbey. There is a monument to John Thornton by Sir Richard Westmacott; a mural tablet with medallion portrait to John Jebb, Bishop of Limerick, who died in 1833 (whose body, however, is interred in the family vault of the Thorntons); and another tablet to Dr. John Gillies (d. 1836), the author of a forgotten 'History of Greece' and translator of Aristotle's Ethics.*

Close by the church is the fire-brigade station. The first year in which there is any record of a parish fire-engine is 1750, when the sum of 20s. per annum was allowed for the care of it, 'on condition that it be brought out and worked

* Thorne, 'Environs of London,' vol. i., p. 112.

at least four times in the year.'*   Opposite the fire-brigade
station is the Cock Inn, one of the oldest houses in Clapham.
As an institution it dates back to the sixteenth century, but
the business was conducted not in the present building, which
was formerly a private house, but in the two wooden cottages
at the rear.   Before the Metropolis had swollen to its present
gigantic proportions, when Clapham was an isolated village,
a market was held here, but it is needless to say no such
thing takes place now.   A curious discovery of an officer's
dress-sword was made in the roof some years ago.   It was
inscribed with ' G. R.,' and was supposed from its appearance to have been of German origin, and it is conjectured
with some probability that it belonged to one of the
Hanoverian officers of the Duke of Cumberland's army,
which was encamped on Clapham Common in 1745, the
year of the Pretender's insurrection.†

The common has recently been embellished by the gift
of a handsome drinking-fountain, presented by the United
Kingdom Temperance and General Provident Institution in
1895.   It consists of a granite pedestal surmounted by a
group in bronze representing a Sister of Mercy giving water
to a wounded soldier.   It formerly stood in front of the
offices of the institution in Adelaide Place, London Bridge,
but as it was not used to the fullest extent, it was presented
to the London County Council for placing in one of their
parks and open spaces, the site ultimately selected being at
Clapham Common, in the pathway close to Captain Cook's
tree.

Some interesting items relating to the common can be
gleaned from the vestry minutes of Clapham.‡   In 1693,
bull-baiting was forbidden throughout the parish.   Although
no details can be gathered respecting the extent to which
this barbarous sport was carried on, it is probable that the
arena would be on the common, and it must evidently have
flourished here before the issue of this vestry mandate.

* Parochial minutes.   † J. W. Grover, ' Old Clapham,' pp. 40, 41.
‡ Quoted in ' Clapham : its Common and Environs,' Batten.

The gorse and thick undergrowth of the common furnished a refuge for hedgehogs and polecats, against which a raid was ordered by the vestry, and 10s. was paid in 1722 for killing nine hedgehogs and seven polecats, and further sums in the following year for similar purposes. Hedgehogs appear in the minutes again in 1728, when the sum of 11s. 4d. was paid for exterminating thirty-four of them. In 1781, the fair which had been held for some years at Clapham was abolished as being a great nuisance to the inhabitants. In 1816 the vestry directed that swine should not be allowed loose on the common, and gave their keeper instructions to impound any stray porkers he might find.

The common and the immediate vicinity were the resort of what was known as the 'Clapham Sect,' a name given to the Evangelical party in the Church of England by the Rev. Sydney Smith in the latter part of the eighteenth century. This earnest body of men laboured together for what they considered to be the interests of pure religion, the reformation of manners, and above all the abolition of slavery. Foremost among the sect were William Wilberforce, Zachary Macaulay (father of the historian) and the Rev. W. Romaine. William Wilberforce lived at Broomfield (now called Broomwood) on the south-west side of the common, and there his son, the celebrated Bishop Wilberforce, was born in 1805. Adjacent to his property was that of Henry Thornton, another famous name in connection with this movement. The meetings were held for the most part in the oval saloon which William Pitt, 'dismissing for a moment his budgets and his subsidies, planned to be added to Henry Thornton's newly purchased residence. It arose at his bidding, and yet remains, perhaps a solitary monument of the architectural skill of that imperial mind. Lofty and symmetrical, it was curiously wainscoted with books on every side, except where it opened on a far extended lawn, reposing beneath the giant arms of aged elms and massive tulip-trees.'* It was in this saloon that there met,

* Sir James Stephen, 'Essays in Ecclesiastical Biography.'

after their long years of effort had been crowned with success, Wilberforce, Clarkson, Granville Sharp, Zachary Macaulay and the other members of the society 'in joy and thanksgiving and mutual gratulation' over the abolition of the African slave-trade.* We find notices of the visits of royalty to this house in the years 1807, 1808 and 1809, when H.R.H. the Duchess of Brunswick, Queen Charlotte, and the Princesses Augusta and Elizabeth visited the house and grounds of Mr. Robert Thornton for public breakfasts.

Another far-reaching society whose influence is felt all over the globe also emanated from Clapham. This is the British and Foreign Bible Society who have been instrumental in translating the Bible into nearly 250 languages. The man to whom may be ascribed the credit of this vast undertaking was its first and greatest President—Lord Teignmouth. After a safe career in India, culminating in his being appointed Governor-General, he returned to England and took up his residence at Clapham Common, where the founders and the earliest secretaries of the society had their meeting-place. What a change has fallen upon his house, for it is now a Roman Catholic nunnery!

Although many of the historical houses skirting the common have been pulled down to make way for the all-devouring march of bricks and mortar, there are yet some which stand out as memorials of a vanished past. Surely no house on the common deserves more respect than Cavendish House, at the corner of Cavendish Road, the residence of that eccentric philosopher, the Hon. Henry Cavendish, who has been justly styled the 'Newton of Chemistry.' The house has undergone considerable alterations, and has been refronted, but it still remains as the sole memorial to this man of talent. In his time very little of the house was devoted to domestic purposes. His whole heart was in his science, and the comforts of life had to give way to that. The upper rooms formed an astronomical observatory, and on the lawn was a large wooden stage which gave access to

* Thorne, 'Environs of London,' vol. i., pp. 111, 112.

the top of a lofty tree, from which the philosopher could the better conduct his researches. The present drawing-room of the house was formerly his laboratory, and near it stood a forge. The scientific discovery which principally makes

The Eagle Pond, Clapham Common.

the name of Cavendish famous is that of finding out the earth's density. 'The man who weighed the earth' found out that it is five and a half times heavier than water of the same bulk. He was most eccentric in all his ways. He would never see or allow himself to be seen by a female servant, and Lord Brougham relates that he used to order his dinner daily by a note, which he left at a certain hour on the hall-table, whence the housekeeper was to take it.* He had a magnificent library which was placed at such a distance from the house that those who came to consult it might not disturb him.

Whilst travelling in England in 1790 with George Foster, Humboldt obtained permission to make use of this library, on condition, however, that he was on no account to presume so far as to speak to, or even greet the proud and aristocratic owner, should he happen to encounter him. Humboldt states this in a letter to Bunsen, adding sarcastically: 'Cavendish little suspected at that time that it was I who in 1810 was to be his successor at the Academy of Sciences.'†

Cavendish was distinguished like many other philosophers by his entire disregard of money, although he had enough of that necessary article. On one occasion the bankers with whom he kept his account, finding his balance had accumulated to over £80,000, commissioned one of the partners to wait upon him to ask what should be done with it. On reaching the house, after considerable difficulty and delay, he succeeded in obtaining an audience with the abstracted chemist, who instead of being pleased with the attention, exclaimed, on hearing of the amount of his balance:

'Oh, if it is any trouble to you, I will take it out of your hands. Do not come to plague me about money.'

'It is not,' replied the banker, 'any trouble to us, but we thought you might like some of it turned to account and invested.'

'Well, what do you want to do?'

'Perhaps you would like to have £40,000 invested?'

'Yes; do so, if you like; but do not come here to trouble me any more, or I will remove my balance.'‡

He very seldom invited friends to dinner, and when he

* Thorne, 'Environs of London,' vol. i., p. 111.

† Bruhn, 'Life of Humboldt,' English translation, 1873, vol. ii., p. 68.

‡ Quoted in ' Old Clapham,' J. W. Grover, p. 49.

did they had to put up with the simplest fare. Once his housekeeper ventured to remind him that, as five friends were coming to dinner, one leg of mutton would not be enough. 'Well then, have two,' said the philosopher. When he died in 1810 he left a fortune of £1,300,000.

To turn now to the other side of the common we shall find the Gauden estate upon part of which was built a house once the property of the ubiquitous Pepys. It had a history before he came there. It is probable that a portion of the common was included in this estate of Sir Dennis Gauden, who flourished in the seventeenth century. Dr. Gauden, the divine, was the reputed author of the ' Εἰκὼν βασιλικὴ, or the Portraiture of his Most Sacred Majesty in his Solitude and Sufferings.' The treatise, which professes to contain meditations and prayers, composed by Charles I. in his captivity, was published in 1649, a few days after the execution of that monarch, and produced an extraordinary sympathy in his behalf. So eagerly was it read that it passed through fifty editions in a single year. Dr. Gauden, who was first of all Bishop of Exeter, and then of Worcester, died of a disease which it is said was caused or aggravated by his not receiving the appointment of Bishop of Winchester. In view of this promised promotion he had had this large house built in his brother's name, which together with the grounds occupied some 430 acres. The Terrace and Victoria Road now mark the site of his mansion, which was pulled down in 1762.* The house and grounds subsequently were purchased by William Hewer, one of the Commissioners of the Navy to James II., and treasurer for the garrison of Tangier. Pepys, who had been Hewer's master at the Admiralty, seems to have resided here with him, and he frequently mentions in his letters absence from town in order to take the air of Clapham. Pepys collected a magnificent library which he afterwards gave to Magdalen College, Cambridge. We may gather a good description of the house from the diary of John Evelyn. He says:

* Thorne, ' Environs of London,' vol. i., p. 111.

'I went to Mr. Hewer's at Clapham, where he has an excellent, usefull and capacious house on the Common built by Sir Dennis Gauden, and by him sold to Mr. Hewer, who got a very considerable estate in the navy, in which, from being Mr. Pepys's clerk, he came to be one of the principal officers, but was put out of all employment on the Revolution, as were all the best officers, on suspicion of being no friends to the change, such were put in their places as were most shamefully ignorant and unfit. Mr. Hewer lives very handsomely and friendly to everybody.'*

Some eight years later his friend Pepys was living here, and Evelyn writes :

'I went to visite Mr. Pepys at Clapham, where he has a very noble and wonderfully well-furnished house, especially with India and Chinese curiosities. The offices and gardens well accommodated for pleasure and retirement.'†

It was in 1703 that Pepys died, when Evelyn summed him up as 'a very worthy, industrious, and curious person ; none in England exceeding him in knowledge of the navy, in which he passed thro' all the most considerable offices, Clerk of the Acts and Secretary to the Admiralty, all which he performed with great integrity. When King James 2nd went out of England, he laid down his office and would serve no more ; but withdrawing himselfe from all public affaires, he liv'd at Clapham with his partner, Mr. Hewer, formerly his clerk, in a very noble house and sweete place, where he enjoy'd the fruite of his labour in greate prosperity.'‡

Turning from the entries of Evelyn to those of his brother diarist, we find that Pepys first visited his future home on July 25, 1663, on the day when he was to have gone to Banstead Downs 'to see a famous race,' which was, however, postponed because the House of Lords was sitting. He continues :

'After some debate, Creed and I resolved to go to

* 'Diary,' June 25, 1692.   † *Ibid.*, September 23, 1700.
‡ *Ibid.*, May 26, 1703.

Clapham, to Mr. Gauden's. When I came there, the first thing was to show me his house, which is almost built. I find it very regular and finely contrived, and the gardens and offices about it as convenient and as full of good variety as ever I saw in my life. It is true he hath been censured

The Cricket Fields, Clapham Common.

for laying out so much money; but he tells me that he built it for his brother, who is since dead (the Bishop), who when he should come to be Bishop of Winchester, which he was promised (to which bishopric at present there is no house), he did intend to dwell here.'

8

At the time when this was written Pepys little thought that this house was afterwards to be his home.

Adjoining Gauden House, on the site of what is now Cedars Road, stood another fine mansion, The Cedars, pulled down in 1864. This was said to have been designed either by Sir Christopher Wren or Inigo Jones, and was very beautifully decorated.

Elms House, at the corner of the Chase, was once the home of the Barclay family, famous for their wealth and their power in using it well. This became afterwards the home of Sir Charles Barry, the famous architect of the Houses of Parliament, who spent the last ten years of his life and died here.

Close by is a row of old houses, built, it is said, by Sir C. Wren, known as Church Buildings. The date 1720 on the arch of one of them tells us when the last house was finished. The house clustering round this arch, and the next on the west side, were formerly one, and here resided at the close of the last century Granville Sharp, whose name we have before mentioned in connection with the abolition of the slave trade. He formed here a school for the education of negroes of high degree from Sierra Leone. The air of Clapham, although beneficial to Mr. Pepys, was not suited to these frequenters of warmer climes, so the school was given up. But Mr. William Greaves, who superintended the establishment, by a stroke of policy, opened it again as a school for the youth of Clapham, and here received his education from 1807 to 1812 no less a personage than the great Lord Macaulay.* It was at Clapham, then, that he acquired the rudiments of English history which was afterwards to make him so famous. Passing through the archway from the common, on the left hand can be seen the school door which formed the entrance for the boys. Zachary Macaulay, his father, had a house in what is now called The Pavement, where young Tom passed a quiet and happy childhood. The common, with all its resources, was at his

---

* G. O. Trevelyan, 'Life of Macaulay.'

command. A graphic description of his connection with it is given by Sir George Trevelyan, in his life of his uncle :

'That delightful wilderness of gorse bushes, and poplar groves, and gravel-pits, and ponds, great and small, was to little Tom Macaulay a region of inexhaustible romance and mystery. He explored its recesses, he composed and almost believed its legends, he invented for its different features a nomenclature which has been faithfully preserved by two generations of children. A slight ridge intersected by deep ditches towards the west of the common, the very existence of which no one above eight years old would notice, was dignified by the title of the Alps ; while the elevated island covered with shrubs that gives a name to the Mount Pond was regarded with infinite awe as being the nearest approach within the circuit of his observation to the majesty of Sinai.'

Macaulay has left his name in the nearest road to his old school leading off the common, which is called after him Macaulay Road.

Upon the western side of the common, near Broomwood Road, is Beechwood, at one time the residence of Field-Marshal Sir George Pollock. He was distinguished as being the only officer of the Indian Army to rise to the rank of Field-Marshal. After a brilliant career in Afghanistan and India, he succeeded Sir John Burgoyne as Constable of the Tower, and was accorded a burial-place in Westminster Abbey, where there are memorials to the following distinguished Claphamites : William Wilberforce, Zachary and Lord Macaulay, Sir J. Mackintosh, and Granville Sharp.*

And now we must leave the common and its varied associations. Changes are inevitable in a city which is always growing, and though the common has not changed much of late years, its surroundings have. Though it would not, perhaps, furnish at the present day the most fitting retreat for the pious reflections of a Wilberforce, it is still dear to the thousands who are cooped up in narrow streets and stuffy courts, for whom it affords the only glimpse of the beauties of Nature.

* J. W. Grover, 'Old Clapham,' p. 61.

8—2

CHAPTER VI.

HILLY FIELDS, DEPTFORD PARK, AND TELEGRAPH HILL.

HILLY FIELDS.

W E next come to a group of three enclosed grounds opened to the public in three consecutive years. The first of these, Hilly Fields, 45½ acres in extent, is situated on the borders of the parish of Lewisham in the county of London. It is on the edge of the crowded parish of Deptford, one of the poorest in London. A writer in the *Times,** in an able appeal for the preservation of this open space, pointed out how unfavourably this district compared, in respect of open spaces, with others in London in proportion to its population. Taking the statistics compiled with such care by Mr. Charles Booth, it will come as a surprise to many to learn that in the county of London there is more poverty south than north of the Thames, while the district lying along the river from Greenwich to Rotherhithe is the second poorest in London. It remains true, no doubt, that in the east of London there is the greatest mass of poverty, the largest area of closely packed houses, because instead of the comparatively open lands which border on the other parts of London, in the East End there is the large poor and populous area of Stratford and West Ham, outside the county. But if only the London division is taken, it appears

* January 4, 1892.

that there is more poverty in proportion to population in the
district lying between Blackfriars and Woolwich than at the
East End, while the highest percentage of poverty in small
areas is also to be found south of the Thames. Thus, deal-

General View of Deptford from Brockley, 1815.

ing with blocks of about 30,000 inhabitants, Mr. Booth finds
that the two poorest are situate, one between Blackfriars and
London Bridge, and the other by the riverside at Greenwich,
which includes Deptford. The first with a population of
33,000 has 68 per cent. of poor; the second with 31,000,

65 per cent., whilst there is no similar block in East London which has more than 59 per cent. In this poor and crowded district, with which we are concerned, 213 persons live on every acre. The time has long since passed when Deptford had its country mansions, and its market gardens get fewer each year. Apart from Deptford Park, there are no open spaces of any size in the parish. All the greater importance therefore attaches to the preservation of such a fine open space as this, which is only a mile distant from some of the most congested quarters. But Deptford is not the only district to reap benefit, for the more immediate neighbourhoods of Brockley and Lewisham, although comparatively open at present, must ere long be covered with buildings to accommodate the ever-increasing population of this mighty London. Moreover, the character of the Hilly Fields gives a wide range to their influence upon the health of the Metropolis. It has long been recognised that it is especially important to keep the hill-tops round London free from buildings, so that the purity of the air blowing in from the country may thus be preserved.

The cost of acquiring Hilly Fields was just under £45,000, towards which the London County Council contributed more than half, the Greenwich District Board £7,000, while substantial sums were given by the Lewisham District Board, the London Parochial Charities Trustees, the Kyrle Society, and by several of the City companies and other bodies. The character of the ground was not materially changed in laying it out for public use. Part of the site which had been used for brick-making had to be levelled, other swampy portions were drained, and trees were planted on the whole. A bandstand has been erected for musical performances, footpaths have been formed, and a plain but substantial railing surrounds the whole area. As the name of the place implies, it is too hilly to provide much scope for cricket and football, but a space has been levelled for the former game, and the remainder is open for children's sports and general recreation. When the works of laying-out were completed, the Fields were dedicated to the public on May 16, 1896, by Sir Arthur Arnold, then Chairman of the London County Council.

The immediate locality in which the Hilly Fields are situate is of too recent growth to boast of many historical associations. A portion purchased from the Corporation was part of the Bridge House estates, which have a curious origin. In evidence given before the Commissioners in 1854 it was stated that the fund was created by ancient grants from the Crown, gifts from different individuals, and purchases of property arising from the saving of the revenues at different periods. On the first building of London Bridge, a sort of crusade was preached by Peter of Colechurch. He went about the country with a brief, and was enabled to collect funds towards the erection of the bridge. Sometimes money was given, and sometimes land, for the purpose of building London Bridge; at other times grants have been made from the Crown, and purchases have been made out of the surplus rents and profits of the Bridge House estates. Property also has been given by will. The result is a fund held in trust for the maintenance and support of London Bridge. Although held originally for the maintenance of this bridge alone, the estates were afterwards charged under the authority of subsequent Acts of Parliament with the cost of keeping up Southwark and Blackfriars Bridges, and of carrying out other works.

Ladywell, or Bridge House Farm, of which the Corporation land formed a part, was an estate of nearly 97 acres. The farm-house was on the south side of Brockley Lane. Game-shooting was quite common here some fifty years ago, but the agricultural lease was determined in 1876, when the character of the estate changed. That portion of the farm now merged in Hilly Fields was then laid out for building purposes. By a lease dated November, 1890, between the Corporation and the New Land Development Association, the company undertook before June, 1895, to erect 'good and substantial brick or stone tenements of the value of not

less than £425 each,' so that the negotiations for the preservation of this open space were only just in time.

In the neighbourhood of Hilly Fields there existed in the
twelfth century a monastery belonging to the Order of the
Premonstratensians. The monks of this Order were first
settled in this country at Ottham, in Sussex, but finding this
place very inconvenient for them, they quitted it for Brockley,
where they were granted land by Countess Juliana, wife of
Walkelin de Maminot, who owned the Manor of Brockley in

The Construction of the London and Brighton Railway   (From a water-colour
drawing dated 1839.)

the latter end of the reign of Henry II.   They did not
remain here long, for they quickly removed to Begham, in
Sussex, and the lands at Brockley were confirmed to them
till the abbey was dissolved with other monasteries in 1526.
The monks were also called Norbertins, after their founder,
St. Norbert, Confessor and Archbishop of Magdebourg.   He
was born at Santen, in the duchy of Cleves, in 1080, died
in 1134, and was canonized by Pope Gregory XII. in 1582.
He built a monastery in the lonely valley of Premontre, in

the forest of Coucy, in 1121. In its primitive institution the rules of the Order were very severe. The monks never wore linen, and observed a perpetual abstinence from flesh, and a yearly rigorous fast of many months. The exact site of the monastery cannot now be located, but it was probably close to the fine church of St. Peter's, Brockley.*

As recently as 1850 there was a fine open space to the immediate north of Hilly Fields, viz., Deptford Common, now all built upon. Previously to 1839, when the London

The Croydon Canal. (From an engraving dated 1815.)

and Croydon Railway was opened (now part of the London, Brighton and South Coast Railway system), the whole of this district consisted of agricultural estates, the farm-houses dotted about amongst the fields being the only buildings. The railway was built upon the site of the Croydon Canal, which connected that town with the Grand Surrey Canal. The nearest building on the west of the recreation ground was the lock-keeper's cottage, which stood close to where Brockley Station is now. The interesting old hostelry, the

* Dew, 'History of Deptford,' p. 49.

' Brockley Jack,' with its traditional associations with Dick Turpin, is one of these lock-houses which still survives, although it is shortly to be rebuilt.

The buildings surmounting the Hilly Fields are those of the West Kent Grammar School, which is one of the twenty-seven schools belonging to the Church Schools Company, Dean's Yard, Westminster. The building is about thirteen years old, and at present affords accommodation for 150 boys.

The Locks on the Croydon Canal, looking South. (From an engraving dated 1815.)

DEPTFORD PARK.

A year after the Hilly Fields were opened, Deptford had a park of its own, dedicated to the public amidst much local enthusiasm on Whit Monday, 1897, by Dr. Collins, J.P., D.L., Chairman of the London County Council in the Jubilee year. The land forming the park, which is 17 acres in extent, was formerly a part of the Deptford estates of the Evelyn family, and had been let as market-gardens. The purchase-money was fixed at £2,100 per acre, or a total of about £36,000, which was less than the market value of the land. Mr. Evelyn, in addition to selling the site at this low figure, contributed £2,000 towards the purchase-money, and also presented two strips of land in order to increase the width of the entrance from Lower Road, Deptford. The rest of the money was raised by votes from the London County Council £24,000, Board of Works for the Greenwich District £8,250, and private subscriptions £1,750.

Perhaps the title of 'park' for this place is too grandiloquent. It is really a recreation-ground of simple design, consisting principally of a central playground, surrounded by a broad walk for promenade, with well-planted margins. Altogether about £7,500 was expended in laying out, the largest item of which was for the boundary railings.

The changes through which the neighbourhood of Deptford has passed even in the present century are many and varied. One hundred years ago a recreation-ground would scarcely have been needed, for at that time there were 500 acres of market-gardens in the parish, chiefly cultivated for the onions, celery and asparagus for which Deptford has long been famous.* A meadow-flower, the *Caryophyllus pratensis*, was named by old botanists the 'Deptford pink,' because of the abundance in which it grew in the fields here. The change in the character of the town from an agricultural to a manufacturing centre has rendered the little land that still remains unenclosed quite unsuitable for market-gardens. It seems to be the opinion of antiquaries that in the time of Chaucer the whole of this district between Shooter's Hill and London was a stretch of woodland and common, covered with gorse and brushwood. For centuries after this the place was nothing but an insignificant fishing village, in many respects like Woolwich, except that it was less fashionable. Henry VIII. was the first monarch to raise it to fame by the establishment of the royal dockyard here in 1513. So rapid was its rise to importance that in less than forty years it came to be the chief English dockyard. Many of the earliest expeditions despatched from this country on voyages of discovery were fitted out here, including those of such men as Frobisher, Drake, Sir Walter Raleigh, Captains

* Lysons, 'Environs of London,' 1811, vol. i., part ii., p. 468.

Cook and Vancouver. The list of famous 'wooden walls' built in the Deptford dockyard would fill many pages, and so great was the fame of its master shipwrights that Peter the Great, Czar of Russia, worked for some time as a ship's carpenter in this yard in order to perfect himself in that art.* Deptford Dockyard was found unsuitable for the construction of the present class of war-vessels that have supplanted those by which England won her naval supremacy, and it was therefore closed in 1869. The greater portion of the site is now occupied by the Corporation's foreign cattle-market.

Deptford derives its name from the *deep ford* by which the river Ravensbourne was crossed before the erection of Deptford Bridge. The first record of any bridge across the river dates back to 1395; but, although the necessity for a ford has been done away with since this time, the old name has still clung to the town which afterwards sprang up around it.

The Manor of Deptford or West Greenwich, of which these lands formed part, was bestowed by William the Conqueror upon Gilbert de Magminot or Maminot, one of the eight barons associated with John de Fiennes for the defence of Dover Castle. No less than fifty-six knights' fees were given him for this purpose, and he was instructed to distribute these among other trustworthy persons who should assist him in this important work. These eight barons had to provide between them 112 soldiers, twenty-five of whom were always to be on duty within the castle, and the rest to be ready for any emergency. Gilbert de Maminot's share of the lands amounted to twenty-four knights' fees, which together made up the Barony of Maminot, held at Deptford as the head of the barony. Maminot built a castle for himself at Deptford, of which all traces have now disappeared; but from the remains of some ancient foundations which have been discovered, it is now conjectured that its site was on the brow of the Thames in the neighbourhood of Sayes Court, near the mast-dock. The grandson of Gilbert

* Dew, 'History of Deptford,' 1883, p. 87.

de Maminot, named Walkelin, held Dover Castle against King Stephen, although he afterwards surrendered it to his Queen. In 1145 Walkelin gave half his estates in Deptford to the Monastery of Bermondsey. His heiress, Alice Maminot, married Geoffrey de Say, and brought to him the lands of the barony. He granted this manor, together with the advowson of the church and other appurtenances, to the Knights Templars; but his son, also named Geoffrey,

The Manor-House, Sayes Court. (From a copy of the original sketch by John Evelyn.)

regained possession by exchanging the Manor of Sedlescomb in Sussex for it. The manor continued in the Say family by direct descent till the reign of Richard II., when Elizabeth de Say, to whom it belonged, married first of all Sir John de Fallesle (Falwesle), and afterwards Sir William Heron. The first-named knight married the heiress without the license of the Crown, and King Richard II. seized her lands, but, upon appeal, Parliament decided in the knight's favour. Her second husband died without issue, and this manor fell to the share of Otho Watlyaton. The next owner of historical importance is William de la Pole, Earl of Suffolk, created Duke of Suffolk in 1448. He was charged with the loss of Anjou and Normandy, and, being impeached by the Commons, was sentenced to banishment for five years; but he was waylaid on his way to France, and murdered in 1450. His infant son, however, was restored to the title, and another descendant and owner of the manor was involved in political troubles; for in Henry VII.'s reign he entered into the plot to place Lambert Simnel on the throne, which cost him his life at the Battle of Stoke, near Newark-upon-Trent, to which place he was marching. His forfeited estates were at once granted by the King to his uncle, Oliver St. John, but Henry VIII. in 1514 granted unreservedly to Charles Brandon, Duke of Suffolk, all the estates that formerly belonged to the De la Poles, apparently prejudicing the rights of St. John, whose representative was then but ten years of age, and unable to protect himself. Charles Brandon secretly married Mary, Queen of France, and upon the advice of Cardinal Wolsey propitiated King Henry VIII. with certain payments out of her dowry. In return for his services as mediator, the Duke bestowed this manor upon the Cardinal for the term of his natural life. He died in 1530, and five years later the Duke gave the manor to the King in exchange. The grandson of Oliver St. John then came forward and petitioned that the estates granted to his ancestor might be restored to him, and in this he was successful.

Before the year 1538 this manor seems to have reverted to the King, who granted it to one of his many wives, Katherine Seymour. On her death it again came to his possession, and he bestowed it on Sir Richard Long, of the Privy Chamber, for the term of his natural life. The next King, Edward VI., granted to Sir Thomas Speke for the term of his life the office of stewardship of his lordships and manors of Sayes Court and West Greenwich (*i.e.*, Deptford), and when he died the same offices were held by Sir Thomas Darcy, K.G.,

Lord Darcy of Chiche. The manor was retained in the hands of the Crown during the reigns of James I. and Charles I.; but Sayes Court, the mansion-house, was leased

Portrait of John Evelyn. (After an oil painting by Sir Godfrey Kneller.)

for a term of forty-one years to Christopher Browne, who had been resident bailiff here, and during his term of office had spent considerable sums in repairing the buildings. This lease was subsequently renewed, and the remainder was devised to his grandson Richard, afterwards knighted, whose only daughter and heiress, Mary, married John Evelyn, who took possession of Sayes Court in 1648. Five years later he bought the mansion for the sum of £3,500, and at once commenced laying out the famous garden there. Charles II. confirmed the same to him in 1663 for ninety-nine years at 22s. 6d. rent, including about 64 acres of land. In the same year the King, in consideration of £3,896, expended by Sir Richard Browne during his residence in France, demised to him for a term of thirty-one years, at an annual rent of 40s., certain other lands adjoining these. This was surrendered in 1672 for a new patent, extending the term to ninety-nine years. We have already seen that his son-in-law Evelyn had been for some time previous to this in residence at Sayes Court, whither he had been sent from Paris to endeavour to 'compound with the soldiers,' and so save something in the general wreck caused by the confiscations of the Commonwealth. The grandson of Evelyn, also John, afterwards became entitled to both leases. He petitioned, therefore, for a grant in fee after payment of such consideration as should be determined by the officers of the Crown, and this petition was allowed. His descendants are still the owners of these lands, called the Evelyn estate in Deptford.

To return now to the manor. Upon the death of Charles I. all the royal estates were seized in order to be surveyed and sold to supply the necessities of the State. The manor and residue were sold to Thomas Buckner for himself and others for the sum of £12,583; but on the restoration of Charles II., in 1660, the manor and demesnes undemised by the Crown returned to the royal revenue, part of which the manor itself continues.*

An inundation of unparalleled magnitude swept over not only the site of Deptford Park, but also the greater part of the town, in 1651. About 2 p.m. on New Year's Day the storm became so violent that the waves forced down the

---

* These particulars relating to the manor are condensed from Hasted's 'History of Kent,' edited by H. H. Drake, 1886, pp. 2-9.

piles of wood, and entered the shipping-yards, removing great trees and baulks of timber that twenty horses could scarcely move. By half-past two there were seven feet of water in the lower town, which had increased to ten very soon. The inhabitants fled to the upper town, leaving all their property 'to the mercy of the merciless waves,' as one writer described it. Those who were not sufficiently prompt to effect their escape in time had to be rescued by

View from the Site of Deptford Park in 1840.

boats from the upper windows of their dwellings, and some are said to have been drowned. Fortunately the waters began to abate by four o'clock, but not till enormous damage had been done. In addition to the havoc wrought in the dockyards and dwelling-houses, more than 200 head of cattle were drowned in the meadows of Deptford and the adjacent fields. It seems that three black clouds were seen in the firmament on the evening preceding the day of this great flood, so that an old chronicler gives warning 'that when you discern the sun to be eclipsed and the appearing of three black clouds, then expect great inundations, loss of cattel, changes and dreadful revolutions, even as a signal from heaven, to purge nations and commonwealths from oppression and tyranny, and to restore to the freeborn their just freedom and liberty, that so peace may abound within the walls of Sion, and each man enjoy their own again.'*

The Grand Surrey Canal to the south of the park, is the property of the Surrey Commercial Dock Company. It commences at a point nearly opposite the eastern entrance of the London Docks, and runs as far as the Camberwell Road, with a branch towards Peckham. It was this canal which connected the old Croydon canal with the river. The land on the other side of the canal, in the direction of Greenwich, now entirely covered with houses, was originally known as Black Horse Fields. Upon this land was a windmill, which was burnt down in 1854 while grinding stores for the use of the Government during the Crimean War.†

TELEGRAPH HILL.

This recreation-ground consists of two detached plots, 9½ acres in all, lying on the upper portions of the slope of Telegraph Hill, the highest point being upwards of 160 feet above sea-level. The first step to secure this land for public use was taken by Mr. Livesey, the managing director of the South Metropolitan Gas Company, who wished to devote to this object a sum of money which he had received as a testimonial to the energy and resources by which he had maintained the supply of gas during the severe strike of gasworkers. He communicated with the Greenwich District Board of Works, offering the sum of £2,000, and asking for their co-operation in the object he had in view. They warmly took up the offer, voting an amount of £2,000 towards the object, and they also applied to the London County Council for assistance, and to the Haberdashers'

* Quoted in Dew's 'History of Deptford,' pp. 248, 249.
† Sturdee, 'Old Deptford,' p. 49.

Company, who owned the property, to sell it on favourable terms. Both these bodies responded to the appeal: the Council promised a contribution of £2,000, and the company, who estimated the value of the land at £8,000, agreed to sell it for this special purpose for £6,000, thus practically giving £2,000 towards the acquisition, which was thus satisfactorily accomplished.

Owing to the steep slopes and the rough nature of the ground, some difficulties were experienced in laying it out to the best advantage. On the larger plot is a small ornamental lake in two sections at different levels, from which paths lead to a gravel promenade at a high elevation, in the centre of which is a bandstand. From this plateau a good view of the lake and grounds generally can be obtained. This plot also contains a handsome drinking-fountain, the gift of Mr. Livesey. The smaller plot, which crowns the summit of the hill, is more level, and upon this lawn-tennis and children's games are practicable. Altogether, a sum of about £7,500 has been expended in fencing and laying out the ground. When these works were completed, Telegraph Hill was opened to the public on April 6, 1895, by Mr. (now Sir) Arthur Arnold.

Telegraph Hill owes its present name to the fact that its highest point was formerly one of the stations on the line of semaphores which were used by the Board of Admiralty before the discovery of the electric telegraph. The invention of the system of semaphore telegraphy is usually attributed to Richard Lovell Edgeworth in 1767, although the idea had occurred to several other inventors in other countries. At any rate, they were in regular use by the French in 1794, a year before they were introduced into England by Lord George Murray. The credit for the invention is given in France to the brothers Chappé, who in their younger days were sent to different schools a mile and a half apart. As they were not allowed to communicate with one another, they ingeniously set to work and devised a means of signalling by means of pieces of wood exhibited at their respective back - windows. In after - years they improved upon their simple device, but the ignorance and superstition of the French prevented it being put to any real use. Being fortunate, however, in 1793 in telegraphing the news of a

Exterior of the Semaphore-Station, Telegraph Hill. (From a sepia sketch dated 1836.)

victory from the frontier to Paris, the utility of the system became at once apparent, and semaphore-stations began to be generally established, not only in France, but all over the Continent. In Russia, particularly, some millions of pounds were spent in building a line of semaphores from the German frontier right through Warsaw to St. Petersburg. This line was only completed in 1858, and no sooner was it at work than the introduction of the electric telegraph made it quite out of date.

The semaphore on Telegraph Hill was on the line between London, Deal, and Dover, communication being established

Interior of the Semaphore-Station, Telegraph Hill.

The men in the centre are working the signals, whilst of those shown with the telescopes one is receiving the message from Shooter's Hill, and the other is seeing that the same is duly sent forward at the next station.

in 1795. Telegraphs were placed at the Admiralty, St. George's-in-the-Fields, Telegraph Hill, Shooter's Hill, and so on down to Deal. The station consisted of a small wooden hut, on the top of which were six shutters, arranged in two frames; by means of opening and shutting these in various ways, sixty-three distinct signals could be formed. Each station was in the charge of a naval officer, usually a Lieutenant, with three men to assist him in receiving the messages and transmitting them to the other stations. It is interesting to know that this was the actual station that communicated to the Metropolis the news of the glorious victory of Waterloo.* The Deal and Plymouth lines fell into disuse soon after the peace of 1815, owing to the cost of maintenance.

It will be gathered from the fact that the Admiralty chose this as a telegraph-station that it must command very extensive views. Situated as it was then in the midst of fields, far away from any houses, there was nothing to impede the view into the surrounding counties. Even now, though the terraces of bricks and mortar considerably mar the prospect, the Tower Bridge, St. Paul's, and Westminster Abbey can be distinctly seen, and on a fine day Alexandra Palace is visible. The finest view is towards the south in the direction of Sevenoaks, the most prominent feature being the well-known clump of trees called Knockholt Beeches.

The growth of this neighbourhood has been very rapid. The Ordnance map of 1873 is a perfect blank as far as houses are concerned, the district being shown as a series of fields. Some of the finest nursery-grounds and market-gardens in the south of London were situated here. When the school adjoining the ground was opened in 1875, there was not a paved road or a house within some hundreds of yards of it, and the party of the Haberdashers' Company at the opening ceremony had to dismount from their carriages and walk along planks placed across the mud, which was too deep for the carriages to traverse. This development has, of course, added considerably to the value. At the time of the purchase of the manor by the Haberdashers' Company, it was assessed for land-tax at a little over £100, now it must be worth a fabulous amount.

Before the telegraph-station here gave it its present name, this hill was known as Plow'd-Garlic-Hill. The derivation of this name is involved in some obscurity. One conjecture is that a member of the Garlic family, of whom there are

* T. Sturdee, 'Reminiscences of Old Deptford,' 1895, p. 18.

still some representatives in Deptford, may have held the land as a farm, and given his name to it.

Geographically, Telegraph Hill is in the county of Surrey, although it has in past years been considered as part of Kent. Philipot (1796) in his ' Villare Cantianum,' speaking of the Manor of Hatcham, says :

General View of Telegraph Hill.

' The manor was formerly considered as part of the county of Kent, and its appropriation to either county became a matter of contest until the year 1636, when it was decided judicially to be subject to assessments as belonging to Surrey. This determination was made on the petition of Mr. Randolph Crew, a London merchant, probably lessee of the manor, who, on a levy of ship-money, was taxed for his property here by the assessors of both counties. He did not, like Hampden, question the legality of the tax, but merely objected to the hardship of being compelled to make a double payment, and petitioned the Lords of the Council for redress ; when, being referred to the Judges of Assize for Kent and Surrey, they, after inquisition and examination of witnesses, on May 31, 1636, certified the Lords that the petitioner's Manor of Hatcham lies in Surrey, and not in Kent. The certificate was signed by Francis Crawley, Justice of the Common Pleas, and Richard Weston, Baron of the Exchequer.'*

The Manor of Hatcham-Bavant, or Hatcham-Barnes, in which the recreation-ground is situated, is an offshoot of the Manor of Hatcham. This parent manor was at the time of the Doomsday survey in the hands of the Bishop of Lisieux. The entry runs as follows :

' In Brixton Hundred, the Bishop of Lisieux holds of the Bishop of Bayeux Hachesham, which Brixi held of King Edward. It was then assessed at three hides as it is now, the arable land amounts to three caracutes. There are nine villanes and three bordars with three caracutes, and there are six acres of meadow ; the wood yields three swine ; from the time of King Edward (the Confessor) it has been valued at forty shillings.'†

From the description given in this entry, together with the old Saxon name of Deptford-Meretone—*i.e.*, the town in the marshes—we can easily gather that in early times this district consisted of well-wooded marsh-land. Owing to its proximity to the Thames, it must have been covered with swamps and creeks. Traces of Roman occupation have been discovered from time to time. In 1735, so Hasted, the Kentish historian, informs us, there were unearthed in a garden near the road at New Cross a simpulum (sacrificial cup), two urns, and ' five or six of those viols usually called lachrymatories.'‡

* Quoted in Dew's ' History of Deptford,' 1883, p. 15.   † *Ibid.*, p. 44.
‡ Hasted's ' History of Kent,' by Streatfield and Larking, edited by H. H. Drake.

In Henry II.'s time Hatcham gave name to a family, one of whom, Gilbert de Hatchesham, accounted for four knights' fees of the barony of Walkelin de Maminot. In the reign of Richard I., two knights' fees in Hatcham and Camberwell were held of the Earl of Hereford by William de Say (from whom Sayes Court is named) and the heirs of Richard de Vabadun. Roger de Bavant, who had married the daughter and heiress of De Vabadun, owned the manor in the time of Henry III., and accounted for two knights' fees of the above-mentioned barony. The tithes of Hatcham were given to the monks of Bermondsey in 1173, and in 1274 a composition was made between the Prior of Bermondseye and the Abbot of Begham concerning the tithes of Hacchesham, in the parish of West Greenwich, let to the said Abbot for 13s. 4d. per annum.

To pass on now to 1285, we find that Adam de Bavant, son of Roger, had free warren for his lands here; but he alienated a portion of the estate directly afterwards to Gregory de Rokesley, citizen of London, formerly Lord Mayor: In the same year he obtained a faculty from the Abbot and Convent of Begham for his oratory, which he had built for the use of himself and family here at Hechesan, in their parish of West Greenwiche, saving to themselves all oblations and other rights. The portion retained by Adam de Bavant, with which we are more immediately concerned, was distinguished from the remainder by the name of Hatcham-Barnes. It was afterwards conveyed, together with other properties, by Roger de Bavant to King Edward III., and he, by letters patent dated July 20, 1371, granted the manor to the Prioress and Convent of Dartford, which he had founded. It remained in their hands till Henry VIII. confiscated the property of all the monasteries, and it was held by the Crown till the time of Philip and Mary. Ann, widow of George Seymour, Duke of Somerset, then had a life interest in the manor assigned to her.*

* Hasted's 'History of Kent,' by Streatfield and Larking, edited by H. H. Drake, p. 19.

On its reverting again to the Crown, James I. in 1610 granted it, together with other lands formerly belonging to the Convent of Dartford, to George Salter and John Williams. They sold the estate to Peter Vanlore, and he in turn to a person named Brookes, who conveyed it to Sir John Gerrard and Sir Thomas Lowe, Aldermen of London, Robert Offley and Martin Bond, citizens of London and haberdashers, for a consideration amounting to £9,000. These funds had been bequeathed by a Mr. William Jones,

Hatcham House. (From a water-colour sketch dated 1841.)

a native of Newland, near Monmouth, to be held by the Haberdashers' Company in trust for founding and support-ing an almshouse and free grammar-school at Monmouth. This William Jones is described as a pedlar or travelling haberdasher, and, as was then the custom, he was a member of the Company of Haberdashers, formerly called mercers or merchants. Having become rich, he left his wealth, like many others of his class, for the improvement of the condi-tion of his less-fortunate brethren, and for the education of children of future generations. The manor - house was advertised to be let in February, 1775, and was described then as being surrounded by a moat well stocked with fish. Hatcham House, as it was called, with its moat and park, has long since disappeared, the site being covered with rows of cottages, although it has given its name to the locality known as Hatcham Park.

The space between the southern plot and Pepys Road is occupied by the modern Church of St. Catherine, Hatcham, built from the designs of Mr. Henry Stock, A.R.I.B.A., and consecrated on October 10, 1894, by the Bishop of Rochester. It is another standing memorial to the liberality of the Haberdashers' Company, the patrons of the living, who built and endowed it at a cost of over £22,000. It is cruciform in shape, constructed of Kentish ragstone, and accommodates 900 people. The total internal length is 127 feet 6 inches, with a maximum breadth of 57 feet 6 inches. The five-aisled arrangement of the transept is a special feature of the interior, which gives an appearance of great size, and keeps the perspective of the aisles unbroken. The pulpit and reredos are of stone, and are enriched with mosaics depicting Scripture scenes. When the tower and spire are added, the church, owing to its elevated site, will be a conspicuous landmark for miles around.

The adjoining Aske's Schools, on the opposite side of Pepys Road, built by the Haberdashers' Company in 1875, have served the purpose for which they were placed in the middle of an estate ready for development. They were founded under a scheme of the Charity Commissioners dated 1873, utilizing funds bequeathed in 1688 by Robert Aske, citizen and haberdasher, for the maintenance of almshouses and the education of twenty boys, sons or grandsons of freemen. This Robert Aske, who lived towards the end of the seventeenth century, was a grandson of the celebrated Robert Aske, of the old Yorkshire family, who headed the insurrection—known in history as the Pilgrimage of Grace, 1536— against Henry VIII.'s arbitrary policy in Church matters, and especially the dissolution of the smaller monasteries. The name Aske is another form of Ash, and refers to the

strong, straight, and useful ash-tree. The amount of the
bequest realized £20,000, and it was the subject of a special
Act of Parliament, December 20, 1690. The trustees—the
Haberdashers' Company — expended this money in the
purchase of 21 acres in Hoxton, and other lands near
Ashford, in Kent. The total amount held reaches nearly
2,000 acres, but whereas the value of the Hoxton lands has
increased enormously during the 200 years they have been
in the company's possession, the depression in agriculture
has seriously lessened the income of the Kentish land. The
4 acres of land on which the school-buildings stand were
purchased in 1873 for £3,200 from Jones' Charity, by consent
of the Charity Commissioners. The schools were opened in
1875, one for boys to accommodate 300, the other for 200
girls. This accommodation proved insufficient as the schools
filled, and a new school was built for 400 girls at the bottom
of Jerningham Road, and opened in January, 1891. The
building up to that time occupied by the girls was then
handed over to the boys' school.

CHAPTER VII.

KENNINGTON PARK.

I T is not often we find, in tracing the past history of any open space which has been left unenclosed for any considerable length of time, that the last state is better than the first; yet such is the case with Kennington Park. So many of the commons in the south of London have been liable to the pilferings of wealthy land-owners, through the laxity of those authorities to whom their care has been entrusted, that it is a relief to turn to a place which has ameliorated in the lapse of time. In these years this attractive and compact little park, as we shall have to relate further on in this chapter, has passed through many vicissitudes, no less strange than the whole district of Kennington itself.

Those who are best competent to form an opinion on such points affirm that in ancient days there was a vast bay here, which covered the whole of the low-lying district. Traces of this can be found in the fossil remains brought to light during excavations. The earliest recorded fact, too, in the history of Kennington is of a maritime nature, namely, the appearance of a fleet of warships in 1016, when Canute the Dane came through it by means of a canal on his way to capture London from the Saxon King Ethelred the Unready.* Another fact in the early history of Kennington, also connected with the Danes, accounts for the etymology of the name, which means King's Town. Canute's son, Hardi-

* Tanswell, 'History of Lambeth,' 1858, p. 197.

canute, was present at the marriage-feast of one of his lords, and drank so heavily that he died here some days afterwards.

The land now enclosed as Kennington Park, together with the outlying portions, were part of the Manor of Kennington, which is the property of the Prince of Wales in right of his Duchy of Cornwall. The entry in Doomsday Book with regard to this manor states that ' Teodric, the goldsmith, holds of the King Chenintune. He held it of King Edward. Then it was taxed for five hides, now for one hide and three virgates. The arable land consists of two caracutes and a half. In demesne there is one caracute and one villan, and one bordar with two caracutes. There is one villan in gross and four acres of meadow. It was worth and is worth £3.'

In 1189 King Richard I. granted to Sir Robert Percy the custody of all his demesne lands in this manor, together with all the profits, in return for the payment of 20 marcs a year. He was also appointed steward of the lordship of Kennington, and keeper of the manor-house there, for which the King remunerated him.

In 43 Henry III. the custody of this manor was granted by the King to Richard de Freemantell. Without entering into the various grants of the manor, which are of little interest, we can come to the time of Edward III., when it was in the possession of Edward the Black Prince, probably by grant from the King. After his death in 1377 it descended to his son Richard (afterwards King Richard II.), who was living at the Palace of Kennington with his mother when his father died.

King James I. in his eighth year settled the Manors of Kennington and Vauxhall, together with a messuage in Lambeth and Newington, on Henry, Prince of Wales ; and subsequently, on his death in 1612, upon Prince Charles, and the Manor of Kennington has ever since remained part of the estate of the Prince of Wales.*

* Allen, ' Lambeth,' pp. 255-259.

Kennington Common and Church in 1830.

It is only of late years that this place has been exalted to the dignity of a park. Before 1852 it was simply Kennington Common, and probably formed a continuous piece of ground with St. George's Fields at a time when they were fields in reality as well as in name. The common was uncared for, the wooden railings which divided it from the public road were allowed to fall into decay, and all kinds of shows and stalls took up their stand there without any interference. Its surface was covered with hillocks, ponds and ditches, and it was divided into sections by the coach-road. Some houses facing the common in Manley Place had back-gates on to it, some of which remain even now. The common, like most other village greens, had no particular notoriety till the middle of last century when it was the scene of several gruesome executions. It must have been a particularly uninviting spot to judge from a description written in 1794 : ' At present it is common to all cattle, without stint, belonging to those parishioners who reside within the Prince of Wales's liberty, whose property it is, who pay a certain stipend per head ; the sum goes towards defraying those expenses which the keeping up of the fence, etc., necessarily incurs. It is shut during the winter six months and opens again in spring ; but it is no sooner opened than the number of the cattle turned in is so great, that the herbage is soon devoured, and it remains entirely bare the rest of the season.' * The result would be much the same now, if the public were allowed to roam freely all over the park, to such an extent is it patronized. On a recent Whit Monday a census of visitors was taken, and it was found that no less than 40,000 entered, giving an average of 2,051 per acre. It is found necessary to close and open alternately the two large grass enclosures in the centre of the park which are specially devoted to the children. If this were not done, there would hardly be a square yard of grass left by the end of July.

It is hard to realize that this little oasis was the scene of

* Malcom's Report, 4to., 1794.

the butcheries we have incidentally referred to. In 1745 Charles Edward Stuart, better known as the Young Pretender, attempted to regain the throne of England for the Jacobites, and made an advance upon Carlisle, where some of his adherents had been left, but was forced to retire. The Duke of Cumberland thereupon laid siege to the town and captured it, taking several prisoners. Among these were several officers of the Manchester Regiment, upon whom was vented all the wrath of the Duke. Jesse, in his 'Memoirs of the Pretenders,' tells us of the fate they met with on Kennington Common. Their names were Francis Townly, who commanded the regiment, Fletcher, Chadwick, Dawson, Deacon, Berwick, Blood, Syddal and Morgan, who were tried in the court-house of St. Margaret, Southwark, on July 15, 1745, and the three following days, and were all ordered for execution. Eight of their brother officers, who were condemned at the same time, received reprieves. The whole of these gallant but ill-fated men met their end with the greatest firmness, remaining true to their principles to the last. About eleven o'clock on July 30 they were conveyed in three hurdles from the new gaol, Southwark, to Kennington Common attended by a strong guard of soldiers. In the first hurdle or sledge were Colonels Blood and Berwick, the executioner sitting by them holding a drawn sword. All the horrors which had been contrived in a barbarous age as a punishment for high treason were actually carried out on this occasion in their most terrible shape. Near the gallows were placed a block and a large heap of faggots; the former to assist the hangman in his bloody task of disembowelling and beheading the prisoners, and the latter for burning their hearts and entrails. While the prisoners were being transferred from their several sledges into the cart from which they were to be turned off, the faggots were set on fire, and the soldiers then formed a circle round the place of execution.

Though unattended by a clergyman, they spent about an hour in devotion, Morgan taking upon himself the task of

10

reading prayers, to which the others calmly but fervently responded. On rising from their knees, they threw some written papers among the spectators, which were afterwards found to contain the most ardent professions of attachment to the cause for which they suffered, and a declaration that they continued true to their principles to the last. They also severally delivered papers of a similar import to the Sheriffs, and then, throwing down their gold-laced hats, they submitted themselves to the tender mercies of the hangman. Having hung about three minutes Colonel Townly was the first to be cut down, and having been stripped of his clothes was laid on the block, and his head severed from his body. The executioner then extracted his heart and entrails, which he threw into the fire, and in this manner, one by one, proceeded to the disgusting task of beheading and disembowelling the bodies of the remaining eight.

Three more of the officers captured at Carlisle were also executed three weeks later at Kennington Common—James Nicholson, Walter Ogilvie, and Donald Macdonald. Being Scotchmen, they came to the place of execution dressed in full Highland costume, and were subjected to the same tortures as the English officers. Once more in November of the same year five more persons were executed—John Hamilton, Governor of Carlisle, who had signed its capitulation; Sir John Wedderburn, Bart., who had taken charge of the Excise in the time of the insurrection; and three others. When in prison they were not informed of their fate till the morning of the day fixed for their execution, when they were conducted to the common. It may be noted here in connection with these executions that one of the titles of the Duke of Cumberland was ' Earl of Kennington,' and this may possibly account for their taking place on the common and not at Tyburn.

An old record of 1678 gives an account of another execution which was quite in accord with the primitive ideas of justice of those times. It is as follows: ' Warning for bad wives, or the manner of the burning of Sarah Elston, who was burnt to death at the stake of Kennington Common for the murder of her husband. On the day of the execution Sarah Elston was dressed all in white, with a vast multitude of people attending her, and after very solemn prayers offered on the said occasion, the fire was kindled, and giving two or three lamentable shrieks, she was deprived both of voice and life, and so burnt to ashes.'*

Flower-Beds in Kennington Park.

The last person executed on the common was a man named Badger, who was convicted of forgery in the early part of this century. He was a man moving in good society, and lived in a large house near Camberwell Green, where he frequently gave musical parties. The circumstances connected with his arrest reveal the humorous side of his character. On this particular night he was giving a musical evening, and when the guests were all assembled the servant

* Quoted in Montgomery's 'History of Kennington,' p. 32.

came up to the master of the house and announced the fact
that there were two gentlemen in the hall who wished to
speak with him. Badger descended to the front-door, and
there found that the two gentlemen were Bow Street officers
sent to arrest him on a charge of forgery. Badger informed
them that he had a large party of friends in the house at that
moment, and represented that it would be most inconvenient
to himself, and a great shock to his family and his friends,
were he to be arrested at once. He asked if they would
consent to remain in the house until the party broke up,
when he would at once accompany them. The two officers
consented to wait till the close of the party on condition
that Badger did not leave their presence on any account.
They were then actually introduced to the company as two
City friends of the host, who had unexpectedly come to the
house, and whom he had prevailed upon to join the party.
Badger kept up his spirits and did his duty so well that no
one suspected the truth; he sang songs at the very piano
which concealed the little closet in the wall where he had
secreted the evidences of his guilt. When the guests were
all gone, the family were informed who these two strange
gentlemen were, and Badger, taking leave of his relatives,
was conveyed away. He was convicted, and sentenced to
be hanged, and his execution was the last which took place
on Kennington Common.* As an example of the irony of
fate, it may be mentioned that Kennington Church now
occupies the site of these executions, so that in a very literal
sense the blood of the martyrs becomes here the seed of the
church. In preparing the foundations, the site of the gibbet
was discovered, and a curious piece of iron, which probably
was the swivel attached to the head of the unfortunate
criminal, was also found.†

This church, dedicated to St. Mark, was one of those built
out of the funds voted by Parliament in 1818 for the building
of churches in London and in the great provincial towns.

---

* Montgomery, 'History of Kennington,' p. 33.
† Allen, 'History of Lambeth,' 1827, p. 386.

After the Battle of Waterloo in 1815 a resolution was passed in the House of Commons 'That it would be necessary and becoming to make some great demonstration of thankfulness to Almighty God for the return of peace by promoting the building of churches.' The sum of a million and a half sterling was the outcome of this resolution, and among others were built the four churches so similar in external appearance: St. Matthew, Brixton; St. Mark, Kennington; St. Luke, Norwood; and St. John, Waterloo Road. The foundation stone of St. Mark's was laid in 1822, and it was consecrated in 1824, the cost of building being £15,274.

The road which divides the church from the main portion of Kennington Park is on the site of the celebrated Roman road, Watling Street, which commenced at Richborough in Kent, and passing through Canterbury, Rochester, and Dartford came to Kennington on its way to Chester and Carnarvon. Old prints of Kennington Common show us the turnpike gate which existed till recent years. Though the gate has disappeared, the name still remains.

The common was favoured with as many preachers as the church, and perhaps there has been no piece of ground so much used for preaching of all kinds. Whitefield sometimes preached here to congregations of 40,000 people, and he records in his diary an account of a farewell sermon he delivered on the common before one of his trips to America. The entry runs as follows:

'*Friday, August* 3, 1739.—Having spent the day in completing my affairs and taking leave of dear friends, I preached in the evening to near 20,000 people at Kennington Common. I chose to discourse on St. Paul's parting speech to the elders at Ephesus, at which the people were exceedingly affected, and almost prevented my making my application. Many tears were shed when I talked of leaving them. I concluded with a suitable hymn, but could scarce get to the coach for the people thronging me to take me by the hand and give me a parting blessing.'

The doubtful quality of some of the itinerant preachers

led to their being caricatured in a play called ' The Hypocrite,' which was produced at Drury Lane in 1769.

' *Lady Lambert.*—Did you ever preach in public ?

' *Mawworm.*—I got up on Kennington Common the last review day; but the boys threw brickbats at me, and pinned crackers to my tail; and I have been afraid to mount, your ladyship, ever since.'*

Father Mathew, the celebrated Irish temperance advocate, who claimed to have made more than a million converts, was also here in 1843. At the meetings which he held on the common on August 7, 8, and 9 of that year, 8,000 persons took the pledge. A former proprietor of the Horns, who was by no means a disinterested party, in entering an account of this event in his diary, very dryly adds : ' Plenty of business in the house.'†

The common was the scene of the first nomination of members of Parliament for the borough of Lambeth. At the time of the passing of the Reform Bill of 1832, Lambeth had 4,768 registered electors, but no member. The Reform Bill made Lambeth a Parliamentary borough with two seats, and when the voting took place, one of the chief polling-booths was on the common.

The next elections on the common were in 1834, 1837, 1841, and 1847.

But although the passing of the Reform Bill had given satisfaction in Lambeth, there were many people—especially among the lower classes—who felt themselves aggrieved by its provisions. They banded themselves together in various parts of the country under the name of Chartists, from their demanding the People's Charter, the six points of which were :

> Universal suffrage,
> Vote by ballot,
> Annual Parliaments,
> Payment of the members,
> Abolition of property qualifications, and
> Equal electoral districts.

* ' Hypocrite,' Act II., Scene 1.
† Quoted in Montgomery's ' Kennington,' p. 56.

The movement did not make much headway till 1838, when the malcontents assembled at different towns, armed with guns, pikes, and other weapons, and carrying torches and flags. A proclamation was issued against them in December of that year, and in 1839 their petition was presented to Parliament by Mr. T. Attwood. Soon after this they committed great outrages at Birmingham and Newport. For some time they held a sort of Parliament called the National Convention, the leading spirits being Feargus O'Connor, Henry Vincent, and Mr. Stephens. After a break of some years, there occurred in 1848 the meeting on Kennington Common which would have made it famous if nothing else had ever happened there. The meeting was fixed for April 10. On the day before there was a feeling of terror throughout the Metropolis. A vague sense of approaching evil seemed to haunt the minds of the populace. But there was one man who was fully equal to the occasion, and by his firmness, coolness, and tact in a great measure allayed the general feeling of unrest. This was the Duke of Wellington. Thousands of soldiers and police were called out for the maintenance of order, and an army of special constables was enrolled, computed at 150,000 for the Metropolis, among them being Louis Napoleon, afterwards Napoleon III. The great Duke himself paid a visit to Kennington early on the morning of the 10th, to see that all the arrangements were complete. The Chartists proposed to hold a meeting of 200,000 people on the common, and thence to march in procession to the Houses of Parliament to present a monster petition embodying their views. They were informed by the police that the monster petition would be allowed to be taken to the House, but that no procession through the streets would be permitted. The proceedings on the common commenced with a few denunciatory speeches, and then the mob became restless, so that no one could obtain a hearing, and the chairman found it necessary to dissolve the meeting. The crowds then melted away without any order, and the bundles forming the petition were ingloriously conveyed to the House in cabs. The numbers on the common fell very far short of the boasted 200,000, some authorities putting them as low as 15,000. About an hour afterwards only 100 persons were to be seen on the common, and by a few hours, anyone who had not been present would have been quite ignorant that anything unusual had happened. So ended this great fiasco, and from this time the proceedings of the Chartists became insignificant.

Soon after the Chartist meeting, Kennington Common was transformed into a park. The credit of this belongs to certain local gentry, chief among whom was Mr. Oliver Davis. With him were associated Mr. Adam and the Vicar of the parish, the Rev. Charlton Lane. It was a long and arduous fight for them, and it was not till 1852 that they succeeded in their object. There were some 200 copy-holders who had rights in the common ; in many cases they were not to be found, and when their consent was asked to the change, it was not granted. This will give some idea of the difficulties that had to be encountered.* Finally an Act of Parliament was obtained, and the Government agreed to form the park on condition that the inhabitants paid the cost of the railing round it, estimated at £1,000. The Prince of Wales, who, as owner of the Duchy of Cornwall estates, has large interests in the neighbourhood, subscribed £200, and the remaining £800 was raised by local subscrip- tion. In the laying out, an amalgamation of the plain geometrical and the English styles has been adopted. It is furnished with a gymnasium and playground, which in this populous neighbourhood is in constant use. Around the lodge there is an effective arrangement of common garden flowers in sunk panels of turf.

Although the park is a small one, being only 19½ acres in extent, the work of laying out has been carried out most effectively. The shrubs are very fine, particularly the belt of flowering varieties just inside the railings. The

* Montgomery, 'History of Kennington,' p. 48.

control of the park was transferred to the late Metropolitan Board of Works in 1887, under the London Parks and Works Act, and two years later to their successors, the London County Council. It has undergone various improvements under their ownership. A strip of land facing South Place was acquired under Parliamentary powers in 1888, and added to the park. There was also a small strip at the junction of Brixton Road and Camberwell New Road, called ' No Man's Land,' used as a store-yard, upon which stood the remains of the old pound and a few large trees. As this proved quite an eye-sore, and could not be put to any profitable purpose, it was laid out, planted, and fenced, and now forms a kind of ' anteroom ' to the park itself.

The laying-out of the park led to another event of almost national importance, namely the formation of the Oval cricket ground, the scene of so many keen struggles. The Prince Consort said that as the inhabitants were losing a place where they played cricket, the Duchy of Cornwall would lease the Oval, which was then a market-garden, to any proper authorities who would encourage the sport. This led to the formation of the Surrey Club, as the consequence of a meeting held at the Horns, and ever since the Oval has been leased to them at a low rental* (£750 at present).

The park contains an unpretentious memorial to the late Prince Consort. In the ivy-covered lodge may be recognised the model lodging-house which was designed by him for the Exhibition of 1851. A story† is told of the first inhabitants of this lodge, which led to the fountain being erected in the park. These people had been dubbed ' Adam and Eve.' One sultry afternoon a gentleman nicknamed ' Young Slade ' went into this house to ask for a glass of water. It may be mentioned that ' Young Slade ' was only eighty, but as his father had always been known as ' Old Slade,' the distinctive name of ' Young ' was given to his son. A glass of tepid water in a dirty tumbler was given him, for which

* Montgomery, ' History of Kennington,' p. 48. † *Ibid.*

he was charged threepence. Slade went home and brooded over this. The result of the brooding was a fountain presented to the park of the value of 500 guineas. It is of polished granite, surmounted by a bronze casting, which represents Hagar and Ishmael at the well, after the design of Mr. C. H. Driver, F.R.I.B.A.

The Drinking Fountain, Kennington Park.

The handsome terra-cotta fountain in the centre of the park between the gymnasium and the lodge was the gift of Sir Henry Doulton to the park, and was erected in 1869. The subject represented at the top is 'The Pilgrimage of Life,' and was suggested to the sculptor, Mr. George Tinworth, by a German original. The fountain is interesting chiefly as being the first piece of Mr. Tinworth's figure modelling after his connection with the Lambeth Art School. The design was carried out under the supervision of Mr. Sparkes, the Principal of the City and Guilds Institute, Kennington Park Road. The figure was not exhibited previous to its erection, and there is no replica of the fountain in existence.*

To leave the park for a moment, we may consider some of the buildings overlooking it which are of interest. The three houses adjoining the park at the corner of South Place and Kennington Road were taken by Dr. Randall Davidson as a temporary episcopal palace for the See of Rochester, and a permanent building was commenced in the grounds from the designs of R. Norman Shaw, R.A. This was completed and inaugurated in 1896.

On the opposite side near St. Mark's Church may be seen the dome of the Oval Station on the City and South London Electric Railway. The lines are carried in two sunk tunnels at some distance from the ground-level, which is reached by hydraulic lifts. The railway, which extends from the Monument to Stockwell, was opened by the Prince of Wales in November, 1890.

Overlooking the principal entrance to the park is the celebrated Horns Tavern, with its assembly-rooms. This handsome pile of buildings was erected in 1887. In the old inn on the same site, which had tea-gardens at the rear, died an eccentric individual, Joseph Capper. He had a very humble beginning in life, coming from Cheshire to London to be a grocer's apprentice. He was very energetic in business, and started on his own account as soon as his apprenticeship was finished. Having been very successful, he amassed sufficient property to retire. For several days he walked about the vicinity of London searching for lodgings, without being able to please himself. Being one day much fatigued, he called at the Horns, took a chop and spent the day, and asked for a bed in his usual blunt manner, when he was answered in the same churlish style by the landlord that he could not have one. Mr. Capper was resolved to stop if he could all his life, to plague the growl-

* Communicated by Messrs. Doulton and Co.

ing fellow, and refused to retire. After some further alterca-
tion, however, he was accommodated with a bed, and never
slept out of it for twenty-five years. During that time he
made no agreement for lodging or eating, but wished to be
considered a customer only for the day. He lived in a most
mechanical way, and his bill for a fortnight amounted
regularly to £4 18s. His conduct to his relations was
extremely capricious; he never would see any of them. As

View of Flower-Beds and Tinworth Statuette, Kennington Park.

they were chiefly in indigent circumstances, he had frequent
applications from them to borrow money. 'Are they industrious?' he would inquire, when, being answered in the
affirmative, he would add: 'Tell them I have been deceived
already, and never will advance a sixpence by way of loan;
but I will give them the sum they want, and if ever I hear
that they make known the circumstance, I will cut them off
with a shilling.' Soon after a Mr. Townsend became proprietor of the Horns, and being in want of ready money, he

applied to Mr. Capper for a temporary loan. ' I wish,' said Capper, ' to serve you, Townsend—you are an industrious fellow—but how is it to be done, Mr. Townsend ? I have sworn never to lend ; I must therefore give it thee,' which he accordingly did the following day. He died in October, 1804, leaving the bulk of his property (upwards of £30,000) among his poor relations, and was buried in a vault under Aldgate Church.*

Opposite to the Horns Tavern on the other side of the park stands St. Agnes' Church, designed by Sir G. Gilbert Scott in the English Middle Pointed style of architecture. It depends mainly for its effect upon its loftiness, the height from the floor to the external ridge being 75 feet. Inside the church the chief features are the stained-glass windows and the chancel screen. It occupies part of the site of what were once very offensive vitriol works. These were the property of Thomas Farmer, and the streets which have been built on the site have been named after him Farmer's Road and Thomas Street. ' Kennington Common,' wrote Thomas Miller, in his ' Picturesque Sketches in London,' published in 1852, ' is but a name for a small grassless square, surrounded with houses, and poisoned by the stench of vitriol works and by black, open, sluggish ditches ; what it will be when the promised alterations are completed we have yet to see.' We feel quite safe in leaving the decision in those of our readers' hands who have visited the park in its present state.

* Allen, ' History of Lambeth,' pp. 385, 386.

CHAPTER VIII.

LADYWELL RECREATION-GROUND—MARYON PARK.

LADYWELL RECREATION-GROUND.

THIS recreation-ground consists of a series of meadows stretching along the banks of the river Ravensbourne for a distance of nearly a mile between Catford Bridge and Ladywell railway-stations. There are two other grounds in the district which are apt to be confused with this, namely, the Ravensbourne Recreation-Ground and the Sydenham Recreation-Ground, but these are maintained locally.

Ladywell takes its name from the old well in Ladywell Road, at the two old cottages near the entrance to Ladywell Cemetery. In ancient times, no doubt, some miraculous cures took place at this spot, which would be connected with Our Lady by the faithful, and so the name has been handed down.

This recreation-ground was formerly part of the Manor of Lewisham, which means the village of pastures. This manor was given by Ælthruda, niece of Alfred, about the year 900 to the Abbey of St. Peter in Ghent, and it is thus described in Doomsday Book: 'The Abbot of Gand holds of the king, Levesham. And he held it of King Edward. And then, and now, it answers for two sulings. There is the arable land of fourteen teams. In demesne there are two teams. And fifty villans, with nine bordars, have seventeen teams. From the produce of the port forty shillings. Thirty acres of meadow there. The entire manor, in the time of King Edward, was worth sixteen pounds. And afterwards twelve pounds. Now thirty pounds.' As St. Peter's of Ghent was a foreign abbey, a cell known as Lewisham Priory was established here in accordance with the usual custom. The manor with its appendages remained in the possession of the abbey till 1414, when the alien priories throughout England were suppressed, and their lands forfeited to the Crown. King Henry V. then granted the manor for the support of his new-founded house, or Carthusian priory, of Bethlehem, of Shene. It reverted to the Crown in 1538 together with

Sketch of the Lady Well. (From Knight's ' Journey through Kent,' 1842.)

other conventual property throughout the country, and ten years later was granted for life to Thomas, Lord Seymour. In 1550 it was in the hands of John, Earl of Warwick, eldest son of the Duke of Northumberland; but on his attainder in 1553 it again reverted to the Crown, where it remained till 1563, when Queen Elizabeth regranted it to the Earl's brother, Sir Ambrose Dudley, for life. James I. gave the manor in 1624 to John Ramsay, Earl of Holderness. John Ramsay, when a page attending James I. at the house of Earl Gowry at Perth, had the good fortune to discover and frustrate the attempt of the Earl and his

brother on the King's life. For this service he was created
Viscount Haddington, and afterwards Earl of Holderness.
In 1664 it was sold for £1,500 to Reginald Grahame, who
in turn conveyed it to Admiral George Legge, afterwards
created Baron Dartmouth. As Admiral of the Fleet, he was
sent to demolish Tangier, and had a grant of £10,000 for his
services. His son, William, was in 1711 created Viscount
Lewisham and Earl of Dartmouth, and the property has
been ever since in the hands of his descendants.*

The land forming the present recreation-ground of 46¼
acres was purchased under Parliamentary power in 1889 at
a cost of £21,880, to which the Lewisham District Board
contributed half. Two years later an addition of 3 acres
was made, being the net gain in an exchange with the
Shortlands and Nunhead Railway Company. In 1894 another
small addition of a quarter of an acre was made by the
acquisition of a plot of land on the east side of the Ravensbourne, near Lewisham Church.

Considerable works of improvement were necessary before
the ground could be suitable for a place of public recreation.
The lands were low-lying and subject to floods, with the
exception of the south-western portion, which is of fair
elevation and commands some good views. This flooding
was due to the peculiar nature of the river Ravensbourne,
which drains a watershed area above Ladywell Bridge of
about 30,000 acres, or nearly 47 square miles. The stream
is tortuous in its course, and in places the channel is
extremely confined, and quite inadequate to the proper discharge of the water brought down from the drainage area
after excessive rainfall. As a consequence, the valley has
always been liable to floods of more or less severity.
Records of that of 1878 show that the low-lying lands here
must have been in parts 4 feet under water. Improvement
works, executed below Ladywell Bridge since that time,
have, however, mitigated the evil to some extent ; but it was
necessary, in order to secure a free and rapid discharge of

* Thorne, 'Environs of London,' vol. ii., p. 417.

the flood waters, to cut off the sharp bends of the stream, to enlarge the channel generally, and otherwise clear the bed of its former obstructions. Several new cuts for the channel were also made, and where the old bed was left islands were formed and planted, thus adding considerable attractiveness and picturesqueness to the ground. In dry summers the river suffered from the opposite extreme,

The Source of the Ravensbourne, Keston Heath. (From an old woodcut.)

namely, an absence of water. To prevent the bed of the river from becoming dry weirs were constructed of moderate height, and the miniature waterfalls so formed have added a decided charm to its appearance. Six rustic wooden bridges have been thrown across the river, and the land generally levelled and drained, in order to give the utmost possibilities for recreation. What was formerly therefore a mere swamp has been converted into a pleasant garden, whilst the proximity of the river always insures a freshness and brightness for the carpet of turf.

11

The river Ravensbourne is probably the chief feature of this open space. There is hardly a river in England without some legendary history to spread a halo of romance around it. The romance of the Ravensbourne lies in its name, of which the following account is given. It is said that Julius Cæsar, on his invasion of Britain, was encamped with all his force a few miles distant from its source. The army was suffering a good deal from want of water, and detachments had been sent out in all directions to find a supply, but without success. Cæsar, however, fortunately observed that a raven frequently alighted near the camp, and conjecturing that it came to drink, he ordered its arrival to be carefully noted. This demand was obeyed, and the visits of the raven were found to be to a small clear spring on Keston Heath. The wants of the army were supplied, and the spring and rivulet have ever since been called the Raven's Well, and Ravensbourne.* Hasted, in his ' History of Kent,' gives a view of the Roman entrenchments on Holwood Hill, and figures the ancient road to the spring of the Ravensbourne as running down to it from where Holwood gates now stand ; he also figures the spring with twelve trees planted round it.† The story is such a pretty one that it would be too cruel to point out how unlikely it is.

A good description of its course has been put into verse :

> ' On Keston Heath wells up the Ravensbourne,
> A crystal rillet, scarce a palm in width,
> Till creeping to a bed, outspread by art,
> It sheets itself across, reposing there ;
> Thence, through a thicket, sinuous it flows,
> And crossing meads and footpaths, gath'ring tribute,
> Due to its elder birth from younger branches,
> Wanders in Hayes and Bromley, Beckenham Vale,
> And straggling Lewisham, to where Deptford Bridge
> Uprises in obeisance to its flood,

---

* Dew, ' History of Deptford,' p. 55.
† Hasted ' History of Kent,' folio, vol. i., p. 129.

Whence, with large increase, it rolls on to swell
The master current of the "mighty heart"
Of England.'*

The Ravensbourne forms the east boundary of Deptford, and in 1849 it also became the boundary of the first Metropolitan Commissioners of Sewers. The place where it empties itself into the Thames is known as Deptford Creek, which has played an important part in the history of Deptford. The Danes in centuries ago had moored here, and it was to this point that Sir Francis Drake returned after circumnavigating the globe. The skeleton of his ship *The Golden Hind*, was laid up in the creek by command of Queen Elizabeth, where she had come herself before to visit Drake and knight him on board his vessel. The ship was broken up shortly afterwards, but a chair was made of the timber, and presented to Oxford University, where it is still preserved in the Bodleian Library.†

The quiet and unpretending Ravensbourne is not without its place in history. 'More than one tumultuous multitude has encamped upon its banks, shouting loud defiance to their lawful rulers. Blackheath, its near neighbour, was overrun by Wat Tyler, and the angry thousands that followed in his train ; and in the Ravensbourne, perchance, many of those worthy artisans stooped down to drink its then limpid waters, when, inflamed by revenge and by the hope of plunder and of absolute power, they prepared to march on London. Jack Cade and his multitudes in their turn encamped about the self-same spot ; and the Ravensbourne, after an interval of eighty years, saw its quiet shores disturbed by men who met there for the same purposes, and threatening bloodshed against the peaceful citizens of London, because, feeling the scourge of oppression, they knew no wiser means of obtaining relief, and were unable to distinguish between law and tyranny on the one hand, and freedom and licentiousness on the other.'‡

* Hone, ' Table-book,' p. 642.
† Lysons, ' Environs of London,' vol. i., p. 466.
‡ Charles Mackay, ' The Thames and its Tributaries.'

Mills have existed on the Ravensbourne from very early times. In the reign of Edward I. we find records relating to the purchase of a sixth part of a mill in Lewisham. These watermills were extremely valuable property, because they were held to be free of tithe, and the lord of the manor could compel all his feudal tenants to have their corn ground at his mill, for which they paid toll. If, however, the lord had leased his mill, he was entitled to have his own corn ground free of charge.* Evelyn refers to one of these mills in his diary under date April 28, 1688. ' To London, about the purchase of the Ravensbourne mills and land round it in Upper Deptford.' Lambarde, in his ' Perambulation of Kent,' written in 1570, speaks of the ' Brooke called Ravensbourne, which riseth not farre off, and setting on woorke some corne milles, and one for the glasing of armour, slippeth by this towne (Deptford) into the Thamyse.' John How, an eminent cutler of Saffron Hill, and county magistrate, died in 1737, leaving a fortune of £40,000 made from his cutlery mill on the Ravensbourne.† It was converted into a flour-mill, and afterwards taken down in 1865 for the introduction of steam power. The foundations of some of the workmen's houses were found to be composed of old grindstones.

At the north-west corner of the recreation-ground, from which it is only separated by the Ravensbourne and its churchyard wall, is the parish church of St. Mary, Lewisham. Part of the glebe land of the vicarage was purchased for this open space. The greater part of the present building dates back to 1774, when the former church standing on this site was taken down owing to its ruinous condition. The present church is a plain oblong structure of stone, with a portico of four Corinthian columns on the south side, and a tower, the lower portion of which (erected between 1470 and 1512) formed part of the old building. The chancel, which was formerly a shallow semicircular recess, was enlarged to its present size in 1882. An unfortunate fire which occurred in

* Kennett, ' Parochial Antiquities,' p. 236.

† Hasted's ' History of Kent,' by Streatfield and Larking, edited by H. H. Drake, p. 253.

1830, through the overheating of the warming apparatus, nearly destroyed the whole of the interior, and consumed the earliest parish registers, dating from 1550.

In the churchyard is a monument to the unfortunate young poet, Thomas Dermody, who was buried here July 20, 1802.

At the south-east of the recreation-ground there formerly stood an old farm-house called Priory Farm, which marked

St. Mary's Church, Lewisham.

the site of the Priory of Lewisham to which we have already referred. The farm-house was built partly with the old material of the priory, and surrounded on nearly all sides by the ancient moat, which was used for watering the farm cattle until it was filled in when the house was pulled down, about 1877. The Priory of Lewisham paid 40s. a year to the Abbey of Ghent as its superior.

MARYON PARK.

Maryon Park, the 'lung' of Charlton, which is about 12 acres in extent, was presented to the London County Council by the late Sir Spencer Maryon Maryon-Wilson, in 1891. It was the site of an old chalk or gravel pit, and in the centre is a large mound, well covered with undergrowth, locally known as Cox's Mount. From the plateau on the top extensive views of the river can be obtained, whilst the southern and western boundaries of the park are picturesque high banks largely covered with brushwood. With the exception of the mound and banks referred to, the park is maintained as a grass area on which lawn-tennis is played. A handsome bandstand has been erected, upon which performances are given once a week during the season, which attract a very large number of children. After his first gift, the late Sir Spencer generously presented sufficient ground for the formation of a gymnasium for children, and for an additional entrance from Woolwich Lower Road. The name of the park was given to it at the request of the donor, who opened it to the public on October 25, 1890. Since this time over £4,000 has been spent upon various works connected with the park and its approaches.

Although the park abuts on Mount Street, Woolwich, it is within the parish and manor of Charlton, of which the Maryon-Wilson family are lords, as well as that of Hampstead. The Manor of Charlton was given by William I. to his half-brother Odo, Bishop of Bayeux; from him it passed to Robert Bloet, Bishop of Lincoln, who gave it, somewhere about 1093, to the Priory of St. Saviour's, Bermondsey. The Dissolution, which deprived so many of the monasteries of their landed estates, resulted in the manor reverting to the King, Henry VIII. Although some leases were granted, the fee remained in the possession of the Crown till James I. granted it to one of his Northern adherents, John, Earl of Mar. King James I. dates several of his warrants and edicts from Charlton, and although there are twenty-one other places of the same name in England, the Wilson family religiously believe that he lived at Charlton House, their family seat, where the ceiling of the principal saloon is adorned with the royal arms of England. The manor was then sold by the Earl of Mar in 1606 for £2,000 to Sir James Erskine, who parted with it in the following year for £4,500 to Sir Adam Newton, who built the present manor-house after the design of Inigo Jones, and also in great measure rebuilt the church. This Sir Adam Newton is famous as having been the tutor of Prince Henry, who died at the early age of seventeen, when Sir Adam was entrusted with the education of Prince Charles. His influence at Court gained him many lucrative appointments, the best being that of Clerk of the Council at a salary of £2,000 a year. In 1659 the manor again changed hands, passing to Sir William Ducie, the banker of Charles I., afterwards Viscount Downe, who died here in 1679. By successive stages it passed through the families of Sir John Conyers, Bart., William Langhorne-Games, and then to the Rev. John Maryon, Rector of White Roding, and finally by marriage to the Wilson family.* The manor-house, Charlton House, built of red brick and stone, is a fine example of the florid Jacobean type.

Maryon Park forms part of Hanging Wood, which has been considerably curtailed in size. It originally stretched from Woolwich Common to the Lower Road, Charlton, and formed a secure retreat for the highwaymen who plied their trade on Shooter's Hill and Blackheath. The Assize Rolls of the thirty-ninth year of Henry III. (A.D. 1255) contain a reference to the 'wood of Woolwych,' and we may reasonably suppose that Hanging Wood is the place referred to. It is as follows : ' John, son of Henry Juventus, was found dead without any mark upon him in the wood of Woolwych. The first finder and the dead man's four next neighbours were attached and not suspected. Verdict, mischance.'†

* Lysons, 'Environs of London,' vol. i., part ii., pp. 431, 432.

† Quoted in Vincent's ' Records of the Woolwich District,' vol. ii. p. 696.

The town of Woolwich was fined on this occasion for not holding an inquest.

Hanging Wood would now be out of the way of travellers on the main road ; but some years ago there was a road— which is now extinct, or nearly so—leading from the cross-roads at Shooter's Hill (near the present police-station) to the Lower Road through the wood. This lane is thus described by Lysons : ' There are in Charlton about ninety acres of woodland called Hanging Wood, belonging to the lord of the manor, through which there is a very pleasant walk to Woolwich. The wood, the variety of uneven ground, and the occasional views of the river, contribute to make the neighbourhood remarkably picturesque.'* A newspaper extract of 1761 tells us that the right kind of hypericum for the cure of worms grows wild in Charlton Wood, near Woolwich.

In olden times the unhappy traveller through the wood along this road would have run considerable risk of being robbed by footpads. The following extracts from contemporary newspapers give us some idea of these ' good old times.' Under 1732 we read : ' On Sunday morning the Rev. Mr. Richardson, going from Lewisham to preach at Woolwich, was attacked by a footpad in Hanging Wood, who robbed him of a guinea (leaving him but twopence) and then made off.' In January, 1762, it was recorded that ' several people have been robbed this week in Hanging Wood, near Woolwich.' Once again, in 1782, ' Three men robbed a boatswain of a man-of-war, near Hanging Wood, of his watch and ten guineas, but some gentleman coming up, they took to the wood.'† A more distinguished resident, Lieutenant-General Sir William Congreve, who projected and completed the Repository, was returning in his carriage from London to Charlton, when he was attacked by two footpads, who issued from a pit on the side of the road crossing Blackheath, called the Devil's Punchbowl. He fired his

* Lysons, ' Environs of London,' vol. i., part ii., p. 430.
† Quoted in Vincent's ' Records of the Woolwich District,' p. 697.

pistol at his assailants, who fled after encountering this un-expected opposition, and although they were tracked to Hanging Wood, they succeeded in making their escape. The *Newgate Calendar* records the trial of two highwaymen who were pursued by the whole garrison of Woolwich, and eventually captured in the wood, where they had imitated the example of hunted foxes, and had gone to earth in an old drain.*

The Ordnance map shows the ground to the immediate south-west of Maryon Park as the site of an ancient Roman camp. There is reason to suppose that the sites of all the towns south of the river were occupied as villages, camps, or fortifications by the succeeding conquerors of Britain. The commanding position of this portion of the wood would certainly be a favourable one from a military point of view, and there are sure proofs of the site having been occupied as a fortified camp. Other portions of Hanging Wood have similar, though less apparent, indications of these fortified camps.

The present appearance of Maryon Park, with its central eminence, was probably caused by the excavation of sand, for which the whole of this district has been famous. This used to be in great request for the sanding of the parlour floors before carpets became the fashion, and also for use in engineering foundries. An extract from a letter of 1762 states: 'The captain, contrary to his intention or desire, being obliged to call at Woolwich, we walked thither. . . . Having spent ye evening very temperately, con-sidering the manner of such partings, we went on board the captain's six-oared boat at ye west end of the yard, where almost all ye sand used by the housewives of London is put on board barges from carts, which bring it down from ye neighbouring hill.'† The site of the park, including the whole of Mount Street extending to the old Toll Gate, was known as the Charlton Sandpits. These were worked for many years by a person of the name of Blight.

* Vincent, 'Records of the Woolwich District,' p. 413.
† Quoted in Vincent's 'Records of the Woolwich District,' p. 23.

The area of the park is too restricted to allow of cricket being played, with the one exception in favour of the boys of the training-ship *Warspite*, which lies at anchor in the Thames off Charlton. This was made one of the conditions in the deed of gift. A peculiar interest attaches to this man-of-war, chiefly centering around the figure-head of the old craft. This is a carved representation of the Duke of Wellington, which was appropriately affixed to the ship when she bore her first name of *Waterloo*. She had been specially fitted at no small expense to go out as flagship of the Mediterranean squadron in 1852, but at the last moment

Cox's Mount, Maryon Park.

the British Cabinet intervened, and put pressure on the Admiralty to keep back the *Waterloo* and send out another ship in her place. It was just after the *coup d'état* in France, and it was thought in high quarters that the name of Waterloo might hurt the feelings of the new master of France, Louis Napoleon. So the *Waterloo* remained in home waters, and was rechristened the *Conqueror* some time afterwards in a very quiet way; but it was not unnoticed, for a protest was raised against it in the very quarter where it was desired to avoid offence. The French naval Attaché to England, the story goes, on learning of the change of name, angrily took exception to it. ' *Waterloo ! Conqueror !*' he declared—' *Conqueror !* Mon Dieu! zat is ten tousand time varse !' The second name of the ship had to stand, however, and the selection was somehow explained away. When the *Conqueror* became a boys' school ship she took over the name *Warspite* from her predecessor at Charlton, the old *Warspite* of 1807, which had been allotted to the Marine Society in the first place for their institution.*

The highest part of Maryon Park,. viz., the Mount, was formerly used as a semaphore-station in connection with that at Shooter's Hill. This mound and its approaches were rented in 1838 from the Lord of the Manor by a gentleman named Cox, who resided at No. 5, Charlton Terrace, for the purposes of cultivation and recreation. He planted the mound round with poplars, and built a large summerhouse in the centre where he entertained his friends. It thus acquired its local name of Cox's Mount, which it has retained to the present day. At the time when the mound was leased the whole of the land in the rear of Charlton Terrace to the main Charlton road was occupied for agricultural purposes.

In addition to being used as a semaphore-station, the mound was rented in 1850 by the Admiralty for the purpose of adjusting ships' compasses. Ten years later they removed their station to the Maryon Road, and built a small observatory for this object.

When the park was being laid out, the workmen came across the foundation of an old kiln in levelling the ground at the base of the mound. This was one of several kilns formerly used for burning red bricks, a large quantity of which were made here. A deep well was sunk by the late Lord of the Manor near the site of the present lodge in order to obtain water to make these bricks.

* *Daily Graphic,* April 16, 1895.

CHAPTER IX.

MYATT'S FIELDS—PECKHAM RYE—PECKHAM RYE PARK
—GOOSE GREEN—NUNHEAD GREEN.

MYATT'S FIELDS.

THE park known by this name consists of 14½ acres of land situate near Camberwell New Road Station. One of the first things to be recorded on the minutes of the newly-formed London County Council in 1889 was the receipt through the Metropolitan Public Gardens Association of an offer of this park, which had been laid out, and was ready for opening. The name of this generous donor was Mr. William Minet, the owner of this and adjoining land, who has been a good friend in other ways to this neighbourhood. Two other tokens of his generosity exist close by—the Minet Free Library and a parochial hall on the opposite corner of the road for the Church of St. James, Camberwell.

The first offer of the ground was made to the Metropolitan Board of Works, who had not then gone out of office, and it was arranged that the park should be taken over when laid out. This the Metropolitan Public Gardens Association undertook to do, but the cost was found to be beyond their means, and so the balance was found from other sources. Before this was completed the London County Council had come into power, and so inherited the offer which had been made to their predecessors.

The park lies a little below the level of Knatchbull Road,

from which it is separated by an open wrought-iron railing, with massive and artistic gates, which are a decided ornament to the park. The principal entrance is through a porch attached to the superintendent's lodge, something after the style of a country lych-gate. The park is tastefully laid out with gravelled walks, flower beds, and grass enclosures, which are large enough to provide room for several tennis-courts. A portion of the ground is used as a gymnasium for boys and girls, the remainder of the buildings comprising a large circular shelter, a bandstand, and the necessary green-houses for the raising of the flowers for decoration.

The history of Myatt's Fields has been very uneventful. It is one of the few places which Pepys has not mentioned in his diary, and which has never been honoured with a visit from Queen Elizabeth. And yet its name is familiar all over London as the place where some of the finest strawberries have been grown which have found their way to Covent Garden Market. These were grown by a celebrated market-gardener of the name of Myatt, who was a tenant of this land from about 1818 to 1869, and in addition to the famous strawberries raised some particularly fine rhubarb here. During his long tenancy the land he occupied came to be generally known as Myatt's Fields, and the name was so general in the neighbourhood that it was perpetuated by being given to the park which covers the site of his holding.

The site of the park is a portion of an estate originally consisting of some 109 acres, purchased in accordance with the provisions of the will of the Right Hon. Thomas Wyndham, Baron Wyndham of Finglass, Ireland, a former Lord High Chancellor of Ireland. Baron Wyndham never married, and when he died in 1745 he bequeathed the residue of his personal estate to Sir Wyndham Knatchbull (who subsequently took the surname of Wyndham) and to other trustees, to be laid out in the purchase of lands in Great Britain. The directions of the will were not carried into effect for some time owing to a Chancery suit which was instituted in connection with some of its provisions. In the meantime, Sir W. K. Wyndham (fifth Baronet), the first beneficiary under the bequest, died in 1749, having purchased

Myatt's Fields Bandstand.

the Camberwell estate in that year. His son, who inherited the property, conveyed it to the trustees of Baron Wyndham's will, and being able to bar the entail, he did so by a recovery in 1762. Upon his death in the following year, the lands were devised to his uncle, Sir Edward Knatchbull

(Wyndham) for an estate tail. This owner again barred the entail by a recovery dated 1764. By this deed, however, he reserved to himself the right to revoke the trusts, and appoint others, a power he exercised in 1768, when he mortgaged the property to Thomas Blackmore, of Briggins, Herts. Two years later, in 1770, the estate was conveyed to Hughes Minet, the mortgages being paid off out of the purchase-money. The property has remained in the Minet family ever since.

Some of the history of the Minet family is retained in the naming of two of the adjacent roads, from which there are entrances to the park. The earliest known member of the family was Ambroise Minet (1605—1675), who was born at Cormont, near Boulogne. He subsequently removed to Calais, and, according to his son's life of him, he 'keept shopp of grocery, druggs, licors,' and built up a considerable business. His son was one of those persecuted Huguenots who were forced to flee from France upon the revocation of the Edict of Nantes.* From these two French towns, which have played so important a part in the family history, we have Cormont Road and Calais Street. The persecution of the Huguenots has resulted in many lasting benefits to England, and Camberwell certainly has received its full share from the hands of the Minet family.

Having now considered the various owners of this estate, it may be interesting to follow its development from the time when it was a farm held under one lease to the present day, when it is a busy colony and an integral part of this mighty London. The property is described in a lease of 1767 as 'all that messuage and tenement and the barn, yard, stables, outhouses, sheds, and other buildings, and all those several pieces or parcells of land arable, meadow or pasture, to the same belonging, containing in the whole, by estimation, one hundred and eight acres.' The principal farm-house and outbuildings stood on the triangular peninsula lying between

* 'Huguenot Family of Minet,' by William Minet, M.A., F.S.A., of the Inner Temple.

Camberwell Green, Camberwell New Road (which did not then exist), and the footway which still connects Camberwell Green with Camberwell New Road. The main portion of the farm extended to the south-west of the farm building as far as the house now known as 62, Knatchbull Road. Through the farm ran a road as far as where the public library now stands. This road, which was formerly known as Myatt's Road, was continued on later, and is now called Knatchbull Road, after the former owners of the estate. South of the main farm in Coldharbour Lane were three outlying fields, which were sold in 1872 as a site for a convict prison; but the project met with so much opposition that it was never carried out.

The first beginning of building on the estate may be dated from the making of the Camberwell New Road in 1819, when part of the property, which was occupied as a stone-mason's yard, was assigned to the trustees of the Surrey and Sussex roads for that purpose. This new road passed close behind the farm-house, which, together with its out-buildings, disappeared early in 1819, and a short time after its place was taken by the houses now known as Nos. 7 to 14, Camberwell Green. The oldest house on the estate dates back to 1815, and stands near the corner of Clarendon Place, on land still used as nursery ground. The opening of the Camberwell New Road provided a large amount of frontage, and it was here accordingly that such building as there was took place. In 1824 and the following year a large number of houses were built, and then for sixteen years nothing more was done. From 1841 building seems to have gone on, though slowly, until 1856, and bricks and mortar swallowed up the fields of strawberries and rhubarb.*

PECKHAM RYE PARK.

Under this heading are included four distinct places—Peckham Rye proper, 64 acres; the park, 49 acres; Goose

* These particulars are taken from a manuscript history of the estate by William Minet, M.A., F.S.A.

Green, 6¼ acres; and Nunhead Green, 1½ acres. They form the greater part of the open spaces of Camberwell, the largest parish in the Metropolis. The Rye has been used as a recreation-ground from 'time immemorial,' and to give some definiteness to this expansive term, we find that mention is made of Peckham Rye in documents of the fourteenth century. A large portion of what was formerly common land has been appropriated to private use and has been built over. A curious discovery was made some years ago which proves this to some extent. A Mr. Weller, who purchased the freehold of the Rectory Nursery or Farm in the Crystal Palace Road, had occasion, soon after taking possession, to take up the stumps of some lime-trees in the corner of the land, in doing which he unearthed a large stone, the top of which had been broken off; but with this exception it was in a good state of preservation. The inscription on the stone, which was still clear, showed that it had been placed there in 1616 to mark the boundary of the land. It is very probable, then, that the common land in past centuries reached as far as this Rectory Farm, including Goose Green as part of the common, and that it spread on the opposite side of the green to the place where the church now stands. St. John's Vicarage occupies the site of the old farm-house. At one time Peckham Rye was covered with trees, and intersected with watercourses in which watercress abounded, so that the physical features as well as the area have undergone considerable change.

There has been a good fight to maintain the people's rights in Peckham Rye. As far back as 1766, and again in 1789, protests were made by the parishioners (which were duly recorded in the vestry minutes) against encroachments on the common land. Great difficulty was experienced before 1869 in preventing the Rye from being privately appropriated. Before this time the Lord of the Manor had granted to a few of the inhabitants in the vicinity leases for twenty-one years, which expired in December, 1866. These lessees usually expended some £100 yearly upon the maintenance of the land, part of which amount was contributed by the local residents. But with the limited power they possessed, they could not prevent objectionable invasions from time to time. In 1864 thirty-two vans of ' Wombwell's wild beast show ' held possession for some time, and this, together with similar misuses, resulted in a meeting of the inhabitants being held the next year to consider the best means to be adopted to prevent building upon the Rye.* The matter was taken up in Parliament, and the deputy-steward of the Lord of the Manor claimed for him before the Committee of the House of Commons, the absolute owner-ship of the Rye, and asserted that he was entitled to the full building value of the land, there being at that time no copy-holders. Evidence to the exact contrary of this was given by the inhabitants, and in 1868 the rights of the Lord of the Manor, whatever they were, were purchased by the Camber-well Vestry. In 1882 the fee simple of Peckham Rye, Nun-head and Goose Greens was acquired from the vestry by the late Metropolitan Board of Works for £1,000. Thus these three open spaces have been secured for the public for ever.

The Rye is very popular with South Londoners—in fact, so much is it patronized for games of all kinds that for many months in the year the greater portion is quite bare of any turf. The performances at the bandstand with its tar-paved promenade attract many thousands of all classes. It was this crowded state of the common, with the attendant dangers from the many playing cricket and football in a confined space, that led the Peckhamites to look round for some extension of their ground for recreation. It was par-ticularly fortunate for them that the owners of the adjoining Homestall Farm were willing to sell land for the purpose of enlarging the common at £1,000 per acre. A committee was formed to try and acquire the land, and after some years of hard work they had the satisfaction of seeing their efforts crowned with success. The cost of the park may be put

* Blanch, ' Camerwell,' p. 353.

down at £51,000, to which the Camberwell Vestry contributed £20,000, the London County Council £18,000, the Charity Commissioners £12,000, Lambeth Vestry £500, and St. Mary, Newington, and St. George the Martyr, South-

The Fountain and Rivulet, Peckham Rye Park.

wark, £250 each. An additional £7,500 was spent in adapting the land to the requirements of a public park, which was opened on Whit-Monday, 1894. The park is well wooded, especially the portion which the vendors have re-

leased from the London County Council for a period of
ninety-nine years from March, 1892. In a secluded hollow
delightfully shaded with trees a lake has been made. It has
an island in the centre, and is fed by a small watercourse
running through the grounds, which has been formed into
a number of pools by artificial dams. This rivulet has its
source in a fountain springing out of rockwork, and thence
meanders through the park, receiving some life when babbling
over some miniature waterfalls before its entrance to the
lake. A veritable lovers' walk has been formed through the
glade on the north side of the tennis-lawns. It runs through
a wide belt of closely-planted trees, whose branches form a
leafy arch, with a luxuriant undergrowth, thus affording a
cool and shady retreat during the hottest time of the summer.
In the spring it is gay with a variety of wild-flowers, daffodils,
anemones, bluebells, and primroses, all adding their colour
to the bright pattern. The avenue through Homestall
Farm is also a favourite walk, affording good views of this
portion of the park. Ample space is provided for games,
2 acres being set apart for tennis (the ground for which
is arranged in a series of terraces), 12 acres for cricket, and
10½ for the two children's playgrounds. These latter are
situated on sloping ground facing the west, and are particularly suitable for the purpose, as they are sheltered from
the wind by the thickly-wooded grounds surrounding this
portion.

The quaint wooden farm-house known as Homestall
Farm, with its out-buildings, is quite 200 years old. It
seems to be the last of the many farms which once surrounded the Rye. The soil must have been particularly
adapted for farming, for there was a time when this part of
Peckham used to furnish melons fit for the King's table.
The old barn, which is exceedingly picturesque with its
red-tiled roof, was very probably used by smugglers to
store away contraband goods, and the old elm-trees
round it which remain must be at least 150 years old.
These, it is surmised, were planted to conceal certain

unlawful kegs which were brought across an old circuitous path, now stopped up, leading across the common to the farm. At one time the chief inhabitants of the barn were owls, but they have entirely disappeared now. The stag's head and antlers which grace its front are fit emblems for this ancient structure, taking us back to the time when Kings used to come to Peckham to hunt the stags in the woods which then covered this spot. A portion of the wood still remained at the beginning of the present century, for we find that a well-known colonial broker named Kymer rented it for pheasant-shooting. Curiously enough, a pheasant may occasionally be seen in the park, though it is doubtful whether it has any connection with the former denizens of the wood.*

Turning our attention now to the manorial history, we find that the parish of Camberwell formerly comprised one manor only, which was held of Edward the Confessor by Norman, and of William the Conqueror by Haimo. The translation of the Doomsday entry runs as follows : 'The land of Haimo, the Viscount or Sheriff. In Brixton Hundred. Haimo himself holds Ca'brewelle. Norman held it of King Edward. It was then taxed for 12 hides. Now for 6 hides and 1 virgate. There are 5 caracutes of arable land. Two are in demesne; and there are 22 villanes, and 7 bordars, with 6 caracutes. There is a church, and there are 63 acres of meadow. The wood yields food for 60 swine. In the time of King Edward it was valued at £12, afterwards at £6, and now at £14.'† There is a separate entry for Peckham, from which it appears that Peckham was part of the Manor of Battersea. 'The Bishop of Lisieux holds of (Odo) the Bishop (of Bayeux) Pecheha' which Alfled held of Herald, in the time of King Edward, when it was included in Patricesy (Battersea).'‡

Soon after the Conquest the parish was divided, and

---

* Communicated by the late Mr. Stevens, the lessee of Homestall Farm.

† Allport, 'Camberwell,' pp. 47, 48.   ‡ *Ibid.*, p. 49.

became several distinct manors. Peckham Rye formed part of two of these—Camberwell Buckingham and Camberwell Friern.

Camberwell Buckingham.—This manor, sometimes called Camberwell and Peckham, was held direct from the King by Robert, Earl of Gloucester, and natural son of Henry I., who had probably received it from his father. For many subsequent generations it was in the hands of the Clares and Audleys, Earls of Gloucester. Margaret, the daughter and heir of Hugh, Earl of Gloucester, who had died in 1347, married Ralph, the first Earl of Stafford, from whom the estate descended to Edward, Duke of Buckingham. He was beheaded in 1521, but ever since this time the manor has been known by the name of Camberwell Buckingham. A tenant of the Duke's, John Scott, who was Baron of the Exchequer in 1529, then obtained a grant of the manor, and it remained in his family till the death of his grandson, Richard Scott, who devised it by will to his five sons. Edgar, one of these, sold his share in 1583 to Edmund Bowyer, from whom it descended to Sir Edmund Bowyer Smyth. The remainder of the estate was purchased by the Cock family. Mrs. Johana Cock was ruined by the South Sea scheme, and the lands were sold under a decree of Chancery in 1776. Two-thirds of this portion were bought by Messrs. John and S. Halliday, and still remain in that family. The other third was purchased by Dr. J. C. Lettsom, whose representative sold it in 1812 to William Whitton, from whom it was afterwards acquired by Sir Edmund Bowyer Smith, so that the two portions became reunited in the same owner.*

Camberwell Friern.—This manor appears to have been formed out of lands granted by Robert, Earl of Gloucester. The Earl gave 100 acres of his wood at Camberwell to Robert de Rothomago, who grubbed it up, and then presented the land to the Priory of Halliwell in Middlesex. Another grant was subsequently made to Reginald de Pointz,

* Allport, ' Camberwell,' pp. 51, 52.

who in turn gave or exchanged part to this same priory, leaving the remainder to his four nephews. One of these, Nicholas Pointz, made a further donation of 10 acres to the priory, and afterwards sold them some more. These successive gifts of land made up the Manor of Camberwell Friern. After the dissolution of the monasteries, the manor was granted July 21, 36 Henry VIII., to Robert Draper, his wife and heirs. This owner was page of the Jewel Office

The Avenue, Peckham Rye Park.

to Henry VIII., and he had other property in Camberwell. It afterwards passed to his son Mathye, but a question as to his title was raised in 10 Elizabeth, when he and his wife were called upon to show by what right they held the manor. He established his claim to the manor, which at that time consisted of 4 messuages, 56 acres of land, 24 of meadow, and 11 of wood in Camberwell and Dulwich. He died without issue, but before his death he conveyed the manor to Sir Edmund Bowyer, whose mother Elizabeth was his youngest sister. From this owner it passed by marriage to Sir Edward Bowyer Smyth.*

We thus see that the two manors, of which the Rye forms part, after passing through many hands, eventually return to the same owner. The manor-house of the Friern Manor was at the south-west of the Rye. Its memory is retained in the names Manor Road, Friern Road, and Friern Place, and it is very probable that Lordship Lane takes its name from the lordship of Friern Manor. Of recent times Friern Manor Farm-house has been known as a dairy farm on a large scale. The farm-house, with its sheds and outbuildings, was sold in about 200 lots in December, 1873. This house, which was not the original one, was built by Lord St. John in 1725, and there is a tradition that Alexander Pope resided here for a time, and wrote some part, if not the whole, of his ' Essay on Man ' beneath its roof.

The derivation of the word ' Peckham ' is involved in some obscurity. To quote the words of Mr. Blanch : ' It certainly is not that which its name at first implies—the village on the hill.' We have seen that in the Doomsday Book the place is called ' Pecheha,' which in all probability was an incorrect description. One theory is that the village of Peckham took its name from its proximity to the hills now known as Forest Hill and Honor Oak Hill, for Peckham Rye is mentioned in documents of great antiquity, and the little *ham* or village under the shadow of the hills above mentioned was evidently a place of some little importance at the time of William the Conqueror.

The word ' Rye,' assuming the above theory to be correct, would then be traced to the Welsh *rhyn*, a projecting piece of land, and Peckham would be the village under the *rhyn* or rye. But in all probability the Rye took its name from a watercourse or river, for before the Roman invasion and the embankment of the Thames the country surrounding the Rye was no doubt partly submerged, and streams more or less rapid abounded.† Peckham Rye is associated with at

---

* Allport, ' Camberwell,' p. 55.   † Blanch, ' Camerwell,' p. 91.

least one literary name. At one time Tom Hood, the well-known author of 'The Song of the Shirt,' lived in a house overlooking the lake on the Rye.*

At the extreme south of the park rises Honor Oak Hill, which tradition links with the name of Queen Elizabeth. It is called on Rocque's map the 'Oak of Arnon,' which is probably a mistake for Oak of Honour, now corrupted to Honor Oak. It is said to have received this name because the good Queen Bess lunched under its shade when returning from one of her excursions to Greenwich. In the Chamberlain's papers for 1602 is the following entry: 'On May Day the Queen (Elizabeth) went a-Maying to Sir Richard Buckley's at Lewisham, some three or four miles off Greenwich.' This house is supposed to have been on the Sydenham side of Lewisham, and it is probable that this is the occasion on which she visited Honor Oak. We must mention another interesting ceremony in connection with this place. On the occasion of beating the bounds in former times, it was customary for those assembled to join in singing Psalm civ. under the shadow of the Oak of Honour Hill.†

The original oak, which is thus said to have been favoured by Royalty, has been blown down, but a successor has been planted to uphold the dignity of the tradition.

At the extreme north-east of the Rye, Heaton Road and the Heaton public-house bring back the memory of a building called Heaton's Folly, which once stood on this site. It was capped with a lofty square tower, giving the building the appearance of a religious edifice. Lysons, writing in 1796, gives the following account of it: 'On the right side of the path leading from Peckham to Nunhead appears this building, environed with wood. It has a singular appearance, and certainly was the effect of a whim. Various tales are related of its founder; but the most feasible appears his desire of giving employment to a number of artificers during a severe dearth. It is related that he employed 500 persons

* Blanch, 'Camerwell,' p. 356.   † Ibid., p. 157.

in this building, and adding to the grounds, which is by no means improbable, as, on entering the premises, a very extensive piece of water appears, embanked,' as he quaintly says, ' by the properties taken from its bosom. In the centre of it is an island, well cultivated ; indeed, the whole ground is now so luxuriantly spread, that I much doubt if

Heaton's Folly in 1804.

such another spot, within a considerable distance from the Metropolis, can boast such a variety and significance. The whole is within a fence, and time having assisted the maturity of the coppice, you are, to appearance, enjoying the effects of a small lake in the centre of a wood. Motives the most laudable, as before observed, induced the founder of this sequestered spot to give bread to many half-starved and wretched families, and to use the phrase of our immortal Shakespeare, "It is like the dew from heaven, and doubly blesses." If from appearance we are to judge of the place, it thrives indeed, and what was meant as assistance to a neighbouring poor, and stragglers, wretched and forlorn, is now, with all propriety, the Paradise of Peckham.'* The building and grounds upon which this eulogium is lavished have now entirely disappeared, and the ' paradise' is covered with bricks and mortar.

Peckham Fair is a very old institution, and several different versions are current to account for its establishment. One of these says that King John, hunting in the woods at Peckham, killed a stag, and was so pleased with the sport that he granted the inhabitants an annual fair to continue for three weeks. The only drawback to the acceptation of this theory is that no charter to this effect has been found. Another version gives the credit to Nell Gwynne, who asked this favour from the Merry Monarch on his return from a day's sport in the neighbourhood to the mansion of Sir Thomas Bond, who was a great favourite of his. At one of the inquiries which was held, evidence was given that the fair had been held in 1715. Whatever the origin, it is certain that early in the last century it had grown into such importance as to originate the proverb, ' All holiday at Peckham.' On May 8, 1823, a vestry of the parish was called for the purpose of inquiring whether the fairs of Camberwell and Peckham were authorized by any grant, charter, prescription, or other lawful or sufficient authority, in order if practicable to their suppression ; but no settlement of the question appears to have been effected. On August 1, 1827, a meeting of the local magistrates took place at the committee-room, Camberwell Workhouse, with regard to Peckham Fair, at which summonses were issued to the representatives of the various Lords of the Manor, directing them to appear at the same place on the 11th, which they did by their attorney, who admitted that he could not show cause for the con-

* Lysons, ' Environs of London.'

tinuance of the fair; and it was accordingly declared unlawful.*

A short walk from the Rye down Nunhead Lane brings us
to Nunhead Green, no longer green, although it retains its
old name, for it has been found necessary to tar-pave it, and
it thus forms a convenient playground for children.   Nunhead is now a very different place to what it was fifty years
ago, when the tea-gardens of the Nun's Head (from which
the district takes its name) were in their prime.   These have
been an institution in the locality for more than 200 years.
Nunhead Hill is mentioned by Hone in his ' Every-day
Book' (1827) as being ' the favourite resort of smoke-dried
London artisans.'   Facing the green is the Asylum of the
Metropolitan Beer and Wine Trade Association, inscribed
with their motto ' Live and Let Live.'   The seven cottages
forming the asylum make a compact building, each tenement
having four rooms and a kitchen, with a plot of garden-
ground in the rear.   The centre one contains three rooms for
the use of the warden, and in addition a large committee and
waiting rooms.   Its institution dates back to 1851, when a
subscription-list was opened at the general meeting of the
society.   The freehold of the site was purchased for £578,
and the buildings, with fittings, cost another £2,500.   The
first stone was laid by Lord Monteagle (a patron of the
society) on June 9, 1852, and the first inmates, numbering
thirteen, were admitted in September, 1853.   There is an
allowance made weekly to the inmates of 6s. to the single
ones, and 9s. for married couples.   A new wing called
Albion Terrace was added in 1872, consisting of eight six-
roomed houses, which are capable of accommodating sixteen
more pensioners.†

The Pyrotechnists' Arms public-house at another corner of
the green is an outcome of one of the staple industries of
Nunhead—firework-making.   Mr. Brock, the great pyrotech-
nist of the Crystal Palace, has large workshops here, where
many hands are employed in making these popular toys.

* Allport,.' Camberwell,' pp. 86-89.   † Blanch, 'Camerwell,' p. 275.

The other piece of common land included in the purchase of Peckham Rye is Goose Green, which was a part of the Manor of Camberwell Friern. It possesses no rusticity beyond its name, although not many years since it was a village green surrounded by a few cottages, and the farmhouse we have before referred to. The geese, who were then the principal frequenters of the green, have disappeared, although they have left their name. Overlooking the green are the Dulwich Public Baths, which reminds us that we are living in the nineteenth century. On the opposite side is the church of St. John the Evangelist, East Dulwich, built in 1865 from the designs of Mr. C. Bailey at a cost of £8,000. It is in the Early English style, and forms a very picturesque feature with its broached tiled spire, recalling to mind the village churches of Surrey and Sussex.

CHAPTER X.

SOUTHWARK PARK — NELSON RECREATION - GROUND —
NEWINGTON RECREATION-GROUND—WALWORTH
RECREATION-GROUND.

SOUTHWARK PARK.

SOUTHWARK PARK, of 63 acres, is situated in
Rotherhithe, immediately west of the Surrey Commercial Docks, with the whole of Bermondsey
between it and Southwark. Its name, therefore, is
as unfortunate as that of Finsbury Park, but the same excuse
may be given in justification, viz., that it was primarily
intended for the inhabitants of the old Parliamentary borough
of Southwark, within the boundaries of which it was situated.
There was formerly a Southwark Park entitled to the name
which was an appendage to Suffolk House in Southwark.
This was a part of the King's manor, and was specially
exempted from the grant of the borough of Southwark to the
city of London in the charter of Edward VI.*

The late Metropolitan Board of Works, impressed with
the importance of securing some place of recreation for the
densely-populated portion of the Metropolis along the southern
side of the river Thames, proceeded in 1864 to obtain from
Parliament the necessary powers for the formation of a park.
About 63 acres of land, then used as market-gardens, were
selected as a suitable site, and the Bill passed through both

---

* Wheatley and Cunningham, 'London : Past and Present,' vol. iii.,
p. 288.

Houses of Parliament without opposition, and received the Royal assent on April 28, 1864. The negotiations for the purchase of the freehold were difficult and prolonged, but were at length concluded in 1865; afterwards, the claims for the yearly tenancies and other minor interests had to be settled, and possession of the whole property obtained. After the park had been laid out and enclosed, it was at length opened to the public on June 19, 1869. The park has proved an undoubted boon to the inhabitants of the crowded parishes of Southwark, Rotherhithe, and Bermondsey. At the time of the formation of the park, the greater portion of the land in this district consisted of fields and market-gardens. Although these are fast disappearing, some still exist in the neighbourhood of Rotherhithe and Deptford. The network of docks, and the consequent difficulty of communication, has in some measure accounted for the less rapid development of Rotherhithe as compared with its crowded neighbour Southwark.

The park boasts of a good cricket-field called the Oval, which, it is hardly necessary to say, is not *the* Oval, the home of the Surrey County Club at Kennington. However, Southwark Park has turned out several good players who have become famous in county cricket, notably 'Bobby' Abel, Surrey's crack batsman. The other features in the laying-out comprise an ornamental lake, an entrance lodge, and a handsome bandstand, purchased from the Exhibition grounds at South Kensington, together with the usual sprinkling of flower-beds interspersed with shrubberies and trees. The lake was not part of the original scheme, but was subsequently added at a cost of about £2,000. It is too small to admit of boating, but it is available in winter for skating, and in the summer-time the water-fowl, of which there are a good selection, seem a constant attraction. Two of the mounds near the entrance from Jamaica Level were formed from some of the earth excavated from the bed of the river in the construction of the Thames Tunnel.* A small decorative house is avail-

* 'Old and New London, vol. vi., p. 136.

able as a winter-garden where flowers cán be exhibited during the cold months. This is also used for the annual show of chrysanthemums, which thrive remarkably well in this atmosphere, which is admitted to be unfavourable for plant-life. It was originally intended, when the park was laid out, to reserve a portion for building purposes, as in the case of Finsbury and Victoria Parks. But here again great opposition was made to the proposal, with a result that it was abandoned.

Entrance and Superintendent's Lodge, Southwark Park.

The carriage-roads formed in the park were broken up and converted into tar-paved footpaths, leaving only one road, viz., that straight across the park from Bermondsey to Rotherhithe, which is left open till midnight.

Beyond the trees in the park can be seen a regular forest of masts belonging to the ships in the adjacent docks. It is worthy of mention that the first commercial docks in London were constructed in Rotherhithe, their origination dating back to about 1660, nearly 150 years before the establishment of the adjoining naval dockyard of Deptford.

The first dock established here was a dry one, belonging to the Howland family. Parliamentary power was obtained in 1696 to build a wet dock, which was completed in 1700, and called the Great Dock. In 1725 the South Sea Company took a lease of the dock, intending to revive the Greenland fishery; it was then called the Greenland Dock. This dock was sold by the Duke of Bedford (the representative of the Howland family) in 1763 to Messrs. John and William Wells, to whom it belonged for many years: they afterwards sold it to Mr. W. Ritchie, from whom it was purchased in 1807 by the dock company. The docks are at present chiefly used for the timber trade.

In the Union Road, opposite the park gate, is situated Christ Church, Rotherhithe, a plain brick structure, erected about 1840. On the external wall is a tablet to the memory of Field-Marshal Sir William Maynard Gomm, G.C.B., Lord of the Manor of Rotherhithe, Constable of the Tower of London, who died March 15, 1875, aged ninety, and from whom the greater portion of the land for the formation of the park was purchased. Gomm Road, on the east side of the park, is named after him.

A few words regarding Rotherhithe, in which, as we have mentioned, the park is situated, may be of interest. The earliest mention of the place is in a charter dating back to 898, in which it is called 'Aetheredes hyd.' Strangely enough, it is not mentioned in Doomsday Book, from which it is conjectured that it was formerly only a hamlet in Bermondsey.* As a separate manor, we find that it was granted in the reign of Richard II., with his permission, by the Abbot of Graces to the Priory of St. Mary Magdalen, Bermondsey. This ancient abbey was situated where the present St. Mary Magdalen Church stands, and existed in Saxon times when it was surrounded by meadows and woodland. At the time of the grant to the abbey, the manor was valued at £20. Among other owners of the manor we may

* Wheatley and Cunningham, 'London : Past and Present;' vol. iii., p. 174.

mention James Cecil, Earl of Salisbury, in whose possession
it was in 1668.

Various interpretations are given of the name Rotherhithe.
The last part of the word, of course, means haven, or harbour.
' Rother' is explained by some as meaning red rose, so
called from some sign of that name. Others prefer to derive
it from a Saxon word meaning sailor, so that Rotherhithe
is the sailor's haven. By the seventeenth century the name
had become corrupted into Redriff. Philip Henslowe informs
us in his diary that he sent his horses ' to grasse in Redreffe '
at a charge in 1600 of 20d. a week.* The greater part of
the parish was formerly a marsh, and in old maps the district
appears as a miniature Holland, being intersected with a
network of dykes and ditches. As a consequence, it has
always been very fertile, and particularly appropriate for
pasture and market-gardens. Thus Lysons in his description (1811) says : ' The land which is not occupied by houses
is principally pasture, of which there are about 470 acres.
The market-gardeners employ about forty.' The marshy
nature of the subsoil has been found to be a serious hindrance
in the erection of buildings within the park. Piles have to
be driven in, and a solid bed of concrete placed upon them
before building operations can be commenced. In some
cases where these precautions have not been taken, although
the result has not been quite so disastrous as in the case
of the house built upon the sand, yet serious subsidences
have taken place.

One incident in connection with the soil is worthy of
mention. Rotherhithe would hardly be chosen as a place
having a soil or situation adapted to the culture of vines.
Yet we learn† that an attempt was made in this parish in
1725 to restore the cultivation of the vine which had once
been successfully followed in England. The particular kind
chosen was the Burgundy grape which ripens early, and so
it was thought that it would be suitable for the English

* Henslowe's ' Diary,' p. 81.
† Hughson, ' History and Survey of London and its Suburbs.'

climate. The attempt was rewarded with success, and the crop amounted to upwards of 100 gallons annually.

This neighbourhood was encircled by two tidal streams—the Mill and the Neckinger. The latter has for centuries contributed to the commercial prosperity of Bermondsey, furnishing the water-power first of all for the various mills of the district, and then when these disappeared supplying the tanners with what their trade required. This stream in ancient days connected the Thames with the Grange, and bore fish and other produce to the monks of Bermondsey Abbey. It was not covered in till about 1850, and though no longer open, it still conveys the Thames water to the Neckinger Mills, the extensive tanneries of Messrs. Bevington.*

The other small river, called the Mill Stream, is more nearly associated with the park. This stream joined the Thames at the bottom of Mill Street. The mill from which this stream took its name is mentioned in one of the oldest records of the district, viz., a deed executed by the Abbot of Bermondsey relating to a mill in a corner of Jacob's Island near St. Saviour's Dock, demised to John Curlew in 1536 for grinding purposes.† The mention of Jacob's Island calls to remembrance Dickens' inimitable description of this fever-stricken and loathsome collection of hovels in ' Oliver Twist,' where it will be remembered the last scene in the life of Bill Sykes was enacted. It was no doubt due to the great novelist's unvarnished and vivid portrayal of the condition of things here that the long-delayed improvement of this spot was carried out, and in 1850 the crazy tenements were pulled down, and the Neckinger and Mill Streams covered in.

On the north-western corner of the park was a district known as the Seven Islands, which comprised a number of tidal ditches, and a large pond, the Mill Pond, all of which were drained and filled up when the park was formed. The inhabitants of the streets round the mill pond were

* Rev. H. Lees Bell, ' History of Bermondsey,' 1880, pp. 37, 38.
† *Ibid.*, p. 39.

dependent till comparatively recent times upon these dirty ditches for their supply of water, which was fetched in pails.*

Before the formation of the park, the marshes and meadows were crossed by a narrow pathway called the 'Halfpenny Hatch,' which extended from Blue Anchor Road to Deptford Lower Road, then past the 'China Hall' tavern to another place of refreshment, the 'Dog and Duck,' near the entrance to the Commercial Docks. The travellers who patronized the 'China Hall' had the privilege of passing along the 'Halfpenny Hatch' without paying the usual toll of a halfpenny. This 'Dog and Duck' reminds us of a barbarous sport, happily now extinct, of hunting ducks by means of spaniels. The fun consisted in seeing the duck dive just at the moment when the spaniel was going to seize it. This cruel pastime of the so-called 'good old times' fortunately fell into disuse at the beginning of this century.

The 'China Hall' still exists in name, but the present tavern just outside the park gates in Deptford Lower Road is quite modern. The ancient tavern was pulled down within the last few years. Pepys, a frequent visitor to Redriffe, as he calls the district, mentions it, but gives us no details. Mr. Larwood, in his 'History of Sign-Boards,' speaking of 'China Hall,' says: 'It is not unlikely that this was the same place which, in the summer of 1777, was opened as a theatre. Whatever its use in former times, it was at that time the warehouse of a paper manufacturer. In those days the West End often visited the entertainments of the East, and the new theatre was sufficiently patronized to enable the proprietor to venture upon some embellishments. The prices were: Boxes, 3s.; pit, 2s.; gallery, 1s.; and the time of commencing varied from half-past six to seven o'clock, according to the season. *The Wonder, Love in a Village, The Comical Courtship*, and *The Lying Valet*, were among the plays performed. The famous Cooke was one of the actors in the season of 1778. In that same year the building suffered the usual fate of all theatres, and was utterly destroyed by fire.'

* 'Old and New London,' vol. vi. p. 135.

The Lake, Southwark Park

The number of places of this kind which at one time
existed within a short radius of the park is indeed remarkable.   In addition to the two already mentioned, there were
the Bermondsey Spa, St. Helena Tavern and Tea-Gardens,
the Blue Anchor, Jamaica House, Halfway House, and
Cherry Garden.   At the time these were flourishing, this
neighbourhood was in the country, and it is no unusual
thing to find an appendix like the following in the advertisements of the entertainments : ' For the security of the public,
the road is lighted and watched by patroles every night at
the sole expense of the proprietor.'*   The existence of these
old places is wisely kept in memory in naming the streets
of this locality.   Thus Jamaica Level and Jamaica Road
are so called from Jamaica House, and Tea-Gardens, which
were formerly at the end of Cherry Garden Street.   Tradition says that Jamaica House was one of the numerous
residences of Oliver Cromwell, but unfortunately there is no
evidence to confirm the statement.   Many of these places
are mentioned by Pepys, who never seemed to miss an
opportunity of enjoying himself.   He was here on April 14,
1667, on which occasion he writes: ' Over the water to
Jamaica House where I never was before, and then the girls
did run for wagers over the bowling-green ; and there with
much pleasure, spent little and so home.'   No doubt this
was not the last visit he paid here.   He has an entry also
regarding Cherry Gardens, which has left its name in Cherry
Garden Street.   ' June 15, 1664.—To Greenwich, and so to
the Cherry Garden, and thence by water, singing finely, to
the bridge, and there landed.'

The last place we shall mention in connection with Pepys
is Halfway House, which he frequently visited on his way to
Deptford Dockyard, to which it was a halfway house.

' May 20, 1662.—Thence to Tower Wharfe, and then took
boat, and we all walked to Halfway House, and there ate
and drank and were pleasant, and so finally home again in
the evening.'

* Quoted in ' The London Pleasure - Gardens of the Eighteenth
Century,' W. Wroth, 1896, p. 234.

'*March* 18, 1662-3. — After dinner by water to Redriffe, my wife and Ashwell with me, and so walked and left them at Halfway House.'

It is supposed by many that this is the same as Jamaica House, but as he says in his entry of 1667 concerning this latter place 'where I never was before,' this must be a mistake. In Corbett's Lane, just off Rotherhithe New Road, is St. Helena Tavern, where were situated St. Helena Gardens, which are of far too modern a date for Pepys. It is not many years since these were in the heyday of their prosperity. There was only one drawback to the visitors, at least to those who had driven from a distance and had exceeded the bounds of moderation, and this arose from the peculiar character of the marshy ground. On each side of the road were ditches of muddy water, and on dark nights, unless the guiding Jehu were particularly sober, it would be no infrequent occurrence for the whole party to be turned into the ditch, and so undergo an involuntary bath, which might have a beneficial effect in bringing them round. These gardens, which were over 5 acres in extent in 1832, were closed in 1881, and their site has since been built over. It is hardly necessary, therefore, to point out the importance of having secured a park in this neighbourhood, which was once so rich in gardens and places of open-air entertainment.

We next come to a group of three recreation-grounds, small in actual area, but situated in localities where open spaces are rare.

NELSON RECREATION-GROUND.

This recreation-ground is found in Kipling, formerly Nelson Street, on the border of Bermondsey, close to Guy's Hospital. Although not of large extent, it forms an important breathing-place in a crowded neighbourhood. It was acquired in 1897 at a cost of £4,600 from Guy's Hospital, or rather ' The President and Governors of the Hospital founded at the sole costs and charges of Thomas Guy, Esq.,' to give the well-known institution its full title. The land had been in the

possession of the hospital authorities since 1789, in which
year it was purchased, together with adjacent property, from
Mr. John Dekewer for the sum of £4,200. It is interesting
to compare the plans of the estate at that time with the maps
of the present day. There are no towering warehouses, no
busy thoroughfares with ceaseless noisy traffic, no iron railroads to be delineated. The plan simply shows a series of
open fields which might be in the heart of the country instead of the very centre of a huge Metropolis. Apart from
the tracks, the only other features to notice are the open
sewers or ditches so characteristic of bygone Southwark and
Bermondsey.

The part of the estate now set apart as a place for recreation was used by the hospital authorities as a burial-ground
down to the year 1855. The burials were confined chiefly to
the poor patients of the hospital, so that no person of note
has found his last resting-place here. Many a gruesome tale
is told in connection with this and other London burial-
grounds of the ' body-snatching ' by the so-called ' resurrec-
tion men,' who secretly exhumed the bodies as soon as
buried. This was a very profitable business if the marauders
could escape detection. The corpses were sold for anatomical
purposes, the coffins for fuel, the nails and coffin-plates had
a marketable value, and there was always the chance of finding rings or other trinkets buried with the bodies. The
passing of the Anatomical Act, 1832, did a good deal to stop
the trading in corpses. Previous to this surgeons and
medical students had been granted the bodies of executed
malefactors for dissection, but owing to the demand exceeding the supply, which led to the terrible traffic just mentioned,
the new statute was introduced, which abated the ignominy
of anatomical research by prohibiting the use of executed
murderers for that purpose, and by permitting, under certain
conditions, the dissection of persons dying in workhouses, etc.

This land, having fallen into disuse as a place of interment,
was let subsequently to a firm of builders as a business yard,
with power to erect temporary buildings thereon, but with

no permission to interfere with the remains beneath. Now that the Disused Burials Act, 1884, has been passed, it cannot of course be built over, and it has been put to the best possible use in being converted into a recreation-ground. The purchase-money, £4,600, was made up as follows: Vestry of Bermondsey, £1,186; St. Olave District Board, £300; Vestry of St. George-the-Martyr, £164; Metropolitan Public Gardens Association, £500; Guinness's Trustees, £150; and the London County Council, £2,300.

A short walk from Nelson Street brings us to

NEWINGTON RECREATION-GROUND.

This recreation-ground, 1½ acres in extent, is situate at the rear of the Sessions House, Newington Causeway. It is approached from the main road by a narrow thoroughfare, Union Road, which widens considerably as the recreation-ground is reached. The ground occupies the site of Horsemonger Lane Gaol, so-called from the former name of Union Road. The gaol and the adjoining Sessions House for the meetings of the Surrey county magistrates were erected in accordance with an Act passed in 1791 'for building a new common gaol and sessions house, with accommodations thereto, for the county of Surrey.' For this purpose 3½ acres of land were purchased, which were then used as market-gardens, and upon them these two buildings were erected at a cost of nearly £40,000. The Sessions House now used is not the original structure.

Horsemonger Lane Gaol was closed in 1878, and when it was pulled down the Surrey justices permitted a portion of the site to be used as a public playground until it might be sold or utilized for other county purposes. This playground, 1¼ acres in extent, was laid out by the Metropolitan Public Gardens Association at a cost of £356, and opened to the public on May 5, 1884. The association also maintained it for a short time, but in the following year the management was transferred to the Vestry of Newington. Subsequently in 1890 the sessions house and the whole of the land were

assigned to the county of London on the apportionment of
the property between the counties of Surrey and London.
The recreation-ground was then enlarged to its present size,
planted and fitted with gymnastic apparatus. The prison

A Band Performance in Newington Recreation-Ground.

walls and gatehouse were pulled down, and the playground
made in every way suitable for public use. Only children are
admitted to the ground, except during the band performances
on Saturdays, when a large adult audience is attracted.

Horsemonger Lane Gaol was built in 1799 on lines

suggested by John Howard, the prison reformer. It was at the time of its erection the largest prison in the county of Surrey, and consisted of two portions, one occupied by debtors, and the other by criminals or persons arrested on criminal charges. It was a substantially-built structure, chiefly of brick, quadrangular in form, and consisted of three stories above the basement, and provided accommodation for 300 prisoners in all. The average duration of imprisonment undergone by each prisoner was not lengthy, as this gaol was not a house of correction. A description of life in the interior (which is none too flattering to the prison authorities) is given by the author of ' London Prisons.' He describes how a visitor must be impressed when entering the wards with the absence of all rule and system in the management. ' He finds himself in a low, long room, dungeon-like, chilly, not very clean, and altogether as uncomfortable as it can conveniently be made. This room is crowded with thirty or forty persons, of all ages and shades of ignorance and guilt, left to themselves, with no officer in sight. Here there is no attempt to enforce discipline. Neither silence nor separation is maintained. . . . In this room we see thirty or forty persons with nothing to do. Many of them know not how to read, and those who do are little encouraged so to improve their time. Some of them clearly prefer their present state of listless idleness. With hands in their pockets, they saunter about their dungeon, or loll upon the floor, listening to the highly-spiced stories of their companions, well content to be fed at the expense of the county—upon a better diet, better cooked, than they are accustomed to at home—without any trouble or exertion on their own part.'* This was written in 1850, some fifteen years before the passing of the Prisons Discipline Act, which changed this and other gaols from being low-class clubs to places where silence was maintained, and where the prisoners passed their sentences in solitary confinement.

Long after the gaol was abolished, the lofty enclosing

* Hepworth Dixon, ' London Prisons,' 1850, p. 286.

wall and the massive gate-house were retained, and the
recreation-ground still presented a very prison-like appearance. Part of the handsome building used as a weights and
measures station is upon the site of this gate-house. Although
this latter structure possessed no architectural beauty, it
was interesting from the fact that the public executions took
place over it. Now that such scenes are hidden from the
curiosity of the public gaze, a large crowd will assemble at
an early hour to see the black flag hoisted, the signal which
tells them that a fellow-creature has paid for his crime with
his life. It can easily be imagined, therefore, how many
would flock to see a public execution with all its gruesome
details enacted before their eyes. Many hours before the
time fixed the roadway before the prison gate was blocked
with these spectators, anxious to obtain a favourable position.
The owners of the cottages opposite used to reap a good
income from those who were able to hire their rooms for
the night, so as to be sure of witnessing the scene in the
early morning. Charles Dickens came here on one occasion
in order to see the crowd, and so gather materials for some
of those vivid pictures of London life which he has so faithfully recorded. It was on November 13, 1849, when the
Mannings were executed. He writes: ' I was a witness of
the execution at Horsemonger Lane this morning. I went
there with the intention of observing the crowd gathered
to behold it, and I had excellent opportunities of doing so
at intervals all through the night, and continuously from
daybreak until after the spectacle was over. I believe that
a sight so inconceivably awful as the wickedness and levity
of the immense crowd collected at that execution could be
imagined by no man, and could be presented in no heathen
land under the sun. The horrors of the gibbet and of the
crime which brought the wretched murderers to it faded in
my mind before the atrocious bearing, looks, and language
of the assembled spectators. When I came upon the scene
at midnight, the shrillness of the cries and howls that were
raised from time to time, denoting that they came from a

concourse of boys and girls already assembled in the best places, made my blood run cold. As the night went on, screeching and laughing and yelling in strong chorus . . . were added to these. When the day dawned, thieves, low prostitutes, ruffians and vagabonds of every kind flocked on to the ground with every variety of offensive and foul behaviour. Fightings, faintings, whistlings, imitations of Punch, brutal jokes, tumultuous demonstrations of indecent delight when swooning women were dragged out of the crowd by the police with their dresses disordered, gave a new zest to the general entertainment. . . . When the two miserable creatures who attracted all this ghastly sight about them were turned quivering into the air, there was no more emotion, no more pity, no more thought that two immortal souls had gone to judgment, no more restraint in any of the previous obscenities than if the name of Christ had never been heard in this world, and there were no belief among men but that they perished like the beasts. I have seen, habitually, some of the worst sources of general contamination and corruption in this country, and I think there are not many phases of London life that could surprise me.'*

So writes Dickens about this scene, which is fortunately one of the things of the past, and it will be a pleasant change to turn to the romance of Horsemonger Lane Gaol if such a thing can be connected with a prison. It was here that Leigh Hunt was confined for two years together with his brother for the serious political offence of calling the 'first gentleman in Europe' an 'Adonis of fifty.' Among his visitors were Thomas Moore and Byron, and this was the occasion when the noble poet met 'the wit in the dungeon' for the first time in his life. The illustrious poets dined with their friend in the prison, who does not seem to have suffered very great hardships, for Moore speaks of 'the luxurious comforts, the trellised flower-garden without, the books, busts, pictures and pianoforte within,'† which are

* Quoted in 'Old and New London,' vol. vi., pp. 254, 255.
† Hepworth Dixon, 'London Prisons,' 1850, pp. 285, 286.

a considerable aid to dispelling the illusion that ' stone walls a prison make.'

One of the ' sights ' that used to be exhibited at this prison was the hurdle upon which Colonel Despard was drawn from the cell in which he was last confined to the place of execution, which was part of the punishment formerly inflicted upon criminals convicted of high treason.   He had been tried together with thirty accomplices by a special commission held in 1803 in the adjoining sessions house, and the Colonel and six of his colleagues were hung and beheaded here.

The last of the group is

WALWORTH RECREATION-GROUND.

This open space, about ⅝ of an acre in extent, is situated in East Street, Walworth, a thoroughfare which connects New Kent Road and the Walworth Road.   East Street in modern times has acquired a more than local celebrity, owing to its connection with the victim of a sensational murder committed on the London and South-Western Railway, the author of which has hitherto succeeded in eluding justice.   Walworth is a place without a history, although its existence can be traced back to Anglo-Saxon times.   It may be that this lack of historical interest has led to the many traditions which connect this place with the illustrious citizen and Lord Mayor, Sir William Walworth, who killed Wat Tyler in Smithfield.   It is a long time back to the reign of Richard II., and in the interim a decided cloud has come over the events of those memorable times.   It is only a conjecture, which has received the support of many historians, that the little isolated hamlet, as Walworth must then have been, was the birthplace of the celebrated Lord Mayor. Another form which the tradition takes is that the Manor of Walworth was bestowed upon Sir William together with the honour of knighthood, as a reward for the part he took in suppressing the formidable rebellion.   This is not borne out by the manorial history which is very simple.

King Edmund gave the Manor of Walworth to his jester, Nithardus, who in the reign of Edward the Confessor, being about to make a pilgrimage to Rome, obtained a license from that monarch to give it to the Church of Canterbury. In the Doomsday record, this manor, called Waleorde, is said to have been held in the time of William the Conqueror by Baynardus of the Archbishop of Canterbury, and to have been appropriated to the support of the monks. It had been valued at 30s. and at 20s., but was then worth £3, and in 1291 was taxed at £10.* The Monastery of Christ Church, Canterbury, was suppressed by Henry VIII. in 1540, who established a Dean and twelve prebendaries in the room of the Prior and monks, and bestowed on them this and other estates, which still belong to the Dean and Chapter of Canterbury.† The statement as to the reward given to William Walworth, taken from Stow's quaint and interesting account of the rebellion runs as follows : ' The rude people being thus dispersed and gone, the King commaunded William Walworth to put a basenet on his head, for feare of that which might follow; and the Maior requested to know for what cause he should so doe, sith all was quieted. The King answered, that he was much bound to him, and therefore he should be made Knight. The Maior againe answered, that hee was not worthy, neither able to take such estate upon him ; for he was but a merchant, and to live by his merchandize. Notwithstanding, at the last, the King made him put on his basenet, and then tooke a sworde with both his hands, and strongly with a good will strake him on the necke. . . . The King gave to Sir William Walworth £100 land, and to the other £40 land, to them and their heires for ever.'

When the manorial history is compared with this account, it will be seen that there are many difficulties in the way of reconciling the tradition with the facts. In the first place, the manor did not belong to Richard II., so that it was not his to bestow. Then we see that the land is valued at £100,

---

* Lysons, 'Environs of London,' vol. i., p. 284.

† Brayley and Britton's 'Surrey,' vol. iii., part ii., p. 401.

which ninety years before had been taxed at £10. And, lastly, we have to account for the transfer to the Dean and Chapter of Canterbury. On the other hand, some colour is given to the tradition by the fact that in the reigns of Edward III. and Richard II., and at subsequent periods, the manor is stated to have been held by a family of the name of Walworth. Margaret de Walworth is mentioned as Lady of the Manor in a register of William de Wykeham, Bishop of Winchester, in 1396, or eleven years after the death of Sir William Walworth, and in 1474 Sir George Walworth died, seised of the manor.* But these persons and others, who are said to have held the manor, were probably lessees under the ecclesiastical lords of the fee, and the identity of the names is simply one of those curious coincidences that often happen.

In the Manor of Walworth two commons were comprised : Lowenmoor Common, of about 19 acres, and Walworth Common, of about 48 acres. It is needless to say that there are no extensive commons in the populous Walworth of the present day, and the provision of this open space, small as it is, will be an inestimable boon in this crowded neighbourhood. It is probably part of the old Walworth Common, and, although now open, has been covered with houses. The ground had already been sold for building purposes, when the London County Council stepped in and made a higher offer. The cost of the site was £5,375, towards which the Council contributed £2,500 ; the Vestry of St. Mary, Newington, a similar sum ; and Mr. James Bailey, M.P. for Walworth, the remainder. Though the web of romance which tradition has woven round the spot has had to be removed, it will not detract in any way from its utility as a recreation-ground for Walworth.

* Brayley and Britton's ' Surrey,' vol. iii., part ii., p. 401.

CHAPTER XI.

TOOTING GRAVENEY COMMON—TOOTING BECK COMMON
—STREATHAM GREEN—STREATHAM COMMON.

Tooting Commons.

T
HE south of London is particularly fortunate in the
possession of so many fine commons, among which
that of Tooting is certainly worthy of holding first
rank. Although split up into three separate areas
by the two branches of the Brighton railway which pass
across it, its golden patches of furze, its noble avenues of
trees, and its ever-green springy turf still mark it as one of
Nature's favoured spots. What is popularly known as Tooting Common is in reality two commons—Tooting Beck and
Tooting Graveney, separated by a majestic avenue, which
faces the keeper's lodge.

The derivation of the name ' Tooting,' or ' Totinges,' as it
is called in Doomsday Book, is enshrined in doubt. Many
authorities assert that it is derived from *theou*, a slave, and
ing, a meadow ; but others assert that it takes its name from
the great Celtic god *Teut*,* or *Teutates*, and *ing*, a meadow.
It is after this Celtic deity that we have Thursday, the day
dedicated to Thor, and many place-names in England, such
as Tot Hill, Tooting, etc., are said to owe their origin to
their being the former site of the worship of this god. At
the most these derivations are but ingenious conjectures.†

* Variously spelt Tuisto, Teut, Teutates, Tot or Thor.
† Arnold, ' Streatham,' 1886, pp. 26, 27.

The Manor of Tooting Graveney is of royal origin. It is mentioned in Doomsday Book as having been held of King Edward, and then of the Abbot of Chertsey. In the year 1285, Bartholomew de Costello had a grant to him and his heirs of free warren in Tooting, and in 1314 Thomas de Lodelaw died, seised of the Manor of Totinge Graveney, and

The Main Avenue, Tooting Common.

a capital messuage, garden, dovecot, 100 acres of land, 12 of meadow, 5 of pasture, 4 of wood, rents of assize, 24s., pleas and perquisites of court. We find that in 1332 Thomas Lodelaw, the son, held his lands in Totinge Graveney of the Abbot of Chertsey, and the Abbot held of the King in like manner. The manor remained in possession of the Monastery of Chertsey until the time of its dissolution in the thirtieth year of King Henry VIII. The earliest court mentioned on the Rolls was held in the thirty-fourth year of that reign: but the name of the Lord of the Manor at that time is not entered. In 1652 Sir John Maynard appears to have been seised of the manor, whose representatives in 1682 sold it to Sir Paul Whitchcote, and in 1695 an Act was passed enabling the Whitchcote family to grant leases of the manor for ninety-nine years. After many changes in ownership, which it would be tedious to enumerate, the manor was in 1861 put up for sale by public auction, when it was purchased by Mr. W. J. Thompson,* from whom the manorial rights were bought by the late Metropolitan Board of Works in 1875, whose successors the London County Council are the present lords of the manor.

Tooting Graveney derives its second name from the De Gravenelle family, in whose possession it was soon after the Conquest, and was at one time held on payment of a rose yearly at the feast of St. John the Baptist.† Here is a splendid example for those who declaim against unearned increments, for the freehold of the two commons cost the late Metropolitan Board of Works £17,771. The manor and the parish of Tooting Graveney are as nearly as possible coterminous, comprising about 565 acres, of which the common claims 63. It is interesting to record that this is the smallest parish in England. The inhabitants of the manor had extensive privileges including pasturage for all beasts, whether commonable or not, a right of cutting furze, bushes, fern and heath, a right to dig gravel, turf, and loam, merely paying the cost of haulage and digging, and lastly the right which has not been taken away from them, viz., that of the use of the common for recreation and village sports. The common has been disfigured in many places by this gravel-digging, but Nature is doing her best to cover over these places with her carpet of green, and so transforming them into lovely dells. The Court Rolls contain many

---

* From evidence prepared for the action of Betts *v.* Thompson.

† Lysons, ' Environs of London,' 1810, vol. i., part i., p. 374.

entries of grants of the common land in consideration of payments made towards building the church or national schools in Church Lane, which are themselves erected on common land. The parish benefited to the extent of £1,417 10s. through these transactions, but a resolution was passed in 1851 denying the right of the Lord of the Manor to alienate in any way the common lands of the parish. If their predecessors had been equally firm the commons would now be of considerably greater area. In 1569 one-fifth part of the two commons of Tooting Beck and Tooting Graveney was enclosed by Robert Lenesey, and it was ordered 'that hereafter he do it not.'* The large district of Streatham Park was originally taken from the common by Mr. Thrale with the permission of the Duke of Bedford.† We have mentioned before that the manor was put up to auction in 1861, and the inhabitants were naturally anxious that the purchaser should be opposed to enclosure. To their great relief they heard that the late Lord of the Manor, Mr. Thompson, was going to bid, and as he was well known to be opposed to enclosure, they resolved not to bid against him. Eventually the manor and some house property were knocked down to him at £3,285. When the lord was in possession, it suddenly occurred to him that some of the land was worth £1,000 an acre for building purposes, and he at once began to enclose portions of it. The inhabitants were up in arms, and the fences were pulled down. The war dragged on for several years till a Mr. Betts, a local butcher, secured an injunction in 1870 restraining the lord from interfering with the rights of pasturage, etc., of the inhabitants.

The Rolls contain several entries with regard to the rights of tenants of the manor to cut furze, and take away gravel. They were extremely strict, and rightly so, with regard to tenants of other manors or foreigners as they are called, exercising these rights, and many are the fines which these trespassers have been called upon to pay.

In 1605 the parishioners themselves were forbidden to cut

* Arnold, 'Streatham,' p. 107.   † *Ibid.*, p. 214.

any more furze, but the order does not seem to have been obeyed, as it is repeated in many subsequent years, but fathers of families are expressly allowed to have 100 bundles a year. The straying of cattle on to the common lands seems to have been a serious matter, so much so that the remuneration of the keeper was settled at the amount of fines he could recover for trespass. It may be interesting to quote the scale of fees which formed the remuneration of John Willson, the keeper appointed in 1660. It is in modern English as follows : ' For every beast within the same manor and parish to be found trespassing upon the common after the first of November as followeth. For every horse, mare, gelding, or cow 4d., and for the same cattle of every stranger 1s. ; for every hog unringed 4d., and for every hog ringed 2d., if they be taken upon the common or elsewhere. For every score of sheep 6d., and so after the same rate for fewer, and double to strangers. And that the same field keep and common keep shall have power and liberty from time to time to impound the same cattle which are there to remain until the same penalties be paid, and the parties trespassed satisfied for the wrongs done to them.'* The succeeding keepers were remunerated at the same rate, and it is hoped this payment by results secured efficient protection for the common.

The remainder of the present common, the Tooting Beck portion, is in the manor of that name, which at the time of the Doomsday survey was held by the Norman Abbey of St. Mary de Bec, from which it derives its name. It left their possession on the dissolution of *alien* monasteries in 1414, and was vested in the Crown. The King gave the manor to his brother, John Plantagenet, Duke of Bedford, and as he had no children, it descended to his nephew, Henry VI. Later on we find the King assigning the manor as part of the endowment of Eton College which he had founded. After he was deposed, his successor, Edward IV., took back several of the grants to the college, the Priory or

* From the Court Rolls.

Manor of ' Totynbeke ' being one of them, and granted it to the Bishop of Durham. It seems then to have reverted to the King again, and he presented it to the Earl of Worcester, who settled it upon a fraternity called St. Mary's Guild for the repose of the soul of Edward IV. At the dissolution of religious houses, the manor was sold in 1553 to John Dudley, Earl of Warwick, and fifty years later it was purchased by Sir Giles Howland. His descendant, John Howland, died seised of the property in 1686, leaving it to his daughter Elizabeth, who conveyed it, by marriage, to Wriothesley, Marquis of Tavistock, afterwards third Duke of Bedford, and Baron Howland of Streatham — a title granted by William III., and since held by the Duke of Bedford. Lord Wriothesley, who was only fifteen at the time, was married by Bishop Burnet of Salisbury at the old manor-house of Streatham. Francis, fifth Duke of Bedford, conveyed the manor to his brother, Lord William Russell, who was murdered by his Swiss valet, Courvoisier, in Norfolk Street, Strand, in 1840.* Before his death he had sold his interests in the estate to R. Borradaile and Richard Rymer. The next Lord of the Manor was Robert Hudson, of Clapham Common, from whom the manorial rights were purchased in 1873 for £10,500 by the late Metropolitan Board of Works, so that their successors, the London County Council, are lords of both the manors of Tooting Graveney and Tooting Beck. The acreage of Tooting Beck Common is 147½, making 201½ for the two.

The chief historical name connected with Tooting Common is that of Dr. Johnson. As the guest of Mrs. Thrale at Thrale Place, he was a most frequent visitor to the common, and made her house a second home. Mr. Thrale, who is usually passed over in any mention of Dr. Johnson, was an opulent brewer of Southwark, and was a person of no mean attainments. It will be sufficient to quote Dr. Johnson's opinion of him. ' I know no man,' said he, ' who is more master of his wife and family than Thrale. If he but holds

* Thorne, 'Environs of London,' 1876, vol. ii., pp. 588. 589.

up a finger, he is obeyed. It is a great mistake to suppose that she is above him in literary attainments. She is more flippant, but he has ten times her learning; he is a regular scholar, but her learning is that of a schoolboy in one of the lower forms.' It may be mentioned that after the death of Mr. Thrale, in 1781, the brewery was put up to auction and bought by Mr. Barclay junior, then the head of the banking

Thrale Place, formerly overlooking Tooting Common.

firm of Barclay and Co. Not knowing much about brewing, he took into partnership Mr. Perkins, who had formerly been manager of the brewery in Mr. Thrale's time, and hence we have the origin of the firm of Barclay and Perkins. Thrale Place estate, which was over 100 acres, faced Tooting Bec Common and extended as far as Streatham Church. The mansion was a large white house of three stories, having a slightly projecting centre and wings with a semicircular end. This house was pulled down in 1863,* and its place has been taken by Streatham Park, a picturesque building estate dotted with red-brick houses. . It would be out of place in this work to give any lengthy account of Dr. Johnson or his work, so we will confine ourselves to a few remarks as to his connection with Tooting. It was not till he was nearly sixty years of age that he took up his residence with Mr. Thrale and his young wife. Up to this time he had had a hard struggle in life. He had to quit his University before obtaining his degree, on account of his poverty, and his first effort in life was as an usher at a school in Market Bosworth ; but the drudgery of this life was unbearable to him, so he tried to earn his bread by translating for a bookseller in Birmingham. In the midst of his troubles he married Mrs. Porter, a widow twenty years older than himself, whose fortune of £800 he attempted to turn to profit by opening a school. This lasted for eighteen months, and during that time he only attracted three pupils, but one of those three was the celebrated actor, David Garrick. Having given up the school he came to London, and contributed largely to the *Gentleman's Magazine.* He now came to be better known, and a proposal was made to him to prepare a dictionary of the English language. This work, with which his name will always be associated, was completed in 1755, and was received with an enthusiasm never bestowed on any similar work before. The success of his dictionary did not at once relieve his pecuniary wants, and it was not till 1762 that his efforts received national re-cognition. But with a pension of £300 a year a complete change came over his career. He was no longer compelled to write for money, and as a consequence his natural indolence revived, and he abandoned himself to talk and tea. It was about this time that he became acquainted with the Thrales.

Tradition couples the name of the burly lexicographer with the old tree we shall have to mention later on, telling us with graphic details how he compiled page after page of his

* Thorne, 'Environs of London,' vol. ii., p. 590.

dictionary under its spreading arms ; but alas for tradition ! his *magnum opus* was completed and published long before he came to Tooting. His favourite resort was the little summer-house in the grounds. One cannot wonder that Dr. Johnson, who had little taste for grand or rugged scenery, should have found in Mr. Thrale's house a spot as congenial in its natural surroundings as in the human companionship it afforded. Here for fifteen years he was a constant visitor ; but during the whole of this time he kept a dingy house at the bottom of Bolt Court, on the face of which is a tablet recording the fact. Although his literary work almost ceased during this time, he is said to have written the greater part of

Summer-house in Mrs Thrale's Garden, the Favourite Resting-place of Dr. Johnson.

his 'Lives of the Poets' at Tooting. Another literary work connected with Thrale Place is the ' Vicar of Wakefield,' for Mrs. Thrale relates that it was from here that Dr. Johnson sallied forth to the help of Oliver Goldsmith, and was the immediate cause of the publication of his masterpiece. In 1781 Johnson lost his best friend, Mr. Thrale, and was one of the executors under his will. The story goes ' that when the sale of Thrale's brewery was going on, Johnson appeared bustling about, with an inkhorn and pen in his buttonhole, like an Excise-man ; and that on being asked what he really considered to be the value of the property which was to be disposed of, answered, " We are not here to sell a parcel of boilers and vats, but the potentiality of growing rich beyond the dreams of avarice.'"* With the death of Mr. Thrale Dr. Johnson's connection with Tooting ceases, and the next year we find him saying farewell to the old place. Many writers attribute this fact to Mrs. Thrale's changed manner towards him. She very soon after the death of her husband married a music-master named Piozzi. Johnson only survived his exile for two years, and was buried in Westminster Abbey.

Another frequent visitor to Thrale Place was Sir Joshua Reynolds, who was commissioned by Mr. Thrale to paint the portraits of the most famous of his guests. This series of portraits, twenty-four in all, was known as the Streatham Gallery, and included Goldsmith, Burke, Reynolds, Chambers, Garrick, besides the host and hostess and the great Doctor himself. When the gallery of pictures was sold by auction in 1816 various prices were realized, ranging from £80 to £378, which latter was for that of Johnson.†

A turnpike gate formerly stood on the highroad between Tooting and Streatham, at the north-west corner of what is now Streatham Park, which was the scene of an amusing incident, in which several important personages figured. The story goes that Lord Thurlow had been dining at Addiscombe with Mr. Jenkinson (afterwards Lord Liverpool) in company with Dundas and the younger Pitt, then Chancellor of the Exchequer. After drinking rather freely of Mr. Jenkinson's champagne, they turned homewards at a late hour of the night. On arriving at this gate the jolly party found it open and dashed through without paying toll. The gate-keeper, who was aroused from his sleep by the noise of their horses' hoofs, got up and dashed into the road, and fired a blunderbuss at them, happily without effect. He had no doubt mistaken them for a gang of highwaymen, who were then infesting the roads.‡

* Quoted in 'Old and New London.'
† Thorne, 'Environs of London,' vol. ii., p. 590.
‡ Arnold, 'Streatham,' pp. 110, 111.

The two approaches to the common from Tooting and
Streatham respectively are both marked by parish churches,
St. Leonard, the parish church of Streatham, being at the
junction of Tooting Bec Road with the main road from
London to Brighton, and at the other end is St. Nicholas,
the parish church of Tooting, at the corner of Church Lane.

Streatham Church at the Commencement of the Nineteenth Century.

There are two lanes which connect the highroad from
Tooting to Mitcham with the common. These are Church
Lane and Back Lane, both bordered with narrow strips
of common land. There are few spots near London which
are so beautiful. Overhead the giant trees meet in a leafy
arch, and the whole scene is rural indeed. Between the
two lanes is a conspicuous red-brick building, originally

called St. Joseph's Academy, a school and training college for Roman Catholics. This is now a workhouse. Continuing by the main road over the common, we shall eventually find ourselves at the parish church of Streatham. The present building only dates from 1831, although the site has been occupied by a church ever since the Norman Conquest. It was the last, not the present, building which was honoured with the presence of Dr. Johnson as a worshipper. The Thrales had a pew there, although the spot cannot now be identified. In 1782 Dr. Johnson came here for the last time. It must have been a great toil for him to come up the hill from Streatham Park. As he left the old church, he kissed the porch, as he records in his diary: 'Sunday, went to church at Streatham. Templo valedixi cum osculo.'* The church possesses two out of the four or five epitaphs which Dr. Johnson is known to have written. One is on the tomb of Mrs. Thrale's mother, to whom he had a particular aversion, and the other on that of Mr. Thrale himself.

Opposite to St. Leonard's is another ecclesiastical building of recent construction, the Roman Catholic Church, built of ragstone with stone facings. This has a tall and graceful spire, and handsome stained-glass windows.

And now leaving the churches, we come back to the common, and among numerous other fine trees will be noticed the relic of a gigantic elm, carefully guarded with railings. This stump was formerly hollow, but as it was chosen by some unhappy being as a suitable place to commit suicide in, it had to be filled up, and a poplar-tree is now planted in it. It is supposed that this noble tree is at least 1,000 years old.† The glory of Tooting Common consists in its trees, and many a story is told about the fine avenues stretching across its wide expanse. One of them is said to have stretched to London, and to have been the favourite drive of Queen Elizabeth. The fact that she came from

* Quoted in Arnold's 'Streatham,' p. 47.
† Arnold, 'Streatham,' pp. 111, 112.

London to see Sir Henry Maynard (secretary to her Minister Lord Burleigh) at his manor-house at Tooting in 1600 may have accounted for this. We find that in the time of the Conqueror, Tooting, like many other places, possessed a wood, and some of the fine old oaks are perhaps a relic of this past grandeur. Many of these trees were sacrificed

Old Tree on Tooting Common.

in the formation of the two arms of the London, Brighton and South Coast Railway across the common. One traverses it from end to end, and leaves it only to intersect the Streatham Park estate, and the other cuts off and completely isolates a corner at the Balham end.

The surroundings of this isolated portion of the common

were once particularly picturesque, and although still rural, it may not be long before the adjacent land will be required for building purposes. Hyde Farm has furnished a subject for many an artist's picture. The land at the rear divided from the common by iron hurdles is let out for cricket and football pitches. On the other side of the leafy lane which forms the approach to the common from Clapham Park is the Telford Park building estate.

At the corner of Tooting Bec Common, close to Bedford Hill Road, is a castellated residence called The Priory, known to fame chiefly through the inquiry into the death of Mr. Bravo, supposed to have been poisoned by his wife, who was, however, acquitted. It is now used as a school for boys. This occupies the site of a real priory, which once belonged to the former owners of the manor, the Abbey of Bec in Normandy. The monastic building was almost totally destroyed through the accidental upsetting of a lamp by one of the monks. Some of the walls of the present building are relics of the older structure. When making some alterations a few years ago, the workmen came across some tiled paving, which possibly may have been used in the refectory of the priory. The tiles were in a good state of preservation, but unfortunately they were all smashed except two by the workmen, who did not know their value. The two that remain are of blue ware, and represent Scriptural incidents, one, 'Christ writing in the dust,' and the other, 'David and Goliath.'*

Another historical personage we shall have to note in connection with the common is Sir Richard Blackmore, physician to King William III. and Queen Anne, who was a very voluminous writer of poetry, medicine, history, and philosophy. His works are more remarkable for their size and good purpose than for their genius. Among other long-winded productions he composed a poem in twelve books called 'King Arthur,' followed by a similar work, 'Eliza,' in ten books. 'It is never mentioned,' says Johnson in his

---

* Arnold, 'Streatham,' pp. 149, 150.

'Lives of the Poets,' 'and was never seen by me till I borrowed it for the present occasion.' A like fate awaited his other 'heroic' effort, in which he sought to enshrine 'King Alfred' in twelve books, 'for,' says Johnson, 'Alfred took his place by Eliza in silence and darkness; Benevolence was ashamed to favour, and Malice was weary of insulting.' Blackmore was repeatedly snubbed by Pope in his poems, chiefly because he had attacked Dryden in one of his works. Blackmore had his country house here for a time.

> 'Blackmore himself, for any grand effort,
> Would drink and doze at Tooting or Earl's Court.'*

Pope gives us his estimate of his poetry as follows:

> 'You limp, like Blackmore on a Lord Mayor's horse.'

But in the face of this adverse criticism we must quote one other authority who was certainly as competent to judge as either Pope or Johnson, and that is Addison. In concluding one of his essays on the poetry of Milton, Addison notices Blackmore's 'Creation.' 'The work,' he says, 'was undertaken with so good an intention, and is executed with so great a mastery, that it deserves to be looked upon as one of the most useful and noble productions in our English verse.'† Though perhaps a man of little genius, he was distinguished by a high and noble purpose, and not a single word could be said against the integrity of his character.

Just in passing we may mention another great personage, the author of 'Robinson Crusoe,' Daniel Defoe, who spent some of his earlier years at Tooting. He founded a Nonconformist church here in 1688, and has left his name in Defoe Road, a turning out of the High Street.

Several improvements have been carried out on the common since it has been in the hands of the London County Council. A few years ago the adjacent lands by

---

* Pope, second epistle of Second Book of Horace.
† *Spectator*, No. 339.

the keeper's lodge were announced to be developed for building purposes. This would have meant that the fine belt of trees overhanging the common would have been lost, so the local residents put their hands in their pockets, and their subscriptions, aided by a Council grant, purchased some 3 acres of close-timbered land as an addition to the common. The ponds, which were in reality disused gravel-pits, have now been considerably enlarged, and form a fine sheet of water. In 1811 John Harwood, a clerk in Woolwich Dockyard, was thrown out of his chaise into one of the disused gravel-pits filled with water, and was drowned. This was probably due to the absence of any proper roads across the common, the only means of communication being tracks varying from 20 to 100 feet in width.* Several of the wet portions have been drained, the surface improved, and trees planted to make an effort to screen the railways. A horse-ride has been formed to save the indiscriminate galloping all over the turf, together with other improvements of minor interest. There must be few, therefore, who would regret the transference of the manorial rights from private to public hands, and the consequent immunity from any likelihood of encroachment upon this favoured spot.

STREATHAM GREEN—STREATHAM COMMON.

Leaving now the commons of Tooting, and passing by St. Leonard's Church, we find ourselves on the highroad to Croydon and Brighton. This is a continuation of the same Roman road we met at Kennington Park. In the days of the Romans, and, in fact, till the first ten years or so of the present century, this Roman road, the Streatham highway, had a narrow patch of common land on either side extending its whole length, where travellers could let their cattle graze. This was fringed with the front ranks of almost pathless woods, amongst which was 'that great wood called Norwood,' which shut Streatham in on all sides, and made travelling

* Arnold, ' Streatham,' p. 108.

exceedingly dangerous. In the latter days of the last century the journey from Streatham to London, *through the country*, was far too dangerous to be attempted by night, unless with a strong escort, and even in the daytime the adventurous traveller had need to be fully armed. It is a fearful and horrible fact that few churchyards have so many people buried within their precincts who have been 'found dead,' foully murdered, in many cases, by the numerous footpads who infested the roads in times gone by, as Streatham churchyard.*

A small portion of land, about 1 acre in extent, lying between the roads to Mitcham and Croydon, known as Streatham Green, forms a connecting - link between the commons of Tooting and Streatham. On one occasion this little place caused a serious hitch between the vestry and the Lord of the Manor. The green was closed April 27, 1794, and the footpath leading to Streatham Church stopped also, whereupon the indignant vestry wrote a letter of protest to the lord's agent. It is satisfactory to note that the green and the path were in the end reopened. Just a word about this vestry. It was certainly not above unbending what Mr. Bumble would call its 'porochial dignity.' In these hours of relaxation they would refresh their inner man at the expense of the ratepayers, but it was found necessary to draw the line somewhere. Hence in 1774 'it was ordered that the church-wardens are not to expend at all the vestry and visitations in the year, any more than ten guineas for entertainments.' As we come across entries like these, we venture to think that the present age has not degenerated so much as pessimists would have us suppose.

Streatham probably derives its Saxon name from the main road we have referred to, being the *ham* or village on the *streat* or street. The name has gone through various changes in spelling. It was variously spelt 'Stretham,' 'Streetham,' or 'Streteham.' It is mentioned in Doomsday Book as 'Estraham,' the Normans having corrupted this along with

* Arnold, 'History of Streatham,' 1886, pp. 19, 20.

many other Saxon names which they could not understand.*
Curiously enough this common, 66 acres in extent, is in the
Manor of Vauxhall, the district known by that name being
some miles away. But this is by no means an isolated case
as regards the dismemberment of manors. The Manor of
Battersea, for instance, includes the district of Penge, a considerable distance away.

The Manor of Fauxhall or Vauxhall was the property of
Baldwin, son of William de Redvers, or de Ripariis, fifth
Earl of Devon, and to whom the Isle of Wight had been
given by Henry I.; whence he was also called de Insula.
Baldwin married Margaret, daughter and heiress of Warine
Fitzgerald, and settled this manor on her as part of her
dower. He died in the reign of King John, in the lifetime of
his father William, leaving by this Margaret a son named
Baldwin, who on the death of his grandfather William
succeeded him, and became the sixth Earl of Devon. In
1240 the second Baldwin was made Earl of the Isle of
Wight, having previously married Amicia, daughter of Gilbert
de Clare, Earl of Gloucester and Hertford. He died when
young in 1244, leaving Baldwin his son and heir, who
became the seventh Earl of Devon, and after having married
Margaretta, a kinswoman of Queen Eleanor, died in 1262,
leaving one child, John, who died whilst an infant. Margaret,
who had married his grandfather, was still living, and held
this estate so settled on her. On the death of her first
husband Baldwin, King John compelled her to marry one of
his favourites, Fulk le Breant, of whose origin we have no
certain account. The monkish historians speak with the
greatest bitterness of him, which, indeed, is not to be
wondered at, as he certainly paid no respect to them, but
apart from their testimony, there are authentic accounts of
his violence and turbulence. In addition to this marriage
the King gave him also the wardship of Baldwin's infant
son, then heir-apparent to the great earldom of Devon.
These wardships were of great value; the grantee, besides

---

* Lysons, 'Environs of London,' 1811, vol. i., part i., p. 361.

receiving the profits of the estate, had the opportunity of marrying his daughter to his ward, and as a proof of their value it may be mentioned that the Earl of Gloucester afterwards gave the King 2,000 marks for this very wardship. Whatever might be the conduct of Fulk in other respects, he remained faithful to King John and his son King Henry, till the commission of that act which brought on his ruin. He had seized some houses and lands in Luton, and to recover these the owners instituted legal proceedings against him. The judges, of whom Henry de Braybrooke was one, in every case decided against him, and imposed fines in addition. This so exasperated Fulk, that, as Braybrooke was going to the council which the King was holding at Northampton, he sent a party of men, seized him and his attendants, and carried them to the castle of Bedford, of which he was then Governor. The indignation of the King and his council was so excited by this deed that they went to Bedford, summoned the Governor to deliver up the prisoners, and surrender the castle; but Fulk had placed his brother as Governor, who refused to comply with these demands, whereupon Fulk was excommunicated, and the castle besieged and taken. The Governor and sixteen of his men were subsequently hanged, and when Fulk was at last prevailed upon to submit himself to judgment, his life was spared in consideration of his past faithful services, but all his estates were forfeited. In the meantime Margaret had not been idle. She had been compelled to marry Fulk against her will, and now succeeded in obtaining a divorce, whereupon she married a third husband, Robert de Aguillon, Lord of Addington. Her son and grandson, and the infant son of the latter having all died in her lifetime, Isabella, the only sister of the grandson, became heiress, she being then the wife of the third Earl of Albemarle.

Isabella had several children by the Earl of Albemarle, all of whom died young, except a daughter named Aveline, married to Edmund Crouchback, second son of Henry III., and afterwards Earl of Lancaster. By him she had no

15—2

children, and died at Stockwell. King Edward had flattered himself that this marriage would bring back the Isle of Wight into the Royal Family ; but his wish being frustrated by the death of Aveline without issue, the King entered into a treaty with Isabella for the purchase of it, together with the Manor of Lambeth and Faukshall, and a conveyance was executed for 20,000 marks, to which 6,000 marks of silver were added for a further deed to rectify a mistake in the first. By the last-mentioned deed in 1293 she conveyed to the King the Isle of Wight, together with other estates in Hants, the Manor of Lambyth (Stockwell), and a manor in Lambyth called La Sale Faukes. This conveyance was executed on her death-bed, and it was hinted that it had been fraudulently obtained inasmuch as Isabella had constantly refused to part with her ancient inheritance. The heir-at-law, Hugh Courtney, Baron of Okehampton, claimed the Isle of Wight, and petitioned King Edward II. that it might be restored to him. The King hereupon directed an inquiry by what means these lands came into the hands of his father. The claimant did not succeed in his suit, and the King retained this manor as well as that of Kennington. About this time a survey was taken of the manor, which shows that land in Streatham was included in the Manor of Vauxhall. It consisted then of 'a capital messuage, 74 acres of arable land, 32 of meadow, a water-mill in Micham, for which the prior of Merton gave 21s. per annum, also in Micham, *Stretham*, and South Lambeth 17 free tenants, 28 customary tenants, etc.' In the same year that this survey was taken the manor was granted with Kennington, to Roger Damorie and Elizabeth his wife, and the heirs of the body of Roger, which grant was confirmed in the following year. On the attainder of this Roger the King seized his estates, but ordered them to be delivered to Elizabeth his widow. This order does not seem to have extended to Kennington or Vauxhall, as the former was granted to Spenser, who in 1324 had a grant of Vauxhall. The Spensers died in 1327, after which Elizabeth probably recovered some of her estates, as we find from a record of 1330 that land was held of her as the Lady of the Manor of Faukeshall. Elizabeth de Burgh in 1338 exchanged the manors of Kennington and Vauxhall for those of Ilketesshall and Clopton in Suffolk, which belonged to the King. In the same year the King granted this manor to his son, Edward the Black Prince, and a few years afterwards, viz., in 1354, the Prince granted it to the monks of Canterbury. On the suppression of monasteries Henry VIII. in 1542 gave this manor, with that of Walworth, to the Dean and Chapter of Canterbury.* The manorial rights were still vested in them, or, rather, in the Ecclesiastical Commissioners, when they were purchased, as far as Streatham Common is concerned, by the late Metropolitan Board of Works, whose successors, the London County Council, now have charge of the common. By the provisions of the Metropolitan Commons Supplemental Act, 1884, the Lords of the Manor sold their rights, except as to minerals, for the nominal sum of £5, and they were also empowered to enclose and appropriate a small strip of the common. A movement is now on foot to purchase this strip, which is well timbered, and add it to the common.

The popularity of Streatham Common is not of modern growth. It came into public favour many years ago for a cause which we will proceed to detail. Epsom salts have become a household medicine, but it is a curious fact that 200 years ago Streatham had its mineral springs, which might have made this little village a rival to the fashionable haunts of Bath and Cheltenham. A drawback to their popularity evidently lay in the fact that their nearness to London made them particularly accessible. If they had only been in some distant Alpine village, many gouty old gentlemen and rheumatic ladies might still have resorted hither, but such is the perversity of human nature that it despises that which is near at hand. The first account of these mineral springs is given by Aubrey. Writing in 1673, he says: 'The medicinal springs here are in the

* Allen, 'History of Lambeth,' 1827, pp. 263-270.

ground east of the green (*i.e.*, Streatham Common); they have a mawkish taste; they were first discovered about fourteen years back, and this is the third year they have been commonly drank. It is a cold, weeping, and rushy clay ground. In hot weather it shoots a kind of salt or alum on the clay, as in the sour grounds in North Wiltshire; it turns milk for a possett. Five or six cups is the most they drink; but the common dose is three, which are equivalent to nine of Epsom.'* He was told by a locksmith that, being very ill, his doctor had ordered him to drink the Epsom waters, which he did, receiving no benefit; but after trying those of Streatham, he was restored to health. Aubrey also informs us that 'they are good for the sight, a Taylor having his sight restored by their use.' The story of their discovery is of interest. It appears that while some horses were ploughing on the field in which they were situated, the ground gave way suddenly, and the consequent inspection led to their being found. But the owner apparently objected to the discovery being made known, and some considerable time elapsed before they came into use.† This is not the only well in the district to which healing virtues are attributed. A spring at Vauxhall, known as the Vauxhall Well, was supposed to be very efficacious in the treatment of eye diseases. Its waters, moreover, had never been known to freeze.

The report of these cures was quickly noised about with the result that Streatham Spa soon became a fashionable resort. The learned physicians of the time analyzed the waters, and reported favourably upon them. The natural beauties of Streatham Common and its surroundings added to the attraction, and in spite of the wells changing hands frequently, they came into great favour. Their reputation was at its height at the beginning of the eighteenth century, and by this time the common and the high road had become fashionable promenades where all the leaders of society

---

* Aubrey, 'History of Surrey.'

† Arnold, 'History of Streatham,' pp. 96, 97.

might be met. The road which had before been trodden by Roman legions now groaned under the weight of the cumbrous family coaches bearing their wealthy occupants to this rural Bethesda. For the accommodation of the numerous visitors, the large house facing the common, now known as 'The Rookery,' was rebuilt and enlarged, while the remembrance of the wells is still kept alive by the adjacent 'Wellfield House.' At the present time may be seen in the kitchen garden of 'The Rookery,' the residence of Sir Kingsmill Key, all that remains of these once celebrated wells. A little house enclosing the pump over the well, which is 35 feet deep, is the sole relic of the glorious past. In the height of the season of 1701, concerts were given at the spa twice a week, which made the crowd of visitors as gay and as frivolous as their ailments would allow. The use of the waters was by no means confined to the locality, as they were supplied to many of the leading London hospitals, and to many coffee-houses, as will be seen from the following advertisement :

'The true Streatham waters, fresh every morning, only at Child's Coffee House, in St. Paul's Churchyard ; Nando's Coffee House, near Temple Bar ; the Garter Coffee House, behind the Royal Exchange ; the Salmon ; and at the Two Black Boys in Stock's Market. Whoever buys it at any other place will be imposed upon.

'N.B.—All gentlemen and ladies may find good entertainment at the Wells aforesaid, by Thomas Lambert.'*

These were in the palmy days of Streatham Spa. With the growth of many other springs in more distant parts, Streatham began to fall into decay, and its downfall was hastened by the closing of the house of 'good entertainment.' Slowly but surely the good old days passed away, and though an attempt was made to stem the tide by opening a new well at the bottom of Wells Lane, the spa has passed into oblivion. The advent of railways made

* *Postboy*, June 8, 1717.

these distant waters within easy reach of the Metropolis, and the rumbling of the clumsy coaches on their way to this once famous spa has long since ceased.

The common has been the scene of some petty riots in the same way as its neighbour Tooting, arising from the encroachment of the Lord of the Manor on the rights of the tenants. On one occasion a mob of people met on the common, and set the heath furze on fire. The conflagration was tremendous, but the neighbours rather promoted than lent any assistance for extinguishing it. It seems that the Duke of Bedford used to let the poor have the furze, but this year (1794) he sold it for £80. His agent, by His Grace's orders, took in some ground from the common which was formerly used for the poor people's cattle; and in the evening a hackney coach drove to the spot, when six men, draped in black and crapes over their faces, got out of the carriage, cut down the paled enclosure, returned into the coach, and drove off.*

A racy story is told of a former resident on the common, whose name, of course, we cannot mention. This worthy conceived a unique scheme for escaping His Majesty's dues. He bought up an enormous quantity of kid gloves abroad, and then told his agent to send all the left-handed gloves to England. On their arrival, as no one claimed and paid the duty, they in due time were put up to auction at the docks. Our friend bid for them, and as no one entered the contest for an apparently worthless article of commerce, he got them for almost nothing. The right-handed gloves were sent soon after with the same result, and a magnificent profit resulted.† This is not the only case on record where Streathamites have defrauded the national purse, for there is a tradition that the house adjoining the common known as 'The Rookery,' which we have before mentioned in connection with the mineral springs, was once the residence of another less scheming smuggler, who warehoused his

* *Gentleman's Magazine*, May 4, 1794.
† Arnold, ' History of Streatham,' p. 215.

goods here till he could find a market for them.   The cellars here far exceed the wants of any private inhabitant, and this seems to give a fair amount of truth to the story.*

The common gently slopes from the highroad to a ridge, which affords magnificent views of the charming country round.   It is said that Woolwich, Stanmore, and Windsor are to be seen from it.   The lower part of the common is

A View on Streatham Common in the Eighteenth Century.

open and available for games, and would be comparatively bare but for the trees on the adjoining properties.   The upper portion, formerly called Lime Common, is covered with a dense undergrowth.   Wild roses, brambles, furze and bracken here flourish in profusion, and the giant trees in clumps and avenues, combined with the rural surround-

* Arnold, ' History of Streatham,' p. 98.

ings, make up a typical scene of Surrey beauty. Dr. Johnson's favourite walk was to go from Mrs. Thrale's house across the common to the top, and then down the field-path towards Norbury, and so home again. His most ardent admirer could not call him a lover of Nature, but he certainly made an excellent choice here. It was doubtless owing to its lofty situation that this spot was chosen for one of the series of beacon fires which were lit on the night of June 21, 1887, to celebrate our Queen's Jubilee. These beacon fires were started on the Malvern Hills at 10 p.m., and were seen from Cottington Hill, Hants. From Streatham Common the fires seemed like so many glow-worms shining in the warm night.

Visitors to the common may have noticed a portion of the cricket - ground zealously guarded by iron posts and chains. This is the cricket ground specially reserved for local clubs under the Act, so that the interests of Streathamites are not lost sight of. The common boasts of a horse-ride and a bandstand on the higher ground, and also two small ponds. There was formerly a cage on the common erected between 1740 and 1760 for the confinement of loose and disorderly persons.* The view of the common in the last century which we give also shows a small cottage close to the pond.

Lord Beaconsfield, in one of his books, pointed out that the environs of London were more beautiful than those of any other city in Europe. Among these, Streatham and its common are by no means in the rear rank, and those who are 'cribbed, cabined, and confined' in our smoky London may well be enticed hither by the attractions they afford.

* Arnold, 'History of Streatham,' p. 209.

CHAPTER XII.

WANDSWORTH COMMON is another of those fine open spaces of which South London can boast. It is within easy reach of Clapham and Tooting Commons, and Putney Heath, which together make up an extensive area of open land, which must be very beneficial to the neighbourhoods in which they are situate. Many of the old houses which formerly surrounded this common have been pulled down, and their grounds let out for building estates, so that on many sides whole towns of small houses are brought up to its very edge. Perhaps none of the Metropolitan commons have suffered so much from encroachments as this. It was once the principal tract of waste land in the large Manor of Battersea and Wandsworth, which extended from Clapham to Wimbledon; but the 183 acres which remain are but a fragment of the original common, and they are terribly cut about by railways and highways. This is all the more to be regretted because we can gather from the scattered portions which are left, what the beauties of the whole must have been. Still there are several picturesque little 'bits' of Nature which yet remain in spite of the incursions of the iron road. The small sheet of water known as the Three Island Pond is a little gem, and the furze which covers the common in many places adds much rural charm, whilst the stump of an old windmill in one portion forms an excellent background to an artist's picture. In parts where there is

no furze, and the ground is sufficiently level, games are
allowed, and in frosty weather a large lake is available for
skating in addition to the pond before mentioned.

The common was acquired first of all in 1871 by a body
of conservators from Earl Spencer, the Lord of the Manor.
The facts which chiefly brought this about were the disgraceful and neglected condition of the common, together
with the numerous enclosures, which threatened to swallow

The Three Island Pond, Wandsworth Common.

up the whole of the open space.   After much correspondence
and local agitation, the Wandsworth Common Act, 1871, was
obtained, mainly through the exertions of Sir Henry W.
Peek, Bart., the Parliamentary representative of the division
at that time, who was well supported by an influential committee.   This Act vested the control of the common in a
body of conservators, and not only provided for its future
maintenance, but also dealt with the important matter of

past encroachments. Section 42 provides that 'In order that this Act may be a final settlement of all questions and claims connected with the lands formerly part of Wandsworth Common . . . those lands shall henceforth be and the same are hereby released and discharged from all commonable, customary and other rights and claims of the commoners.' Lord Spencer has been subjected to considerable abuse in connection with Wandsworth Common, and it is only fair to him to record what many of his critics conveniently forget, that he generously surrendered his manorial rights in return for a perpetual annuity of £250. This amount was raised by levying a local rate. In 1887 the duties of the conservators were transferred to the London County Council's predecessors by the Metropolitan Board of Works (Various Powers) Act, 1887, and it is generally admitted that the state of the common has not suffered thereby.

When the common was first placed under public control in 1871, it was more of a nuisance than a place of recreation and enjoyment. Its surface was bare, muddy and sloppy after a little rain, undrained, and almost devoid of trees or seats. It was covered with huge gravel-pits, many of them full of stagnant water, which, in addition to being very offensive, constituted a positive source of danger owing to their great depth and want of protection.

One portion was the resort of gipsy vans and tents, one of whose occupants has been immortalized in a picture by C. R. Leslie, R.A., painted about 1830. The following story is told about this gipsy beauty :

' There is a very small tent about the middle of Wandsworth Common ; it belongs to a lone female whom one frequently meets wandering, seeking an opportunity to *dukker* (tell fortunes) to some credulous servant girl. It is hard that she should have to do so, as she is more than seventy-five years of age, but if she did not she would probably starve. She is very short of stature, being little more than five feet and an inch high, but she is wonderfully strong

built.   Her face is broad, with a good-humoured expression
upon it, and in general with very little vivacity ; at times,
however, it lights up, and then all the gipsy beams forth.
Old as she is, her hair, which is very long, is as black as the
plumage of a crow, and she walks sturdily, and if requested
would take up the heaviest man in Wandsworth and walk
away with him.   She is upon the whole the oddest gipsy
woman ever seen ; see her once and you will never forget
her.   Who is she?   Why, Mrs. Cooper the wife of Jack
Cooper, the fighting gipsy, once the terror of all the light
weights of the English ring, who knocked West Country
Dick to pieces and killed Paddy O'Leary, the " Pot Boy,"
Jack Randall's pet.   Ah it would have been well for Jack if
he had always stuck to his true lawful Romany wife, whom
at one time he was very fond of, and whom he used to dress
in silks and satins, and best scarlet cloth, purchased with the
money gained in his fair, gallant battles in the ring.'*

A characteristic song was written on her in the original
Romany of which the translation runs :

> ' Charlotte Cooper is my name,
>   I am a real old Lee ;
> My husband was Jack Cooper,
>   The fighting Romany.
> He left me for a shameful girl
>   Who stole a purse, while he
> Took all the blame, and all the shame,
>   And went beyond the sea.'

A gipsy encampment forms a romantic subject for a picture,
but the reality is quite a different thing, and Wandsworth
is quite willing to sacrifice the romance in losing these
unwelcome visitors.

A large amount of money has been expended in improving
Wandsworth Common; and its present condition will compare favourably with other large open spaces.   Though
lacking the picturesque combination of hill, dale, and thicket
that may be enjoyed on its neighbour, Wimbledon Common,

* Mr. George Borrow in ' Lavo Lil.'

it still possesses beauties of its own. When the bright golden blossoms of the gorse mantle its somewhat scarred surface, we cannot help remembering that Linnæus worshipped that flower when he saw it in full bloom on Putney Heath.

The origin of the name of Wandsworth is very apparent. Its various forms of Wandesore, Wandelesorde are corruptions of Wandlesworth, which is the *worth*, or village, on the river Wandle which passes through it.

In the parish of Wandsworth there are, or were, four manors—Dunsfold, Downe or Downe-Bys, Allfarthing, and Battersea and Wandsworth. The common is the waste land attached to this latter manor.* The other manors have no commonable land remaining, but it is said that a considerable portion of the original common in the Manor of Allfarthing was enclosed by the lord at the beginning of this century. This manor appears to have belonged to the baronial family of Molins, and at different periods afterwards was part of the possessions of the Monastery of Westminster, and then of Hampton Court. It was leased by Henry VIII. in 1534 for sixty years to Thomas, Lord Cromwell, who assigned the lease to Elizabeth Draper, widow. This manor was among the lands settled in 1625 upon Charles I. when Prince of Wales, and was leased for ninety-nine years to Sir Henry Hobart and others, under whom Endymion Porter, Gentleman of the Bed-chamber, and one of the favourite attendants of King Charles, took a lease. He afterwards procured the reversion of the remainder of the original lease, and his descendant had the fee simple of the manor granted to him.† The Mr. Porter, who was Lord of the Manor when the alleged enclosure of the common was made, sold it in 1811 to Rev. Mr. White, who in turn disposed of it to Lord Spencer.‡

Long before the incorporation of a body of conservators to protect the commoners' rights, the question of encroach-

* For descent of this manor, see p. 3.
† Lysons, 'Environs of London,' vol. i., pp. 380, 381.
‡ Brayley and Britton's 'Surrey,' vol. iii., part ii., p. 492.

ments upon the common land had become sufficiently serious to engage the attention of the inhabitants. About 1760 a kind of club was formed by those living near that part of Wandsworth Common which adjoins Garratt Lane, in order to watch over their interests in the waste land of the manor. They were mostly people in humble circumstances, but they agreed at every meeting to contribute a small sum for the defence of their rights. In the event of any attempted filching of the common land, proceedings were taken against the offender, which in most cases were successful. The president of this association was called the ' Mayor of Garratt,' and connected with his installation are some of the most curious of parochial reminiscences.* The ' mayor ' was chosen after each General Election, and the chief requirements in the candidates appear to have been an unlimited capacity for talking, and some personal deformity · or peculiarity. When a party spirit was introduced into the election, and the facetious members of the club turned it into a burlesque of a Parliamentary contest, polling-day became quite an event in all the villages round about. The publicans, with a keen eye to business, subscribed for a purse in order ' to give it character,' and a very bad character it was at the best of times. The fight for the post of ' mayor ' has its own chronicler, who writes that ' none but those who have seen a London mob on any great holiday can form a just idea of these elections. On several occasions 100,000 persons, half of them in carts, in hackney coaches, and on horse and ass back, covered the various roads from London, and choked up all the approaches to the place of election. At the two last elections I was told that the road within a mile of Wandsworth was so blocked up by vehicles, that none could move backward or forward during many hours, and that the candidates dressed like chimney-sweepers on May-day, or in the mock fashion of the period, were brought to the hustings in the carriages of peers drawn

* ' How the Battle of Wandsworth Common was Fought and Won,' p. I. Reprinted from the *Mid-Surrey Gazette.*

by six horses, the owners themselves condescending to become their drivers.'* In addition to this honour conferred upon the candidates, it is said that Foote, Garrick, and Wilkes wrote their addresses. Foote, who was present at the election in 1761, made it the subject of a farce, ' The Mayor of Garratt,' which was produced at the Haymarket. The electors on this occasion were the mob, and the electoral oath was taken on a brickbat. The chosen of the people was dubbed knight and M.P., and as these worthy ' mayors ' are so intimately connected with Wandsworth Common, a passing note on those whose history has been handed down to us may not be out of place. The first was ' Sir ' John Harper, a man of wit, who combined the hawking of brick-dust with the mayoralty of Garratt. He was succeeded by ' Sir ' Jeffrey Dunstan, who held the post during three Parliaments, whose trade was that of buying old wigs—an occupation now obsolete. He was so enthusiastic in his defence of his constituents and his attack on the corruptions of power that he was prosecuted, tried, and imprisoned for using seditious expressions. The next and last ' mayor ' was ' Sir ' Harry Dimsdale, a muffin-seller, who died before he could stand a second time, and the election was suppressed in 1796.†

An attempt was made, though without success, to revive the farce in 1826. Electoral addresses were put forward in favour of ' Sir John Paul Pry,' ' Sir Hugh Allsides ' (a beadle of a neighbouring church), and ' Sir Robert Needale,' described as ' a friend to the ladies who attend Wandsworth Fair '; but the authorities intervened and prevented the election.‡

No doubt as time went on the purpose for which the ' mayor ' was originally chosen became a secondary one, but this is certain : that after the office was abolished, there was no organization to prevent enclosures of common land,

* Sir Richard Phillips, ' A Morning's Walk to Kew,' 1817, p. 81.
† Robert Chambers, ' Book of Days.'
‡ Hone, ' Every-day Book,' vol. ii., col. 819-866.

The Garratt Election (From a drawing by Valentine Green)

and as a consequence Wandsworth Common suffered severely. A sort of tradition lingers that the persons most resolute against these encroachments, after vainly trying to prevent others appropriating, at last quieted their own consciences by enclosing portions for their own benefit. From 1794-1866 there have been fifty-three enclosures of areas varying from a quarter to 96 acres.*

One of the largest parcels of land enclosed from the common is that upon which the buildings of the Royal Victoria Patriotic Asylum stand. These were erected as part of the scheme for the relief of the families of those who might fall in the Crimean War. A commission was appointed with the Prince Consort at its head in November, 1854, and large sums were collected from this country and the colonies, amounting in the total to £1,460,861. This was made up of contributions from all grades of society. The Commissioners reported that 'artisans, domestic servants, workpeople, labourers, individually and in associations, have felt a patriotic pride and a generous satisfaction in answering their Sovereign's appeal. In one striking instance the inmates of the Reformatory Asylum, Smith Street, Westminster, having literally nothing of their own to give, denied themselves a meal that its value might be offered as their gift. 'We deem it a fact deserving your Majesty's notice, that even the children of the poorer classes have very generally contributed their "mite" to enlarge the amount of the nation's bounty.'† Out of the large sum received £200,000 was appropriated towards founding an asylum for 300 orphan girls, and as a site for its erection 55 acres of Wandsworth Common were purchased from Lord Spencer for £3,700.. The architect was Mr. Rhode Hawkins, and the building is a free imitation of Heriot's Hospital, Edinburgh, with the omission of the ornamental details.‡ The

* Mr. J. C. Buckmaster in the *Daily News*, November 30, 1886.

† First Report of the Royal Commissioners of the Patriotic Fund, 1858, p. 11.

‡ Thorne, 'Environs of London,' part ii., p. 665.

first stone was laid by Her Majesty in person on July 11,
1857. A similar institution for boys was commenced in
1871 upon part of the land of the girls' school. It was
intended to teach them gardening, and to cultivate the
ground so as to produce sufficient vegetables for the boys
and girls; but as the cultivation resulted in a loss, and the
number of legitimate candidates for admission fell off, the
boys' school and grounds were sold for £30,000 under the
provisions of the Patriotic Fund Act, 1881. This is still
continued as a boarding and day school under the name
of Emmanuel School.

The remaining land of the girls' school was let to a
contractor, who conducted an extensive vehicular traffic over
a portion of the common in order to gain access to public
roads. It is needless to say that this carting caused grievous
damage to the turf, gorse, and footpaths. Representations
were made to the Patriotic Fund Commissioners as to the
damage and annoyance occasioned by their tenant; but as
these resulted in nothing, a lawsuit was commenced against
them by the Wandsworth Common Conservators. Before
this action could be heard, Parliament transferred the control
of the common to the late Metropolitan Board of Works.
That Board, after considerable delay and some legal difficulties, revived the action, and it was still pending when
the London County Council came into office. After dragging
on for some years a compromise was at length arrived at,
by which the carting across the common was stopped, and
a sum of £200 was paid by the London County Council
towards the cost of forming a new road over other land.

Another huge block of buildings adjacent to the common
constitutes the Surrey House of Correction, built in 1851.
This brick and stone erection makes provision for 1,000
convicted criminals, with all appliances for ensuring order
and discipline among the inmates.* Some seventy years
before this was built, there had been a proposal to establish
a prison at Wandsworth, but in view of the opposition it

* Thorne, 'Environs of London,' part ii., p. 666.

met with, the scheme was entirely laid aside. In front of the prison are 10 acres of common land, acquired in 1861 by the justices of Surrey to prevent its being built upon. The London County Council have secured the manorial rights in this and another plot of land at the southern end of the common.

The land close to Wandsworth Common Station, to the south of the common is part of an enclosure of 20 acres made in 1846 for the purpose of the industrial schools of St. James's, Westminster. Much of this has unfortunately been built upon, and there is no chance of its being restored to the common.

At a short distance south of the prison on land part of the common which has since been restored, the Rev. John Craig built a gigantic telescope, 85 feet in length, but very imperfect. This instrument was the largest which had been constructed up to that time, and was completed in 1852. The tube, which could be placed in almost any position for celestial observation, was supported at each end, and was slung at the side of a massive central tower 64 feet high, while the lower end of the tube rested on a support running on a circular railway. Not fulfilling the original expectation of its proprietor, it was taken down and removed some years ago.*

When the control of the common was transferred to the conservators, Lord Spencer reserved to himself a portion at the north adjoining the London and South-Western Railway main line on which was a sheet of ornamental water called the Black Sea, which was studded with thirteen small islands, and beautified with shrubs and flowers. This lake was formed by a Mr. Wilson, the founder of Price's Candle Works, who lived at Black Sea House. For every child added to his family, he constructed a little island in the lake, fringing each with yellow iris. In the Act for the construction of the railway, the Black Sea was called Pond No. 3, and was said to be the property of Battersea and Wands-

* 'Old and New London,' vol. vi., p. 482.

worth, and by this Act the Company were compelled to place apart the sum of £8,000 for the purpose of covering the bed of the pond with clay, and making a drain into it from the cutting, and keeping it in good condition. Shortly after-wards in another Act certain sections enabled the Company to withdraw this sum, but still they were bound to repair and keep intact the pond and drain on both sides of the railway. The Black Sea has now been filled up, and Spencer Park has been built upon its site so that one of the most picturesque and ornamental waters near London has dis-appeared. The old windmill, of which the stump remains (which is utilized as a toolshed by the labourers on the common), was used for the purpose of pumping water into the Black Sea.

At the parting of the roads to Clapham and Vauxhall are the offices of the Board of Works, the site of which was a roadside pond down to about 1863, when it was enclosed and planted with lime-trees. Behind these, nearer the common, is a very ancient but small burial-ground—the Huguenots' Cemetery. When the French Protestants came over to England upon the revocation of the Edict of Nantes, a small community settled at Wandsworth, and maintained themselves by dyeing silk and making hats. For their worship they rented and enlarged the old Presbyterian chapel in High Street, where service was performed in French for over a century. In this graveyard there are remains of many old gravestones of Frenchmen, but on the later ones are many English names, showing that in course of time the Huguenot element became absorbed in the surrounding population.*

Adjoining the St. Mark's portion of the common are the buildings of the Royal Masonic Institution for girls, erected in 1852 from the designs of Mr. Philip Hardwick, R.A. The tall clock-tower forms a very conspicuous object in the landscape. This institution, which is supported entirely by voluntary contributions, was founded on March 25, 1788, at

* Thorne, ' Environs of London,' part ii., p. 664.

the suggestion of the Chevalier Bartholomew Ruspini, surgeon-dentist to George IV. Since its establishment nearly 2,000 girls have been provided with education, clothing, and maintenance. A school-house was erected in 1793, near the Obelisk, St. George's Fields, on ground belonging

The Old Mill - Wandsworth Common.

to the Corporation of London, the lease of which expired in 1851. Subsequently the present site of nearly 3 acres was purchased at a cost of £1,075, and the buildings erected for a further £7,272. Later additions have been made, the principal of which is the 'Alexandra' Hall, named after the Princess of Wales, who was present with the Prince at the inauguration ceremony on March 12, 1891. The funds for this were raised at the centenary festival held in 1888 at the Royal Albert Hall, when the Prince of Wales, the King of Sweden and Norway, and the late Duke of Clarence were amongst those present. The subscription - list on this occasion amounted to over £51,500.*

Although most of the old mansions which formerly overlooked the common have been pulled down, and their grounds turned into building estates, there are a few left here and there to remind us of the time when the surroundings of the common were as rural as parts of the village still are, in spite of all the modern improvements. There is a cluster of these houses on the north side of Wandsworth Common, close by the Huguenots' Cemetery. 'The Gables' is certainly the most picturesque of these, if not the oldest. It is now divided into two, but the division does not take away any of its beauty. Covered from top to bottom with creeper, and approached by an essentially English garden, it forms a lovely picture. Of course tradition has been very busy in trying to weave some romance around this ancient dwelling. It is commonly reported that it was used at some time as a nursery for some of the children of Queen Anne, but although the records of the life of this good lady have been most carefully searched, no trace can be found of her ever having been at Wandsworth, so that this romantic legend must be dismissed altogether. A fact which is much better authenticated is that 'The Gables' was the residence of Francis Grose, F.S.A., the learned and jovial antiquary. The recklessness of his early life stands out in striking contrast to the care and attention required by his learned calling. He was a native of Greenford, in Middlesex, where he was born in 1730 or 1731. He was well provided for by his father, who obtained a position for him in the Heralds' College, where he attained to the dignity of Richmond Herald. In 1763 he sold this office for 600 guineas, and entered the Surrey Militia as Paymaster and Adjutant. Some idea of the way he carried on business may be gathered from his own

* G. B. Abbott, 'History of the Royal Masonic Institution for Girls.'

statement that he kept but two books, 'his right and left hand breeches pocket,' took no vouchers, and gave no receipts. Eventually his accounts showed a serious deficiency, which had to be made good out of his own pocket. This calamity roused his energies, and led him to develop the talent for drawing which he possessed. He devoted himself to the study of the then standing 'Antiquities of England and Wales,' under which title he brought out a costly book, which proved successful and profitable. This was followed by the 'Antiquities of Scotland,' and during his stay in Scotland he made the acquaintance of Robert Burns, who describes him as

> 'A fine fat fodgel wight
> Of stature short, but genius bright.'

Grose produced several other works of standard merit, and he intended to bring out the 'Antiquities of Ireland,' but he died at Dublin in 1791, before he could complete his task.*

Bolingbroke Grove, which forms the eastern boundary of the common for a considerable distance, was formerly known as Five House Lane, from the five houses at its commencement. The only one of these now standing is the Bolingbroke Hospital. In the last of these resided Sir John Gorst, Q.C., before he was knighted, a fact which has not been forgotten in the naming of Gorst Road on its site. The network of roads bordering on this side of the common has taken the place of some large mansions with extensive grounds, but their inhabitants did not acquire fame outside their own particular circle. On the western boundary are, or were, Burntwood House, Lodge and Grange and Collamore. Some of the present and past occupiers of these houses came into prominence with regard to the many legal actions which took place owing to the attempted encroachments upon the common. The conservatories and grounds of Burntwood Grange have been singled out by the author of Bohn's 'Pictorial Hand-book of London' for special description and praise as being some of the most remarkable for their size in and around London.

* 'Imperial Dictionary of Universal Biography,' vol. ii., p. 739.

CHAPTER XIII.

BETHNAL GREEN GARDENS.

THE gardens included under this heading are situated in the Cambridge Road, at the junction of Bethnal Green Road with Green Street. They comprise in all 9 acres—2½ for the northern or museum section, and 6½ for the southern. These lands are the remains of an estate of 15½ acres, originally part of the extensive commonable waste lands of the Manor of Stebonheath, or Stepney, which have now, with few exceptions, been built over. To prevent a similar fate befalling these 15½ acres, a number of persons joined together and purchased them from the Lady of the Manor, Philadelphia, Lady Wentworth, in 1667. As these contributors of the original purchase-money did not live at Bethnal Green, the lands were made the subject of a trust-deed dated December 13, 1690, by which they were conveyed to twenty-seven local trustees, in order that they might the better supervise the administration of the estate, and the proper distribution of the profits. From this deed many interesting particulars can be gathered relating to the estate, the proceeds of which were to be given to the poor, from which fact the lands were identified as the Bethnal Green Poor's Lands. The deed reminds us that 1690 was ' the second yeare of the reigne of our Sovereign Lord and Lady William and Mary, by the Grace of God, of England, Scotland, *France*, and Ireland King and Queen, Defenders of the Faith.' One of the sections gives particulars of the cost and the names of the contributors, which certainly deserve

to be recorded. In considering the amount of their donations, it must be borne in mind that the value of money has
greatly increased since this time:

'The purchase of which said waste ground and inclosure
of the said closes and other incident charges did amount
unto the sums of £332 2s. 10d.,* towards which charges the
said Thomas Rider did freely contribute £20, the said Rogor
Gillingham £20, the said William Sedgewicke £20, the said
Samuel Stanier £20, the said John Goldsborough £10,
Richard Warner, late of London, grocer, £10, the said
Joseph Blissett £10, the said Sussannah Andrewes £10,
David Clarkson, late of London, clerk, £5, Ffrancis Howell
£2, the said Thomas Walton £2, William Gill, late of
Bethnall Green, gent., £2, in all the sum of £131. The
remainder of the charges—being £201 2s. 10d.—was raised
out of the rents, issues and profits of the premises.'

The first yearly distribution was made in the months of
November and December, 1685, and consisted ' of 12 chaldrons
of coals and £12 in money (equally divided) to 24 families of
the said Hamlett (i.e., Bethnal Green) that neither received
pension nor paid to the poor, amounting to £25 4s.,' and
of the balance of the profits for that year ' there was £8
disbursed in providing and setting up the monument and
four dialls upon the watch house of Bethnall Green and
£9 10s. more distributed in coals and money to 19 other poor
familys,' and the balance remainder £4 13s. 2d. was in hand
' towards drawing, engrossing and enrolling these Deeds and
Conveyances.'

The next clause is a very important one, in that it proves that
in laying out the residue of Poor's Lands as a public garden,
the original object of the purchasers has been preserved:
' . . . the said waste grounds were purchased and the greatest
part of them have been since enclosed *for the prevention of any
new buildings thereon*, and to and for the yearly reliefe of the
poor.' It is provided that the profits of leasing the lands,
subject to the payment of the necessary expenses, shall be

* Lysons says £200 of this was for the land.

devoted to the maintenance of the paths, gates, and *stiles,* etc., the cost of leases, etc., and the distribution of coals and money to the poor.

Of the 15½ acres which originally made up the Poor's Lands, 4½ were appropriated for the site of the Bethnal Green Museum. This land was sold by the trustees under Parliamentary powers in 1868 for the sum of £2,000, the conditions including a proviso that all the space not actually covered by the museum buildings should be thrown open to the public as a recreation-ground. This was the first portion of the Poor's Land dedicated to public use, and was maintained by the Government under the name of Bethnal Green Museum Garden. This was subsequently in 1887 transferred with Battersea, Victoria, and Kennington Parks to the late Metropolitan Board of Works under the London Parks and Works Act, 1887. Previously to this a part had been taken as a site for St. John's Church and Vicarage, Bethnal Green. At the beginning of the present century the amount produced from the charity was about £60.* In 1890 this had increased to £517 13s., made up as follows : £430 for the open land which was leased to the proprietors of the asylum at Bethnal House, £3 for a strip of ground forming part of the forecourt of houses in Victoria Park Square, and £84 13s. dividends on stock which had been purchased out of the moneys realized by the sale of land.†

The Poor's Lands were held under this trust-deed of 1690 for nearly 200 years, the proceeds being spent in charitable purposes as we have seen. But changes were proposed by a majority of the trustees, because it was thought that a larger income could be secured by selling the land for building purposes and investing the proceeds ; seeing that the value of land in close proximity to London has increased so enormously. As the trustees were expressly prohibited from building upon the land, it was necessary for them to apply to the Charity Commissioners to frame a new scheme, which was done in

* Lysons, ' Environs of London,' 1811 edition, vol. ii., part i., p. 23.
† Schedule to the scheme of the Charity Commissioners.

February, 1888. The draft scheme formulated by the Commissioners and agreed to by the trustees, if made law, would have empowered the latter body to sell to the local Board the whole of the remaining land as sites for an infirmary, a public hall, and a free library, for a sum of £18,000, out of which a grant of £4,500 might be given toward the cost of the library, a similar condition being imposed to that in the case of the northern section, viz., that any land not used for building purposes should be devoted to the public as a recreation-ground. This proposed scheme was vigorously opposed both by the London County Council and the Metropolitan Gardens Association, and after lengthy negotiations the clause relating to buildings was entirely struck out, and the final scheme established by law, February 27, 1891, provides that ' the trustees shall, with the approval of the Charity Commissioners, and upon terms such as to secure a sufficient benefit to the poor . . . forthwith grant to the County Council of the Administrative County of London, or any other public body, all the land held in trust for the charity, provided that the said land be secured and permanently maintained . . . as a recreation-ground accessible to the inhabitants of the said parish.'* The scheme also provides for the sale of part of the land to the trustees of the neighbouring asylum, upon which they had built a counting-house.

The terms finally agreed upon for the purchase of the recreation-ground were £6,000. Adding to this the amount received for the northern portion £2,000, and the land sold to the asylum trustees £1,000, we have a total of £9,000, as against £200 paid in 1667.

The land, as it came into the possession of the London County Council, consisted of orchard, paddock, kitchen-garden, and pleasure-grounds all in a rough and neglected condition, and it therefore required to be entirely re-modelled for public use. The principal works of laying-out comprised the erection of an ornamental wrought-iron en-

* Section 30 of the scheme.

closing fence ; the formation of broad walks and shrubberies ; a sunk garden with a central fountain flanked by an extensive rockery for the display of alpine and other suitable plants ; the construction of a gymnasium for children, together with other necessary buildings. The amount spent on these improvements was over £5,000, and the gardens were publicly opened on Whit Monday, 1895.

The parish of Bethnal Green, formerly part of the Manor of Stepney, was separated from it in 1743. Till comparatively recent times, it was correctly described as chiefly inhabited by weavers of silk. It is a region of small and mean houses closely huddled together, and the provision of an open space here must be an incalculable boon. The older houses still bear the traces of their former inhabitants in the wide windows of the upper stories, so constructed in order to give light to the weavers' looms.*

The origin of the name of Bethnal Green is not very clear. Most writers adopt the conjecture put forward by Lysons that it is a corruption of *Bathon Hall*, the supposed residence of the family of Bathon or Bathonia, who possessed considerable property at Stepney in the reign of Edward I. One of this family, Alice de Bathon, died seised of a messuage in Stepney in 1274, and her son, who is called John de Bathonia, died in 1291.†

On the north of the Bethnal Green Museum are some houses which are built upon the site of an ancient chapel, called in a survey of the manor in 1703 St. George's Chapel. It is uncertain now whether this was a public place of worship for the inhabitants, or a private chapel, perhaps connected with the adjacent palace of the Bishops of London, who were Lords of the Manor. Stow says : ' In my remembrance another chapel of ease was on the northern part of Bethnal Green, but . . . is now turned into houses.' One of these, Netteswell House, according to a tablet in the wall,

---

* Wheatley and Cunningham, ' London : Past and Present,' vol. i., p. 178.

† Lysons, ' Environs of London,' 1811, vol. ii., part i., p. 17.

was built in 1553, and considerable restorations have been necessary at different times in order to secure its preservation. The Bethnal Green Museum is a branch of that at South Kensington, and the steps which led to its erection are rather interesting. When it was decided to build a permanent structure at South Kensington, the Education Department offered the temporary buildings (popularly known as the Brompton boilers) to any London parochial authorities who would establish a district museum. The only parish to rise to the occasion was Bethnal Green, where a committee was quickly formed, and sufficient funds obtained to purchase the present site from the trustees of the Poor's Lands.. When all the arrangements were complete, the building was erected from the designs of Major-General Scott, C.B., and opened on June 24, 1872, by the Prince and Princess of Wales. The structure is of red brick, with a broad frieze, whilst the interior consists mainly of the 'boiler' part of the temporary exhibit, and comprises a central hall, surrounded by a double gallery. The flooring of this hall is a mosaic pavement formed from refuse chippings of marble, executed by female convicts in Woking Prison. The only permanent collections here are those illustrating food and animal products, and the utilization of so-called 'waste products'; but the temporary exhibits have drawn many visitors at various times. Among the most interesting of these may be classed the exhibition of the Queen's Jubilee presents, the Indian collection of the Prince of Wales, and the portraits now housed in the National Portrait Gallery. In front of the museum is the celebrated St. George Fountain, executed in majolica by Mintons, which was one of the features of the exhibition of 1862.

Between the north and south portions of the gardens are the church and vicarage of St. John's Church, Bethnal Green. The former is a solid-looking building in the classic style, erected 1824-25 from the designs of Sir John Soane, R.A., the architect of the Bank of England. This was the first church consecrated by Bishop Blomfield, as the results of his efforts to supply the East End with more church accommodation. It contains sittings for 1,600.*

On the southern portion of the gardens, the chief historical association is with the adjacent asylum, Bethnal House, called in the survey of 1703 Bethnal Green House. This part of the Poor's Lands was formerly leased to the trustees of the asylum as a recreation-ground for the unfortunate inmates. The first owner of the house was John Kirby, a merchant of London, and it consequently received the name of Kirby's Castle. Fleetwood, the Recorder of London, in a letter to the Lord Treasurer, Sir William Cecil, written about 1758, mentions the death of '*John Kirby, who built the fair house upon Bethnal Green*, which house, lofty like a castle, occasioned certain rhymes abusive of him and some other city builders of great houses who had prejudiced themselves thereby.'†

After John Kirby, the next occupant was Sir Hugh Platt, a prolific writer, who is best remembered as the author of ' The Garden of Eden.'‡

After this versatile author, we find Sir William Ryder, an eminent citizen, well versed in naval affairs, was the next resident. He was Deputy-Master of the Trinity House, and a Commissioner for Tangier. Having been knighted on March 12, 1660-1, he bought Bethnal House the same year. From the Restoration till his death he was a person of much importance, and in 1666 he obtained a license from the Lord of the Manor to drive in his coach across Mile-end Common on his way to Stepney Church.§ He was evidently a great friend of Pepys, and it is from the diary of this ubiquitous secretary of the Admiralty that we gain our best knowledge of Sir W. Ryder, and also some glimpses of life

---

* Wheatley and Cunningham, 'London : Past and Present,' vol. i., p 178.

† Stow, 'Survey of London,' 1755 edition, vol. i., p. 47.

‡ Lysons, 'Environs of London,' 1811, vol. ii., part i., p. 18.

§ Hill and Frere, ' Memoirs of Stepney Parish, p. 244 note.

at Bethnal Green in the seventeenth century. Pepys seems to have attended many gay dinner-parties at Bethnal House, the entry under June 26, 1663, telling us he went ' By coach to Bednall Green to Sir W. Rider's to dinner. A fine merry walk with the ladies alone after dinner in the garden ; the greatest quantity of strawberries I ever saw, and good.

Kirby Castle, Bethnal Green (the Blind Beggar's House).

This very house was built by the Blind Beggar of Bednall Green, so much talked of and sung in ballads ; but they say it was only some of the outhouses of it.' We shall have something more to say about this legend of the blind beggar, but this reference to strawberries in Bethnal Green is particularly remarkable. To continue the record : ' At table, discoursing of thunder and lightning, Sir W. Rider did tell

17

a story of his own knowledge, that a Genoese galley in Leghorn Roads was struck by thunder, so as the mast was broke a pieces, and the shackle upon one of the slaves was melted clear off his leg without hurting his leg. Sir William went on board the vessel, and would have contributed towards the release of the slave whom Heaven had thus set free; but he could not compass it, and so he was brought to his fetters again.'*

Bethnal House again comes into prominence in connection with the Fire of London. Pepys appears to have received his friend's permission to bestow his valuables here for safety, so on September 3, 1666, to quote his own words, 'about four o'clock in the morning my Lady Batten sent me a cart to carry away all my money and plate and best things to Sir W. Rider's, at Bednall Green, which I did, riding myself in my nightgown in the cart; and Lord! to see how the streets and the highways are crowded with people running and riding, and getting of carts at any rate to fetch away things. I find Sir W. Rider tired with being called up all night and receiving things from several friends. His house full of goods, and much of Sir W. Batten's and Sir W. Pen's. I am eased at my heart to have my treasure so well secured.'† Another treasure was safely kept here, being none other than the diary itself, for on September 8 he went to 'Bednall Green by coach, my brother with me, and saw all well there, and fetched away my journal-book to enter for five days past.' However much we may smile at Pepys for the entries in his diary, it would have been an undoubted loss to the literature of England if his work had perished in the Great Fire, giving us as it does so complete an insight into contemporary history. Two days later, while at Sir W. Batten's, he hears 'that Sir W. Rider says that the town is full of the report of the wealth that is in his house, and he would be glad that his friends would provide for the safety of their goods there. This made me get a cart, and thither, and there brought my money all away.'‡

        * Pepys' 'Diary,' Cassell's edition, pp. 160, 161.
        † Ibid., p. 148.   ‡ Ibid., p. 161.

Sir W. Ryder died in this house in 1669, and after his death the property passed through many hands till about the end of the last century it was purchased by Dr. Warburton for a private lunatic asylum,* and it has remained as such ever since.

Having now dealt with the history of this house, some account must be given of the very romantic legend attached to it, preserved in the ballad called 'The Beggar's Daughter of Bednall Green,'† which dates back to the reign of Elizabeth. The heroine of the story is 'pretty Bessee,' the daughter of a blind beggar. Being very beautiful, she had many a gallant suitor; but when they found out that she was only a beggar's daughter, and that her dowry would be *nil*, their ardour cooled considerably. So she left home to seek out her fortune, and eventually came to Romford, where she very quickly had four lovers, to all of whom she gave the same answer, that they must obtain her father's consent before she would be married. This they joyfully said they would do, and asked where he lived. Then the sad truth had to come out that he was the 'blind beggar of Bednall Greene,' and only one, a knight, was willing to marry her then. So the happy pair went from Romford to Bethnal Green, pursued by the disappointed swains, whom the favoured suitor had to fight in the lists. At the end of the fray, his relations chide the 'pretty Bessee' with her poverty; but to their surprise her father offers to give as her dowry as many angels as the bridegroom's friends can produce; and when he had given 'full three thousand pound,' and their funds were exhausted, he added one hundred pounds more to buy her a gown. This ends the first part of the ballad. The second describes the marriage feast to which a great number of nobles came, when the blind beggar declares himself to be Henry, the eldest son of Simon de Montfort, who was felled by a blow at the Battle of Evesham, which deprived him of his sight. Whilst lying amongst the heap of slain, he was found by a

* Lysons, 'Environs of London,' vol. ii., part i., p. 18.
† Printed in the collection known as the 'Percy Reliques.'

baron's daughter, who removed him from the field, and eventually married him; but in order to conceal his identity, he disguised himself as a beggar. Upon hearing this story the company embraced the knight's bride as being of noble birth, and so the romantic legend ends. Unfortunately this account clashes with history, which most emphatically declares that the younger De Montfort was slain at Evesham. Lysons very tersely observes that although the writer might have fixed upon any other spot with equal propriety for the residence of his beggar, the story has gained much credit in Bethnal Green, 'where it decorates not only the signposts of the publicans, but the staff of the beadle; and so convinced are some of the inhabitants of its truth, that they show an ancient house upon the green as the palace of the blind beggar, and point to two turrets at the extremities of the court wall as the places where he deposited his gains.'* These two turrets, which have now disappeared, stood on the site of the present approach road to the asylum.

We have already mentioned some of the past inhabitants of Bethnal Green, who are more particularly identified with these gardens; but there are others who have helped to bring honour to the neighbourhood, whose residences cannot be fixed, who are yet worthy of a passing notice. Among these may be included Sir Richard Gresham, father of the more celebrated Sir Thomas Gresham. It was at Bethnal Green that Sir Balthazar Gerbier, by profession a painter and an architect, founded in 1649 an academy in imitation of the Museum Minervæ.† Here he delivered weekly public lectures, where he professed to teach among other subjects, 'astronomy, navigation, architecture, perspective, drawing, limning, engraving, fortification, fireworks, military discipline, the art of well-speaking and civil conversation, history, constitutions and maxims of state, and particular dispositions of nations, riding the great horse, scenes, exercises, and magnificent shows.' The inclusive terms for this compre-

* Lysons, 'Environs of London,' 1811, vol. ii., part i., p. 18.
† J. N. Brewer, 'London, Westminster and Middlesex,' vol. iv., p. 279.

hensive repertoire were £6 a month, half of which was for riding the great horse.*

Another personage as peculiar as this master of all the arts was Roger Crab, who resided at Bethnal Green at the time of his decease. He wrote a life of himself called 'The English Hermit; or, The Wonder of the Age,' from which it appears that he served seven years in the Parliamentary army, and had his skull cloven to the brain in their service, for which he was so ill requited that he was once sentenced to death by Cromwell, and afterwards suffered two years' imprisonment. When he had obtained his release he set up a shop at Chesham, and began to imbibe strong notions against eating meat, and he adopted for his food bran, herbs, roots, dock-leaves, grasses, washed down with water for his only drink. He was buried September 14, 1680, and a very handsome tomb, which has now decayed and been removed, was erected to his memory at Stepney Church.†

Lysons mentions two other eminent inhabitants—Ainsworth, the compiler of the dictionary bearing his name, who kept an academy at Bethnal Green, perhaps on a less ambitious scale than that of Sir Balthazar Gerbier, and William Caslon, the eminent letter-founder, who died here in 1766, some years after he had retired from business.

* Lysons, 'Environs of London,' 1811, vol. ii., part i., p. 19.
† *Ibid.*, pp. 697, 698.

CHAPTER XIV.

THE EMBANKMENT GARDENS.

WHEN the late Metropolitan Board of Works passed out of existence, *Punch* wrote an epitaph to be placed on a proposed memorial-stone to that body. Its virtues were summed up in one sentence : ' It drained London and gave an embankment to the Thames.' To this latter gigantic undertaking the Embankment gardens owe their existence, for they form part of the land which was reclaimed from the river during the work. The Embankments are three in number—the Victoria, which extends from Blackfriars Bridge to the Houses of Parliament ; the Albert, on the opposite side of the river from Westminster Bridge to Vauxhall; and the Chelsea, from Battersea Bridge to Chelsea Bridge.

VICTORIA EMBANKMENT GARDENS.

Taking these in chronological order, we must deal first with the Victoria Embankment, commenced in 1864, and opened by the Prince of Wales as the representative of the Queen on July 13, 1870. The conception of the formation of a continuous embankment on the northern shore of the Thames appears to have originated with Sir Christopher Wren, who incorporated it as a part of his scheme for re-building London after the Great Fire in 1666. Since that time it has been recommended by several eminent men, among whom may be mentioned William Paterson, founder

General View of Victoria Embankment Gardens (Whitehall Section).

of the Bank of England, about 1694; Gwynne, 1767; Sir
Frederick Eden, 1798; Sir Frederick Trench, 1824; James
Walker, 1840; the Duke of Newcastle, 1844; and John
Martin, the painter, 1856. The late Board in deciding to
form this Embankment had another object in view besides
the improvement of the banks of the Thames, viz., to enable
them to complete the main drainage without interference
with the traffic through the Strand. The line laid down by
Mr. Walker was approved and recommended by various
Parliamentary Committees and Royal Commissions, until at
length it received the sanction of Parliament. An Act for
the formation of an embankment to this line, entrusting its
execution to the late Board of Works, was accordingly passed
in 1862. The length of the roadway is about a mile and a
quarter, and the total area of the land reclaimed from the
river has been 37¼ acres, 19 of which are occupied by the
carriage-road and footways; 7½ acres have been conveyed
under Act of Parliament to the Crown, the Societies of the
Inner and Middle Temples, and other adjacent land-owners,
and the remainder has been devoted to the public gardens.
Many difficulties were encountered during the progress of
the work; that which entailed the most delay being the
arrangement, authorized by Parliament, for the construction
of the Metropolitan District Railway in connection with the
roadway. The railway construction proceeded very slowly
in consequence of the difficulty of raising sufficient capital.

The Embankment roadway, planted with trees, and
ornamented with so many noble buildings, is recognised as
one of the finest in the world. It is exempted, under an
Act passed in 1872, from the operation of the general law
which vests roads and streets in the local authorities of the
parishes or districts in which they are situate, and is maintained out of the rates by the London County Council. The
total cost of the Embankment and works connected therewith was £1,156,981. It is very appropriate that Her
Majesty's name should be associated with the accomplish-
ment of a work of construction which, in the opinion of

Englishmen and foreigners alike, is one of the greatest of which any city can boast.

The Victoria Embankment Gardens consist of three sections, named from the adjacent properties, the Temple, Villiers Street, and Whitehall Gardens. The first-named must not be confounded with the celebrated Temple Gardens, famous among other things for chrysanthemum shows, and which were made by Shakespeare the place where the rival houses of York and Lancaster first assumed the distinctive badges of the white and red rose.*   The Temple section of

The Press Band, Victoria Embankment Gardens (Temple Section).

the Embankment gardens as first laid out was but a narrow strip intersected by a tar-paved path, with flower-beds on each side. In 1895 an alteration of the garden was made in order to find room for a bandstand in the centre. This is the outcome of a successful venture originated by Mr. Thomas of the *Graphic.* These gardens are close to the printing-offices of many of the leading papers, and a band performance is given during the printers' dinner-hour on every day of the week except Saturday and Sunday during the summer months. The expense of this was in the first case borne by

* First Part of ' King Henry VI.'

subscriptions from the press. Originally the band performed on the grass, and a vast audience crowded together as best they could on the narrow path. The erection of the band-stand and the alteration of the paths have now added greatly to the comfort of this large mid-day audience.

In the gardens is a statue, with granite base, with the following inscription :

'WILLIAM EDWARD FORSTER,
BORN JULY 11, 1818,
DIED APRIL 5, 1886.
TO HIS WISDOM AND COURAGE ENGLAND OWES
THE ESTABLISHMENT THROUGHOUT THE LAND OF A NATIONAL
SYSTEM OF ELEMENTARY EDUCATION.'

Round the sides of the base :

'WILLIAM EDWARD·FORSTER,
FOR TWENTY-FIVE YEARS MEMBER OF PARLIAMENT FOR
BRADFORD, 1861-1886.'

This statue is appropriately placed in front of the School Board for London offices, established in pursuance of the Elementary Education Act 1870. This handsome building is of red brick in the Queen Anne style from the designs of Mr. E. R. Robson. At the other end of the garden is a statue of John Stuart Mill, adorned with no other inscription than his name. Close by at the entrance are two figures presented by Mr. Buxton, L.C.C., representing ' The Wrestlers ' from Herculaneum, and a fountain has recently been erected as a memorial of the temperance work of Lady Henry Somerset.

At the time when the Strand was the favourite dwelling-place of the aristocracy, and the site of these gardens was washed by the Thames, there were two mansions adjacent to this spot, whose grounds extended to the water's edge— Essex House and Arundel House. On their sites have been built Essex and Arundel Streets, and Milford Lane.

When the Order of Knights Templars was dissolved, this portion of their property was bestowed on the Prior and Canons of the Church of the Holy Sepulchre, who disposed of it in 1324 to Walter, Bishop of Exeter, who erected

thereon the stately edifice called Exeter House.* It was subsequently alienated from the Bishops, and came into the possession of several noble families. At the Reformation it was
called Paget House, because William, Lord Paget enlarged
and owned it. Subsequently it was known as Leycester House,
after Robert Dudley, Earl of Leicester, and afterwards as
Essex House,† from Queen Elizabeth's favourite, the Earl
of Essex. By a lease dated March 11, 1639, and in consideration of the sum of £1,100, Lord Essex let one half of
his house for a period of ninety-nine years to William Seymour, Earl of Hertford.‡ Pepys paid a visit here in 1669

Essex House.

when Sir Orlando Bridgman, the Lord Keeper, was in
possession. 'By-and-by the King comes out, and so I took
coach and followed his coaches to my Lord Keeper's, at
Essex House, where I never was before, since I saw my old
Lord Essex lie in state when he was dead; a large, but ugly
house.'§ The main portion of the building was pulled down
about 1680, and upon the site of the grounds 'the great

* Rev. J. Nightingale, 'History of London and Middlesex,' vol. iv.,
p. 197.
† Stow, p. 165.
‡ Wheatley and Cunningham, 'London : Past and Present,' vol. ii.,
p. 17.
§ Pepys' 'Diary,' January 24, 1669.

builder,' Nicholas Barbone, built Essex Street.*   The
remaining portion was not taken down till 1777, and in it
the Cottonian Library was kept from 1712 to 1730.

The adjoining mansion, Arundel House, was also originally
an episcopal residence, belonging to the See of Bath and
Wells.   It was disposed of by Edward VI. to his uncle,
Lord Thomas Seymour, High Admiral of England, and by
him named Seymour Place.   It remained in his possession
till his attainder, when it was purchased of the Crown by the
Earl of Arundel, together with several other messuages and
lands in the parish for the sum of £41 6s. 8d.†   The mansion
then passed by marriage into the Howard family, and became
the residence of the Dukes of Norfolk.   It was described at
that time as ' a large and old built house, with a spacious
yard for stabling towards the Strand, and with a gate to
enclose it, where there was the porter's lodge ; and as large
a garden towards the Thames.'   Philip Howard, Earl of
Arundel, was attainted in the reign of Elizabeth, and this
mansion was granted in 1603 to the Earl of Nottingham,
but four years later was transferred to Thomas Howard, son
of the last-mentioned Philip, who was restored to the
earldom of Arundel by James I.‡   It was afterwards
appointed for the residence of the Duke of Sully, who
described it as one of the finest and most commodious in
London, from the great number of apartments on the same
floor.   Although it covered a great deal of ground, the
general appearance was low and mean, but the views from
the gardens were remarkably fine.§   In this mansion was
kept the magnificent collection of works of art formed by
Henry Howard, Earl of Arundel.   When complete this
collection contained 37 statues, 128 busts, and 250 inscribed
marbles, exclusive of many other gems.   Evelyn tells his
own story of his endeavours to obtain a safer keeping for

* Strype, book iv., p. 117.   † *Ibid.*, p. 105.
‡ Wheatley and Cunningham, ' London : Past and Present,' vol. i., p. 73.
§ Rev. J. Nightingale, ' History of London and Middlesex,' vol. iv.,
p. 188.

these works of art. 'To London with Mr. Hen. Howard, of Norfolk, of whom I obtained ye gift of his Arundelian marbles, those celebrated and famous inscriptions, Greek and Latine, gathered with so much cost and industrie from Greece, by his illustrious grandfather, the magnificent Earl of Arundel, my noble friend whilst he liv'd. When I saw these precious monuments miserably neglected and scatter'd up and down about the garden, and other parts of Arundel House, and how exceedingly the corrosive air of London impaired them, I procur'd him to bestow them on the University of Oxford. This he was pleas'd to grant me, and now gave me the key of the gallery, with leave to mark all those stones, urns, altars, etc., and whatever I found had inscriptions on them that were not statues.'*

This owner presented his library to the Royal Society, who met here at the invitation of the Duke, after the Fire of London. Pepys records a visit to this house during the first year the Royal Society was installed here: 'I by water with my Lord Brouncker to Arundell House, to the Royall Society, and there saw the experiment of a dog's being tied through the back, about the spinal artery, and thereby made void of all motion; and the artery being loosened again, the dog recovers.'† When Arundel House was ordered to be pulled down in 1674, the society returned to Gresham College. Some three years before this it was proposed to build a mansion house on the gardens next the river, but although a private Act was obtained for the purpose, the design was never carried out.‡ Arundel Street was built on the site of the old house in 1678, and Gay hits off a description of the street as it appeared in 1716:

'Behold that narrow street which steep descends,
Whose buildings to the slimy shore extends;
Here Arundel's fam'd structure rear'd its frame,
The street alone retains the empty name:

---

* Evelyn's 'Diary,' September 19, 1667.
† Pepys' 'Diary,' July 16, 1668.
‡ 'London : Past and Present,' vol. i., p. 73.

Where Titian's glowing paint the canvas warm'd,
And Raphael's fair design, with judgment, charm'd,
Now hangs the bellman's song, and pasted here
The coloured prints of Overton appear.
Where statues breath'd, the work of Phidias' hands,
A wooden pump, or lonely watch-house stands.'*

These lines of Gay are interesting, not only for the description of the art treasures of Arundel House, but also for that of 'the slimy shore,' which is the exact site of the present gardens.

Leaving now the historical houses which once graced this site, mention must be made of another building facing the gardens, the property of Mr. J. J. Astor, the American millionaire. Although dwarfed by the larger School Board offices, this is a gem of architecture, inside and out.

The main or Villiers Street section of the Victoria Embankment Gardens is perhaps the best known and the most used of all the parks and gardens of London. The area is so limited that every portion of it has to be treated as a garden, and the cost of maintenance is therefore very heavy. The number of flower-beds is extremely large, and from early spring to late autumn they present a mass of bloom. In the gardens is a temporary wooden bandstand, removed from the Naval Exhibition at Chelsea, and the frequent high-class band performances attract numerous crowds. All the paths are tar-paved, which is rendered necessary by the incessant flow of the passers to and fro. Owing to their public character, these gardens have been chosen as sites for the erection of statues to many eminent celebrities. One of these is to Robert Burns (1759—1796), with the following inscription :

'THE POETIC GENIUS OF MY COUNTRY FOUND ME AT THE PLOUGH, AND THREW HER INSPIRING MANTLE OVER ME. SHE BADE ME SING THE LOVES, THE JOYS, THE RURAL SCENES AND RURAL PLEASURES OF MY NATIVE SOIL IN MY NATIVE TONGUE : I TUNED MY WILD, ARTLESS NOTES AS SHE INSPIRED.'

* Gay, ' Trivia,' book ii.

This memorial was the gift of John Gordon Crawford, 1884.

A second is to

'ROBERT RAIKES,

FOUNDER OF SUNDAY SCHOOLS,

1780.

'This statue was erected under the direction of the Sunday-School Union by contributions from teachers and scholars of Sunday-schools in Great Britain. July, 1880.'

There is also a mural fountain and tablet :

'Erected to the memory of Henry Fawcett, by his grateful country-women. A.D. 1886. Fortiter, fideliter, feliciter.'

This tablet has a medallion portrait of the blind Postmaster-General, who did so much for the introduction of female labour into the Post-Office.

The huge monolith erected on the river bank facing the gardens is the celebrated Cleopatra's Needle, which created so great a sensation when first brought to England. The inscriptions on the four sides of the base give a brief account of the romantic story connected with this monument.

'This obelisk, prostrate for centuries on the sands of Alexandria, was presented to the British nation, A.D. 1819, by Mahomed Ali, Viceroy of Egypt—a worthy memorial of our distinguished countrymen, Nelson and Abercromby. This obelisk, quarried at Syene, was erected at On (Heliopolis) by the Pharaoh Thotmes III. about 1500 B.C. Lateral inscriptions were added by Rameses the Great. Removed during the Greek dynasty to Alexandria, the royal city of Cleopatra, it was there erected in the eighth year of Augustus Cæsar, B.C. 23. Through the patriotic zeal of Erasmus Wilson, F.R.S., this obelisk was brought from Alexandria encased in an iron cylinder. It was abandoned during a storm in the Bay of Biscay, recovered, and erected on this spot by John Dixon, C.E., in the forty-second year of the reign of Queen Victoria, 1878.'

The fourth side is a memorial tablet to six sailors who 'perished in a bold attempt to succour the crew of the obelisk ship *Cleopatra* during the storm, October 14, 1877.' The bronze base and sphinxes on each side of it were designed by a former architect of the Metropolitan Board of Works—George Vulliamy.

At the Waterloo Bridge end of this garden we are on the site of the foreshore of the old Savoy Palace, of which only the chapel remains.   This palace, which seems, from the old prints still extant, to have been built literally in the water, dates back to the thirteenth century.   It was the seat of Peter, Earl of Savoy, uncle of Eleanor, the Queen of Henry III.   The Earl bestowed it on the fraternity of

Cleopatra's Needle.

Montjoy, from whom it was bought back by Queen Eleanor for Edmund, Earl of Lancaster.   The palace was restored by Henry Plantagenet, fourth Earl and first Duke of Lancaster.   It was in this palace that John, King of France, was confined after the Battle of Poictiers (1356), and after his release, he died in his old prison on a visit to England. The Savoy was burnt and entirely destroyed by Wat Tyler and his followers in 1381.   It does not seem to have been rebuilt till 1505, when Henry VII. restored it as a hospital of St. John the Baptist for the relief of 100 poor people.   It was suppressed in 1553, but re-endowed by Queen Mary, and was continued as a hospital till 1702, when it was dissolved. The fruitless meeting between the Bishops and eminent Puritans, known as the 'Savoy Conference,' for

Chapel Royal, Savoy.

the revision of the liturgy was held here in 1661.* The river front contained several projections, and two rows of angular mullioned windows. North of this was a court

* Wheatley and Cunningham, 'London : Past and Present,' vol. iii., pp. 217, 218.

18

formed by the walls of the body of the hospital, the ground-plan of which was in the shape of a cross. This was more ornamented than the south front, and had large pointed windows and embattled parapets. At the west end of the hospital was a guard-house, used as a receptacle for deserters, and as quarters for thirty officers and men.* In 1763 the recruits for the East India Service, temporarily confined in the Savoy, made a determined attempt to escape. They overpowered the guard and obtained possession of the keys, but before they could force the outer gate a detachment of soldiers arrived, who disarmed them after a short struggle. Several of the recruits were wounded, and three killed outright.† This gate was embellished with the arms of Henry VII., and the badges of the rose and portcullis. Strype, in his description of the buildings, says : 'In this Savoy, how ruinous soever it is, are divers good houses. First, the King's printing press for proclamations, Acts of Parliament, gazettes, and such-like public papers ; next a prison ; thirdly, a parish church, and three or four of the churches and places for religious assemblies—viz., for the French, for Dutch, for High Germans, and Lutherans, and lastly, for the Protestant dissenters. Here be also harbours for many refugees and poor people.'‡ The ' parish church ' referred to is the Chapel Royal of the Savoy, which once possessed the privilege of sanctuary, and long after this was legally abrogated, it was a refuge for debtors and disorderly persons. After many restorations, the chapel was destroyed by fire in 1864, and rebuilt at the Queen's expense, and re-opened in November, 1865.§ The chapel, with its richly decorated interior, and the burial-ground adjoining, are all that remain of this once famous palace and hospital. All the other ruined buildings were cleared away (1817-19) in the formation of the approaches to Waterloo Bridge from the Strand. The present buildings erected on the site are the Medical Examination Hall (the foundation-stone of which was laid by the Queen

* Malcolm, ' London.'   † Lambert, vol. ii., p. 193.
‡ Strype, book iv., p. 107.   § ' London: Past and Present,' vol. ii., p. 500.

The Fox under the Hill.

on March 24, 1886), and the Savoy Hotel, opened in August 1889.

There was also another building of rather dilapidated appearance which stood close to this spot, which was

18—2

interesting from its associations with Dickens. This was the tavern called The Fox under the Hill. Dickens' biographer says : ' One of his favourite localities was a little public-house by the water-side called the Fox under the Hill, approached by an underground passage, which we once missed in looking for it together ; and he had a vision, which he has mentioned in " Copperfield," of sitting eating something on a bench outside, and looking at some coal-heavers dancing.' The dismal aspect of this tumble-down place was very suggestive of this despairing season in Dickens' childhood.*

Adjoining the Savoy is another mammoth hotel—the Cecil —built upon the Salisbury estate, for which the sum of £200,000 was received by the present Marquis. This property included what are now called Salisbury and Cecil Streets, and occupies the site of old Salisbury House. This was built by Sir Robert Cecil, the first Earl of Salisbury, and like the other large mansions that formerly graced the Strand, had gardens extending down to the river. It is interesting to note that this nobleman caused the high street of the Strand to be levelled and paved for the convenience of passengers. Queen Elizabeth was present at the housewarming, as we learn from Manningham's ' Diary ' : ' On Monday last the Queen dyned at Sir Robert Secil's newe house in the Strand. Shee was verry royally entertained, richeley presented, and marvelous well-contented. . . . His hall was well furnished with choice weapons, which her Majestie took speciall notice of.'† The mansion was subsequently divided into two portions, the one called Great Salisbury House, which was retained by the Earl, and the other; which was still a large house, was leased to various gentlemen under the name of Little Salisbury House. In 1692 this latter was pulled down, and about 1698 Salisbury Street was built on its site. A part next to Great Salisbury House over the long gallery, together with some adjoining

* P. Fitzgerald, ' Picturesque London,' p. 47.
† Manningham's ' Diary,' December 7, 1602.

property, was converted into an exchange, which consisted of a very long and large room with shops on both sides, and extended from the Strand to the river-side, where there were steps down to the water. This was opened, with great eclat, in 1608, in the presence of the Royal Family. It was intended to rival the Royal Exchange, but was allowed to go to decay. The river end subsequently fell into disrepute, and as the shops were all unlet, the Earl pulled down the Exchange and Great Salisbury House, and built Cecil Street on their site.*

Durham House, 1660.

In Little Salisbury House lived William Cavendish, third Earl of Devonshire, father of the first Duke of Devonshire; and at one time Thomas Hobbes, the philosopher, found a home here.†

The next mansion to be noticed, whose grounds extended to the site of the present gardens, is Durham House, the town residence of the Bishops of that see. ' This house . . . was buylded in the time of Henry 3, by one Antonye Becke,

* Strype, book iv., p. 120.
† Wheatley and Cunningham, ' London : Past and Present,' vol. iii. p. 205.

B. of Durham. It is a howse of 300 years antiquitie ; the hall whereof is stately and high, supported with lofty marble pillars. It standeth upon the Thamise veriye pleasantly. Her Matie hath committed the use thereof to Sir Walter Rawleigh.'* So writes Norden in 1593. Before it came into the possession of Sir Walter Raleigh, it had been conveyed to King Henry VIII. by Cuthbert Tunstall, Bishop of Durham. Afterwards it was owned by Queen Elizabeth, but when she died it was restored by Mary to the See of Durham. In the reign of Charles I. the premises were leased to the Earl of Pembroke and Montgomery for the yearly sum of £200.

In 1540 a magnificent feast was given here by the challengers of England, who had caused to be proclaimed in France, Flanders, Scotland, and Spain, a great triumphal tournament for all comers that would accept their challenge. Both challengers and defendants, however, that entered the lists at Westminster were English. At the close of each day, open house was kept at this mansion, where the King and Queen were feasted, together with the Lord Mayor, Aldermen, and knights and burgesses of the House of Commons. The King gave to each of the challengers, and his heirs for ever, 100 marks out of the lands pertaining to the Hospital of St. John of Jerusalem.†

The grounds of Durham House were encroached upon for the erection of Salisbury House, and the stabling was converted into the Exchange we have before referred to. The Adelphi now occupies the site of Durham House. This estate was designed and built by four brothers of the name of Adam, whose Christian names—John, Robert, James, and William —are preserved in the adjoining streets. These brothers were architects of great reputation, patronized by royalty, and among other works designed by them may be mentioned Caen Wood House at Hampstead. Over the wharves which fronted the

* Norden, MS. History of Middlesex.
† Rev. J. Nightingale, 'London, Westminster and Middlesex,' vol. iv., pp. 229, 230.

river they threw a series of arches, and connected the river with the Strand by another archway, so that the streets are built over wide-spreading vaultings. The principal feature in the design is the noble Adelphi Terrace, fronting the river, which at the time of its erection must have had fine views across the Thames, now unfortunately spoilt by the unsightly buildings on the opposite shore. In the centre house of this terrace lived David Garrick from 1772 to 1779, which fact is commemorated by a memorial tablet. The actor died in the back room of the first-floor, and his widow in the same room

Durham House and the Strand in 1660.

in 1822.* Johnson's friend Topham Beauclerk also lived here. Boswell writes : ' He (Johnson) and I walked away together. We stopped a little while by the rails of the Adelphi, looking on the Thames, and I said to him with some emotion, that I was now thinking of two friends we had lost who once lived in the buildings behind us : Beauclerk and Garrick. " Ay, sir," said he tenderly, " and two such friends as cannot be supplied." ' The arches under the Adelphi had a very bad name some years ago, owing to the use to which they were put by thieves and other ruffianly

* 'London : Past and Present,' vol. i., pp. 6, 7.

characters, but they are now enclosed and are used for wine-cellars. At the time of the Chartist scare, a battery of guns was hidden in these vaults, so as to be ready for use if required.

Of the next mansion in the Strand, something more remains than the mere name. This relic of the former grandeur of the river-side buildings is the York Water-gate, now the chief architectural glory of the Embankment gardens. It is interesting apart from its associations, because it shows the point to which the river extended before it was embanked. The terrace walk, planted with trees behind the water-gate, was the former river-side promenade, and is considerably below the level of the Embankment roadway. From the beauty and magnificence of this gate, some idea may be obtained of the splendid mansion of the Buckinghams. The original house was anciently the Bishop of Norwich's inn, but was exchanged in 1535 for the Abbey of St. Bennet Holme, in Norfolk. The next possessor, Charles Brandon, Duke of Suffolk, received it in exchange for Southwark Palace. In the reign of Queen Mary it was purchased by Dr. Heath, Archbishop of York, and called York House. He was the only Archbishop to live in it, and his successors seem to have let it to the Lord Keepers of the Great Seal. In the reign of James I. it was exchanged with the Crown for several manors. Subsequently it was granted to George Villiers, Duke of Buckingham, who rebuilt it in princely style.*

In order to adorn his house, the Duke purchased for 100,000 florins the splendid collection of paintings and works of art which formerly belonged to Rubens. The Duke, who did not live in the mansion, but only used it for state occasions, was assassinated in 1628. Some twenty years later, under the Commonwealth, the mansion was given by Cromwell to General Fairfax, whose daughter and heiress married George Villiers, the second Duke of Buckingham, so that eventually it returned to its rightful owner. When the Pro-

* Rev. J. Nightingale, 'London, Westminster and Middlesex,' vol. iv., p. 244.

tector was told of the marriage, he gave the Duke permission to reside at York House, but not to leave it without consent. Buckingham disobeyed this order, and was promptly sent to the Tower, whereupon his father-in-law went to Cromwell, and pleaded for him without effect.* From references by Pepys in 1661 and 1663, it appears that York House was the residence of foreign Ambassadors:

Adelphi Terrace prior to the Formation of the Embankment.

'*May* 19, 1661 (*Lord's Day*).—I walked in the morning towards Westminster, and, seeing many people at York House, I went down and found them at masse, it being the Spanish Ambassador's; and so I got into one of the gallerys, and there heard two masses done, I think not in so much state as I have seen them heretofore. After that, into the garden, and walked an hour or two, but found it not so fine a place as I always took it for by the outside.'

* Wheatley and Cunningham, ' London : Past and Present,' vol. iii., p. 539.

'*June* 6, 1663.—To York House, where the Russian Ambassador do lie ; and there I saw his people go up and down ; they are all in a great hurry, being to be gone the beginning of next week.   But that that pleased me best, was the remains of the noble soul of the late Duke of Buckingham appearing in his house, in every place, in the door-cases and the windows.'

Pepys refers to the arms of Villiers and Manners—lions and peacocks—with which every room was adorned.

After the Restoration, Buckingham returned to York House, which he sold in 1672 as a building estate, and the streets were named after him—George Street, Villiers Street, Duke Street, Of or Off Alley, Buckingham Street.   In this last street, in the first house from the gardens on the right-hand side, lived Peter the Great, and the house opposite to him was occupied by Pepys, who came here in 1684.   This house was subsequently used as a studio by William Etty, R.A., from 1826 to 1849, and after that by Stanfield, the landscape-painter.*   One of these two houses was the residence of Dickens before he went to Furnival's Inn.   It will be remembered that David Copperfield once lived in Buckingham Street in rooms which are described as being at the top of one of the end houses, with an outlook on the dreary fore-shore of the unembanked Thames.   In this description Dickens was reproducing the story of his own career.†

The York or Buckingham Water-gate is traditionally ascribed to Inigo Jones, and certainly it is worthy of this great man ; but if he designed it, Buckingham must have employed two architects, for the house itself appears to have been the work of Sir Balthazar Gerbier, of Bethnal Green fame.   The water-gate was built by Nicholas Stone, who also claims the design.   In his Works Book, now preserved in Sir John Soane's Museum, is the entry :

' The Watergate at York House hee dessined and built ;

---

* J. T. Smith, ' Book for a Rainy Day,' p. 292.

† ' Notes on some Dickens Places and People,' by Charles Dickens the younger.   *Pall Mall Magazine*, July, 1896.

and yᵉ right hand lion hee did, fronting y Thames. Mʳ Kearne, a Jarman, his brother by marrying, did ye shee lion.'*

On the gardens side of the water-gate are the Villiers' arms, and on the other is the motto 'Fidei coticula crux.' Under an Act passed in the twenty-ninth year of George II., the proprietors and inhabitants of the houses in York Buildings were enabled to levy a rate upon themselves for the purpose of repairing or rebuilding the terrace walk and the water-gate. The late Metropolitan Board of Works made frequent attempts to secure this gate and terrace, which was

York or Buckingham House and York Water-gate (now in the Victoria Embankment Gardens.)

enclosed, and add the land to the gardens, but the difficulties in the way proved too formidable. This object was, however, eventually attained by the London County Council under the London Open Spaces Act, 1893, which vested the site in them. At the time when the Embankment was formed, there was a proposal to transfer the water-gate to the new river-front, and re-erect it as a part of the Embankment wall. In view, however, of the wishes of the architectural profession, this scheme was abandoned, and there is now no

* Quoted in 'London : Past and Present,' vol. iii., p. 542.

prospect of the interesting relic being moved from its present site.

Close by the water-gate, on the site of the photograph shop at the corner of the gardens, was an unsightly tower of wood, octagonal in shape, about 70 feet high, with small round loopholes instead of windows. This belonged to the York Waterworks, a company now defunct, which was established for the purpose of supplying the neighbourhood with water pumped direct from the Thames. The company was incorporated by an Act of Parliament in 1691, and the tower was built between this year and 1695.*

Charing Cross Station and the land on this side adjacent to the gardens is on the site of Hungerford Market. It took its name from the family of Hungerford, whose seat was at Farley, on the borders of Wiltshire and Somersetshire. Sir Edward Hungerford was created Knight of the Bath at the coronation of Charles II., and had a large mansion here, which he converted into a number of small tenements, which together formed the market. Pepys records how the old mansion was destroyed :

' *April* 26, 1669.—A great fire happened in Durham Yard last night, burning the house of one Lady Hungerford, who was to come to town to it this night ; and so the house is burned, new furnished, by carelessness of the girl, sent to take off a candle from a bunch of candles, which she did by burning it off, and left the rest, as is supposed, on fire. The King and Court were here, it seems, and stopped the fire by blowing up the next house.'

The permission to hold a market on the site for three days a week was then granted to Sir Edward Hungerford, known in history as 'the spendthrift.' The market was rebuilt early in the present century from the designs of Mr. Fowler, the architect of Covent Garden Market. The upper part of the market consisted of three avenues, roofed over, with shops on each side, and the principal article sold here was fish,

* Rev. J. Nightingale, ' London, Westminster and Middlesex,' vol. iv., p. 245.

but there were also stalls for fruit, vegetables, and butchers' meat. This building was pulled down in 1862, together with Hungerford Bridge (which was approached through the market), in order to make way for Charing Cross Railway Bridge and Station. The old Suspension Bridge was designed by Brunel, to whom there is a statue on the Embankment in a shrubbery near the Temple Gardens. It was then re-erected as the Clifton Suspension Bridge near Bristol.*

The remaining section of the ornamental grounds on the Victoria Embankment is distinguished by the name of Whitehall Gardens. Under the terms of the Act authorizing the formation of the Embankment, this reclaimed land in front of Crown property was to belong to the Crown. When, however, the wall was about to be erected to enclose this land, considerable feeling was exhibited throughout the Metropolis, as it was deemed unfair that the Crown should reap the benefit of works which had been carried out at the expense of the inhabitants of the Metropolis. The attention of the House of Commons was called to the subject, and the late Mr. W. H. Smith, then M.P. for Westminster, gave notice of his intention to move the following resolution: ' That, in the opinion of this House, it is desirable that so much of the ground reclaimed from the Thames at the cost of the ratepayers of the Metropolis as may be in front of the ancient line of buildings, shall be devoted to the purposes of public recreation and amusement.' This motion was subsequently withdrawn in consequence of the House of Commons having, on the motion of the Prime Minister, appointed a Select Committee ' to inquire whether, having regard to the various rights and interests involved, it is expedient that the land reclaimed from the Thames, and lying between White-hall Gardens and Whitehall Place, should, in whole or in part, be appropriated for the advantage of the inhabitants of the Metropolis, and, in such case, in what manner such appropriation should be effected.' This Committee went into the whole question very fully, and agreed to report to the House

* ' Old and New London,' vol. iii., pp. 131, 132.

that the land should be leased to the late Metropolitan Board of Works for public purposes at a rental calculated after the rate paid by the Crown tenants for adjoining reclaimed land. These terms were agreed to by the late Board, but the initiative rested with the Government, and their decision was awaited with some anxiety. In the following year the Chancellor of the Exchequer introduced a Bill which was to enable the late Board to acquire possession of the land in question on payment of the sum of £40,000. After the previous proposal these terms naturally came as a surprise, and the Bill was petitioned against, and was afterwards dropped. After several compromises had been suggested, the arrangement finally come to was that the land should be conveyed to the Board for the purposes of a public garden in consideration of the payment of £3,270 and the surrender of a small plot of ground on the south side of Whitehall Place. This agreement was ratified by the Thames Embankment (Land) Act, 1873. After the land had been fenced in and ornamentally laid out, it was opened to the public on May 8, 1875, by the late Mr. W. H. Smith, M.P., to whose exertions in the House of Commons the arrangement made for the appropriation of the ground to the public was largely due.

Before the terms of the compromise could be fully carried out, Northumberland House and grounds had to be acquired. The acquisition of this mansion was made almost a necessity if the full benefit of the formation of the Embankment roadway were to be reaped. This provided a wide and convenient thoroughfare from the extreme west of the Metropolis to the heart of the city, but owing to the steep and narrow approaches in the neighbourhood of Charing Cross, it was not utilized to so large an extent as was anticipated. From the first it was felt that a direct approach from Charing Cross was necessary, but unfortunately the only way in which this could be effected seemed to be through the grounds of Northumberland House, the removal of which could not be lightly entertained.

Northumberland House was the last of the palatial residences of the nobility which formerly skirted the banks of the river, and on this account its demolition must be regretted, but beyond the associations which surrounded it there was little to recommend its preservation. It could not boast of any very great antiquity, its existence dating from the reign of James I., and it had been so frequently altered since that date that but little of the original structure remained at the time of its demolition. To architectural beauty it had but little claim, the gateway on the Strand front being indeed the only portion worthy of preservation on this score. Internally the ballroom was an apartment of noble proportions, and the grand staircase was a design of considerable merit, but beyond these there was nothing in its architectural features worthy of any great consideration.

It would appear that a house existed on this spot in the time of Henry VIII., of which little or nothing is known, but in the reign of James I. a mansion was erected by Henry Howard, Earl of Northampton, in 1605, and was then known as Northampton House, but subsequently passing into the hands of the Earl of Suffolk its designation was changed to Suffolk House.

In 1642 it passed by marriage into the possession of Algernon Percy, Earl of Northumberland, and its name was again altered to Northumberland House, a designation which it retained during the remainder of its existence, being in possession of the Dukes of Northumberland to the last.

The house as erected by the Earl of Northampton consisted of only three sides of a quadrangle, one towards the Strand, with a wing on either side, the fourth side, fronting the river, having been subsequently built by the Earl of Suffolk; but this was afterwards reconstructed by the Earl of Northumberland from the designs of Inigo Jones. The first building is said to have had for its architects Bernard Jansen and Gerard Christmas. The greater part of the Strand front was rebuilt about the middle of the eighteenth century, but the central portion, over the gateway, was probably but little interfered with.   It was at this time that the ballroom was erected, forming a western wing projecting towards the river on the garden front, together with a corresponding wing on the eastern side; but this had no architectural pretensions, and was far from adding to the dignity of the mansion.

The proposed demolition of this building was much canvassed at the time the suggestion for the formation of this approach was first brought before the public, but now that the Avenue has been flanked with buildings of good architectural elevation, few, probably, will be found to regret that the old building has given place to a necessary and handsome public thoroughfare.

The Duke naturally felt much reluctance to agree to the destruction of a mansion which had been the residence of his ancestors for two centuries, and consequently when application was first made to Parliament in 1866 by the late Metropolitan Board of Works to obtain power to purchase this property for the formation of Northumberland Avenue, the then Duke offered his strenuous opposition, and the Bill had in consequence to be withdrawn.   The Duke had previously obtained the insertion of clauses in the Thames Embankment Act of 1862 preventing the erection of lofty buildings on the reclaimed ground between his property and the river.

The pressing necessity for an approach to the Embankment roadway from Charing Cross became, however, every year more obvious.   The late Duke having died in August, 1867, representations were made to his successor, pointing out to him the importance of this Metropolitan improvement and the impossibility of effecting it without the removal of Northumberland House.

The late owner, although sharing the feeling of reluctance of his predecessor, was at length induced to waive those objections in favour of so great a public necessity, and with his sanction an Act was passed in July, 1873, authorizing the purchase of the mansion for the sum of £500,000.   Possession was obtained in July of the following year, but before the house was pulled down the public were given an opportunity of inspecting it, and during the days set apart for the purpose the building was visited by many thousands of persons. The sale of the old materials realized £6,376, and the ground was cleared in June, 1875. The lion, which was such a conspicuous object in the Strand, has been carefully re-erected with the arched pedestal on which it stood on the river-front of Syon House, Isleworth.

Northumberland Avenue is well known for its imposing buildings, which now occupy the site of the mansion and grounds. It contains some of the largest hotels in London, the Constitutional Club, and the offices of the Society for Promoting Christian Knowledge.

The Whitehall Gardens form a series of flower-beds grouped round three statues, and together are an excellent frontage to Whitehall Court, one of the handsomest buildings in London. The three statues which add such attraction to the gardens have been erected by public subscription to Sir James Outram (the 'Bayard' of India), Bartle Frere, and a third to William Tyndale, the first translator of the New Testament into English from the Greek.

CHAPTER XV.

THE EMBANKMENT GARDENS (continued).

Albert Embankment Gardens.

THE construction of the Southern or Albert Embankment was commenced in July, 1866, and completed in 1869, at a cost of £1,014,525, so that it was finished before the twin work on the Middlesex shore, although put in hand some years afterwards. The reason for this more rapid progress was owing to the easier nature of the work to be done. The Victoria Embankment was a very complicated piece of engineering, owing partly to the greater depth of water in which it had to be constructed, partly to its combination with the formation of the sewer and subway above, and last, but by no means least, to the execution of the underground railway works on its site. The Embankment wall is of uniform character throughout its length, and is similar in elevation to that on the Middlesex side, having a highly-dressed granite facing, and is surmounted with a moulded parapet and plinth. The mouldings are stopped at frequent intervals against plain pedestals of granite, ornamented with bronze mooring rings and standards for gas-lights. At a point about 800 feet above Lambeth Bridge the wall was constructed in a trench excavated out of the solid ground, and the space between it and the water was excavated and thrown into the river-bed, so as to increase its width, which was formerly very narrow. The footway promenade running alongside the Embankment wall is reached by a broad flight of steps from Westminster Bridge, which thus forms a connecting-link between the two Embankments. The roadway diverges from the foot promenade between Lambeth and Westminster Bridges, and the reclaimed land is situated between the two. On the greater portion of this are the blocks of buildings constituting the St. Thomas's Hospital, which add materially to the architectural embellishment of this Embankment. The governors had been compelled to give up their old hospital in St. Thomas Street, Southwark, to make room for the railway works near London Bridge. The remaining strips of reclaimed land, partly fronting Lambeth Palace, have been enclosed and planted and constitute the Albert Embankment Gardens.

Before the formation of the Embankment the land in the neighbourhood of the water was subject to periodical flooding, and no doubt in ancient times the incursions of the river were more serious, before any attempt was made to confine it within bounds. The extensive district of Lambeth Marsh, which is some distance from the river, bears testimony to the far-reaching effects of the flooding. The part of the foreshore upon which the Embankment has been formed has been celebrated for its boat-building yards as far back as the reign of Charles II. Pepys came ' to Lambeth, and there saw the little pleasure-boat in building by the King, my Lord Brouncker, and the virtuosos of the town, according to new lines, which Mr. Pett cries up mightily ; but how it will prove we shall soon see.'* The boat-building yards have now been removed higher up the Thames, but the venerable palace of Lambeth, which has stood here for centuries, still remains to add dignity and picturesqueness to this bank of the river. It looks more like a fortress than a Bishop's palace, and reminds us of those feudal times when the King, the nobles, and the ecclesiastics were all struggling for the supreme power.

For more than 600 years Lambeth Palace has been

* Pepys' ' Diary,' August 13, 1662.

19—2

the London residence of the Archbishops of Canterbury, and much of the history of the English nation is wrapped up in its walls. In the centre of the river-front is the hall, a room of noble proportions, having a carved roof similar to that of Westminster Hall. This part of the building is the work of Archbishop Juxon, and dates back to 1663. In it is placed the celebrated Lambeth Library, established by Archbishop Bancroft in 1610, which comprises 30,000 volumes, and many valuable manuscripts relating to the see. In this hall was held one of the most famous trials of modern times, that of Dr. King, Bishop of Lincoln, for ritualistic practices. The case was tried before the Archbishop and five other Bishops, and in the end Dr. King was acquitted of all the charges except two.

The oldest part of the palace is the chapel, built in 1245, by Archbishop Boniface. It consists of nave only, 72 feet long, and is divided by a handsomely-carved screen. The only Archbishop buried here is Dr. Parker, who died in 1575, and expressed a wish to be interred at the upper end of the chapel against the Communion-table. During the Civil Wars his monument was destroyed, and his body exhumed and submitted to other desecration. At this troublous time the palace fell to the share of Colonel Scott, who pulled down the hall and turned the chapel into a hall or dancing-room. After the Reformation the remains of Archbishop Parker were discovered and re-interred in their original place. Underneath the chapel is a spacious crypt with finely-groined roof, which probably dates from the middle of the thirteenth century.

The so-called Lollards' Tower (properly the Water Tower) adjoins the west end of the chapel. The exterior of this end of the palace has a fine venerable appearance, and is the only remaining part that is built entirely of stone. It is thought to have received its name from a little prison at the top, in which the Lollards were confined. The chief features inside this prison are the iron rings fastened to the wainscot which lines the walls. Besides the Lollards, other political prisoners have been secured here, including the Earl of Essex, Queen Elizabeth's favourite (1601), Lovelace, the poet (1648), and Sir Thomas Armstrong (1659). On the exterior of the tower are the arms of Archbishop Chicheley, who built this keep in 1434; and there is also a fine Gothic niche, which formerly contained a statue of Thomas à Becket.

Lambeth Palace and Albert Embankment Gardens.

At the opposite end of the palace is the gate-house, rebuilt by Cardinal Morton about 1490. This magnificent building consists of two immense square towers, with a spacious gateway and postern in the centre; the whole embattled and built of red brick with stone dressings. The guard-chamber, another fine hall, which runs parallel with the west side of the library, was formerly the armoury, and contained the weapons for the defence of the palace. These warlike preparations are no longer needed, and the chief orna- ments now are the portraits of the Archbishops since 1533.*

The grounds of the palace are very considerately given up to the public for cricket and football, a boon which is much appreciated in this crowded neighbourhood. One of the 'improvements' made necessary by the formation of the Em- bankment is much to be regretted, viz., the abolishing of the terrace called Bishop's Walk, with its fine elm-trees. A view of the palace by Hollar, dated 1647, shows it almost exactly as it appears now, with the addition of these stately trees and the stairs at the waterside, which formed an important approach to the palace when communication by river was more general.

Chelsea Embankment Gardens.

The third Embankment is at Chelsea, between the Royal Military Hospital and Battersea Bridge. From Millbank to Chelsea Hospital the river was already embanked, the work having been done by the Commissioners of Her Majesty's Works and Public Buildings. It was deemed desirable that the thoroughfare along the river-side from Westminster should be extended to Battersea Bridge, and the formation of an embankment offered the opportunity, which was found so useful in the case of the Victoria Embankment, of con- structing with it a portion of the main sewerage works. The Act of Parliament authorizing the Chelsea Embankment was passed in 1868. The works were begun in July, 1871, and completed in May, 1874. On the 9th of that month the new road was opened in state by the Duke and Duchess of Edinburgh on behalf of the Queen. This was the first public function performed by Their Royal Highnesses since their marriage in the early part of that year. The length of the Chelsea Embankment is rather more than ¾ mile, and the road is 70 feet wide. The execution of these works com-

* The particulars relating to Lambeth Palace are chiefly taken from Allen's 'History of Lambeth.'

Chelsea in 1738 (From a print by Thomas Preist, of Chelsea)

pleted a thoroughfare by the river-side, extending from Black-friars to Battersea Bridge, 4½ miles in length.   The net cost of the undertaking was £269,591.

In the course of making the Embankment an old block of houses, which stood between Battersea Bridge and Chelsea Church, had to be removed, in order to widen the thorough-fare.   The backs of these houses fronted the river, and in some places overhung it.   After passing the church, the width of the reclaimed land is sufficiently great to admit of its being laid out as a garden.   The gardens extend from Church Street to Flood Street, which takes its name from Luke Thomas Flood, a benefactor of the parish.

Chelsea, the ' village of palaces,' the home of Sir Thomas More, Carlyle, Atterbury, Smollett, and many another celebrity, was originally a village which clustered round the silvery Thames, and from the strand of which it takes its name.   Norden says ' it is so called of the nature of the place, whose strand is like the chesel which the sea casteth up of sand and pebble stones, thereof called Cheselsey, briefly Chelsey, as in Chelsey (*i.e.*, Selsey) in Sussex.'* This retired and rustic village formed a very pleasant retreat from London, and at the end of the last century its character had not materially altered.   There is record of a stag-hunt in Chelsea, which took place about 1796.   The stag swam across the river from Battersea, and made for Lord Cremorne's grounds.   Upon being driven from thence, it ran along the waterside as far as the church, and turning up Church Lane (now Church Street) took refuge at last in a barn, where it was easily caught alive.†   The pebbly strand of Chelsea gave way in the course of time to unsightly mud-banks. These in their turn have given way to the Embankment roadway, planted with trees, and in parts relieved by the ornamental gardens.

In the gardens opposite Cheyne Row is a statue by Boehm of Thomas Carlyle, the ' philosopher of Chelsea.'   This is

---

* Norden, ' Spec. Britanniæ ': ' Middlesex,' p. 17.
† Faulkner, ' Chelsea.'

Chelsea Embankment Gardens and Cheyne Walk.

appropriately placed near the house No. 24 (formerly No. 5) which was his residence from 1834 till his death in February, 1881. The house in which he wrote ' The French Revolution,' 'Cromwell,' the 'Latter-Day Pamphlets,' and ' Frederick the Great,' is distinguished by a mural medallion of the sage. In December, 1895, the centenary of his birth was commemorated, and this house was purchased for the sum of £1,750, and will be preserved by a trust committee. It is now open to the public on payment of a small fee. An attempt has been made to collect all the relics which might add interest to this historic house. In the garret-study, which Carlyle made sound-proof because of the ' demon-fowls,' is preserved the great writer's chair, several portraits, and other memorials which have been lent. The inscription on the statue runs as follows :

'THOMAS CARLYLE,

B. Dec. 4, 1795,

AT

Ecclefechan, Dumfriesshire.

D. Feb. 5, 1881,

AT

Great Cheyne Row,

Chelsea.'

To the west of Oakley Street, facing the gardens, is a very new house, styling itself the Ancient Magpie and Stump. The ancient hostelry from which this residence takes its name is one of the old landmarks of Chelsea which has disappeared. It acquired its interest from the fact that it was the house at which, from the time of Charles II., the parish dinner was held. This was provided in past times out of the funds bequeathed by Mr. Leverett, from whom Leverett Street is named.* In front of this noted public-house, which was burnt to the ground, was a notice-board bearing the name of the inn, above which was an iron magpie on an iron stump, with a rusty old weathercock. There may still be traced the top-stone of an old water-staircase,

* Right Hon. Sir C. W. Dilke, 'Chelsea.' p. 3.

embedded in the ground forming the Embankment Garden, which is now on the site of this sign. This staircase is as historical as the York Water-gate in the Victoria Embankment Gardens, for Queen Elizabeth must often have used it when coming to Chelsea to visit her friend and subject, the powerful Earl of Shrewsbury.* His mansion, Shrewsbury House, was hard by the old church on the eastern side, afterwards

Shrewsbury House.

called Alston House. Several Earls of Shrewsbury lived in this house, and also the famous Bess of Hardwicke, wife of four husbands, the second of whom was Sir William Cavendish, ancestor of the Dukes of Devonshire. She built three of the most famous mansions in England—Hardwicke Hall, Chatsworth, and Oldcoates. Lord Shrewsbury could not get on well with her, and writes in his letters of the 'cunning devices of his malicious enemy, his wife.'† From

* B. E. Martin, 'Old Chelsea,' p. 129.
† Right Hon. Sir C. W. Dilke, 'Chelsea,' p. 23.

contemporary views of this house, it seems to have been an irregular brick structure, much gabled, and built about a quadrangle. In one of the rooms was a trap-door, giving entrance to an underground passage, said to have been used by the Jacobites of 1745.*

The next mansion eastward on the river - front was Winchester House. When the ancient palace of the Bishops of Winchester in Southwark was injured in the troubles of the seventeenth century, an Act of Parliament was obtained in 1663 authorizing them to purchase a new brick house at Chelsea, then lately built by James, Duke of Hamilton. This adjoined the manor-house, and was purchased for the sum of £4,250. Winchester House was by this same Act held to be within the Diocese of Winchester.† Externally, this palace displayed little grandeur or magnificence. It was built of red brick, and was two stories in height, forming a quadrangle, with its principal entrance in the south front. On the ground-floor of this side were the great hall, kitchen, and chapel, the latter being of moderate dimensions. A grand staircase at the eastern end led to three large drawing-rooms, which extended the whole of the south front, and were furnished in splendid style—a great contrast to the plain exterior. Among those who lived in this palace must be mentioned Trelawney of the song ' And shall Trelawney die,' Bishop Hoadley, and the Hon. Brownlow North ; but after his death the wife of his successor took a dislike to the house altogether, and the Bishops left Chelsea for ever. A further Act had accordingly to be obtained to enable the episcopal residence to be sold, which was accomplished after some difficulty. After the palace had gone to ruins, the remains were sold by auction, and the whole fabric pulled down, and upon its site have. been built the Pier Hotel and the first houses in Oakley Street.

The manor-house adjoining was built by Henry VIII., who, it is said, formed a very favourable opinion of Chelsea

* B. E. Martin, 'Old Chelsea,' p. 129.
† Faulkner, 'Chelsea,' 1829 edition, vol. i., p. 285.

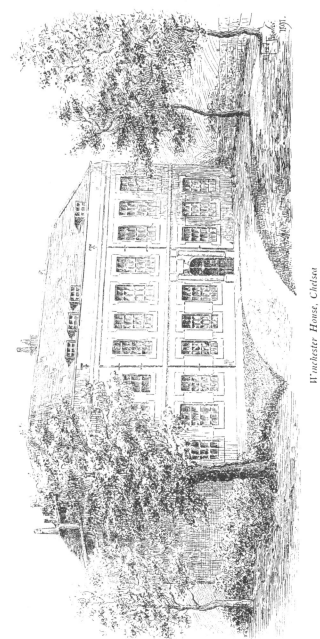

Winchester House, Chelsea

in his numerous visits to Sir Thomas More. He therefore
acquired the manor; but finding that the manor-house was
ancient, and at that time in the possession of the Lawrence
family, he built a new manor-house, which fronted the
Thames between old Winchester House and Don Saltero's
coffee-house.* This building, too, was quadrangular, enclosing a spacious court. From an old view of the manor-

The New Manor-House, or Chelsea Place, built by Henry VIII.

house, it seems to have been enlarged at various times in
varying styles of architecture, giving the whole a very patchy
appearance. Sir Hans Sloane was one of the chief notabilities
connected with this mansion, and during his residence here
he collected the library and other curios which afterwards
formed the nucleus of the British Museum. The manor-
house was pulled down in the middle of last century, after

* Faulkner, 'Chelsea,' 1829 edition, vol. i., p. 311.

Queen's House, from Chelsea Embankment Gardens.

the death of Sir Hans Sloane, and part of Cheyne Walk now occupies the site.

This terrace of red-brick houses, screened with a row of well-established trees, is one of the most interesting and historical spots in London. With their wrought-iron gates and their old - fashioned architecture, they form a most picturesque background to the gardens. Many of the original houses in this row have had to be pulled down, and their places taken by modern ones, but they have been rebuilt in a style which harmonizes with the remainder. Cheyne Walk takes its name from Viscount Cheyne, Lord of the Manor of Chelsea towards the close of the seventeenth century. The finest house in the row, No. 16, which can easily be distinguished by the figure of Mercury with which it is surmounted, was the residence of Dante Gabriel Rossetti, poet and painter. It is known as Queen's House, from its associations with Queen Catherine of Braganza. Legend states that this mansion was built for this unfortunate and long-suffering consort of Charles II., and some go so far as to say that she lived in it.

Inside the gardens opposite his house, a fountain has been erected to the memory of Rossetti. This was designed by John P. Seddon, the architect, and the bronze bust above it is the work of Ford Madox Brown, both intimate friends of Rossetti. At No. 4 in this row, in December, 1880, died George Eliot, the eminent novelist, in the same house where, ten years before, Daniel Maclise, equally celebrated as a painter, had passed away.[*]

On the site of No. 18, Cheyne Walk was Salter's or Don Saltero's coffee-house. Salter was a jack-of-all-trades, poet included, and in an advertisement of his establishment[†] gives a brief biographical notice of himself:

> ' Through various employs I've passed,
> A scraper, virtuoso, projector,
> Tooth-drawer, trimmer, and, at last,
> I'm now a gimcrack whim collector.'

---

[*] B. E. Martin, ' Old Chelsea.'
[†] *Weekly Journal*, June 22, 1723.

He was a trusted servant of Sir Hans Sloane, who gave him a large number of curiosities. With these he furnished his 'Museum Coffee-house,' opened in 1695, and many other celebrities contributed towards his collection, which became

Don Saltero's, Cheyne Walk, 1840.

quite famous in its day. He was christened 'Don Saltero' by Vice-Admiral Munden, who had spent many years on the coast of Spain, and had acquired a fondness for Spanish titles.* When the Don died, the museum was continued by

* *Gentleman's Magazine*, 1799.

20

his daughter, Mrs. Hall, but having lost the personality of its former owner, the receipts fell off, and in 1799 the collection was sold by auction, but only realized a little over £50.

Don Saltero is not the only eccentricity of which Cheyne Walk can boast. At No. 4, John Camden Neild, the miser,

Cheyne Walk, 1750.

dragged out a miserable existence. When he died, in 1852, he left his savings, amounting to half a million sterling, to the Queen, who caused a painted window to be erected to his memory at North Marston, near Aylesbury, where he was buried.*

* Cassell s ‘ Old and New London,’ vol. v., p. 60.

Here also, at No. 6, lived Dominicetti, the Italian quack, who made a great stir with the medicated baths which he established in Cheyne Walk in 1765. At one of Dr. Johnson's chatty evenings the subject of Dominicetti's baths was mentioned. The doctor thought there was no efficacy in the much-boasted system, but someone in the company ventured to disagree with him, when he retorted : ' Well, sir, go to Dominicetti, and get thyself fumigated ; but be sure that the steam be directed to thy head, for that is the peccant part.'* For all this, Dominicetti is said to have had under his care more than 16,000 persons.

One of the chief features of this side of the river is Old Chelsea Church, which is the commencing-point of the gardens. It is composed chiefly of brick, and though adding picturesqueness to the scene, it is not conspicuous for its beauty. Its various parts have been erected at different periods, with no attempt at architectural harmony, so that the whole has rather an incongruous effect. The oldest part of the structure, dating back to the fourteenth century, is a chapel of the Lawrence family, at the eastern end of the north aisle. Another ancient chapel is that built by Sir Thomas More, one of the chief historical personages connected with Chelsea. This chapel is at the east end of the south aisle. There is a mural tablet to him in the south side of the chancel, surmounted by a flat Gothic arch, but it seems certain that his remains are not interred here. The peculiar feature about the More Chapel is that until recently it was private property exempt from the control of the authorities of the Church. It was purchased, however, by a Mr. R. H. Davies, and presented to the Vicar and churchwardens, and has since been restored. At the corner of the churchyard nearest to the gardens is a monument to Sir Hans Sloane, which consists of a large vase of white marble around which serpents are entwined, under a portico supported by four square pillars. The wall of the church facing

* Boswell, ' Life of Johnson,' vol. ii., p. 72.

the river has a number of monuments to various members of the Chamberlayne family.

Lawrence Street, to the east of the church, is famous as being the birthplace of the Chelsea china, which is so eagerly sought after by collectors.*   At the upper end of this street, at the corner of Justice Walk, the china factory was established in 1745, but why it failed, and why the manufacture was discontinued, is one of those mysteries which time does not seem to solve.

There are a few small enclosures close by known as the Pimlico Shrubberies, which were transferred to the late Metropolitan Board of Works in connection with Battersea Park.   They comprise the shrubberies at the end of Vauxhall Bridge (which will probably be absorbed in forming the new bridge), the enclosures at the back of the wharves along Grosvenor Road, the river-banks near the Victoria Suspension Bridge, and a rectangular piece of ground at the end of St. George's Square, which has since been laid out as a garden. At the south-western corner of Putney Bridge there is a small enclosure, not large enough to be opened to the public, which is also maintained by the London County Council.

* B. E. Martin, ' Old Chelsea,' p. 132.

CHAPTER XVI.

FINSBURY PARK—CLISSOLD PARK.

FINSBURY PARK.

FINSBURY PARK is in point of date the second of London's municipal parks, although the Act authorizing its construction is the oldest relating to the municipal parks of the Metropolis. To trace the earliest steps which led to its formation, we must go back nearly fifty years, when a borough meeting of the inhabitants of Finsbury was convened, in 1850, and a committee appointed for the purpose of obtaining a park for the northern suburbs of the Metropolis. The resolution then carried was, ' That a park on the borders of a district so large as the borough of Finsbury, and containing a dense industrial population of nearly half a million, is universally admitted to be a public necessity.' In the same year representations were made to the Government, showing the urgent need for such a park, in consequence of the buildings which had spread over the Shepherd and Shepherdess Fields, the Rosemary Branch Fields, Spa Fields, White Conduit Fields, and other open spaces, which, while available for the recreation of the community at large, were more particularly used by the inhabitants of the city of London and borough of Finsbury. Successive administrations of the Government favourably entertained the scheme, and under that of Lord John Russell surveys were made, plans prepared, and notices served on the occupiers of the land, but the operations were suspended by the succession of Lord Derby as Prime Minister. In the new régime, the proposal was again considered, and the Government once more decided to give the Parliamentary notices, and deposit plans as before.   But another Ministerial change prevented the carrying out of the undertaking, and in Lord Aberdeen's Cabinet it was decided to defer all operations till the passing of the Act which vested in the

Flower-Beds, Finsbury Park

late Metropolitan Board of Works the power of making parks, and of effecting other improvements.   The Finsbury Park committee had submitted to Parliament plans and estimates of the lands proposed to be acquired, and Viscount Palmerston promised to advise Her Majesty's Government to contribute £50,000, or one quarter of the supposed cost.   When the new body, the Metropolitan Board of Works, came into power, the question of parks and open spaces was one of the first to be considered, and, as a consequence, the Finsbury Park Act,

1857, was obtained. By this the late Board were authorized to acquire for the purposes of the park any portion of certain lands, specified in the Act, which contained altogether 250 acres, and it was provided that 20 acres should be reserved for building purposes. The enthusiasm of the Board cooled down, however, especially when the Government negatived the proposed grant of £50,000, and many years passed away without any practical steps being taken. At last a smaller area of 115 acres was purchased at a cost of about £472 per acre, after the scheme had been nearly abandoned altogether, and Finsbury Park was secured for the public for ever. The park was opened by Sir John Thwaites, Chairman of the Board, on August 7, 1869.

The name which Finsbury Park bears has been the subject of much adverse criticism, seeing that it is separated from Finsbury by many large districts ; but when it is considered that the original intention of the committee was to provide a park for the inhabitants of the borough of Finsbury, the name does not seem so much out of place.

The ground slopes gently down on all sides from the lake, which is the highest point, and the eastern boundary is formed by the Seven Sisters Road and Green Lanes. The Great Northern Railway bounds it by a cutting and embankment on the western side, and Finsbury Park Station adjoining makes the park conveniently accessible from the Metropolis. Inside the gates there is provision for all kinds of sport and recreation—cricket, lawn-tennis, and football grounds, gymnasia, a boating lake well stocked with water-fowl, and a bandstand for musical performances. There are a number of flower-beds and ornamental shrubberies, while a roomy decorative house with a varied display of flowering plants, winter and summer alike, forms a permanent attraction. In its season, too, the annual show of chrysanthemums brings a large number of visitors to the park who have an opportunity of seeing blooms uncontaminated with the smoke which spoils those at places nearer London. From the plateau round the lake there is a fine view northwards

towards High Beech and Epping Forest, although the building of houses in the neighbourhood has interfered with the
immediate rural surroundings of the park.   The New River
runs at the foot of the northern slope of the hill, and adds
much to the appearance of the cricket-ground.

It is satisfactory to think that the gloomy prognostications
of those who condemned the scheme at the outset have not
been fulfilled.   One leading journal at the time of opening

Manor-House Entrance, Finsbury Park.

predicted that the park would never become popular during
this century, and perhaps the best answer that could be given
to this prophecy would be to visit the park on some Saturday
afternoon and then judge.   The money spent on the acquisition and formation of Finsbury Park has by no means been
wasted, and it will always be one of the most popular parks
in the north of the Metropolis.

The south-east side of the parish of Hornsey, including
Finsbury Park, is in the sub-manor of Brownswood, which

is the corps of a prebend in St. Paul's Cathedral. By a survey taken in 1649 it appears that this manor had been demised to John Harrington in the year 1569 for ninety-nine years, and that by several mesne assignments it was then the property of Lady Kemp, the reserved rent being £19 per annum. It was sold, together with the Manor of Friern-Barnet, to Richard Utber for the sum of £3,228 4s. 10d. In 1681 Sir Thomas Draper, Bart., was lessee under the prebendary.. His wife bequeathed the benefit of the lease to John Baber, who assigned it in 1750 to John Jennings, and eight years later Richard Saunders, the last owner's sole executor, became lessee. His only surviving son, Thomas, in 1789, sold the lease to John Willan. The manor court was held in the Hornsey Wood tavern. The descent of this manor does not present any features of particular interest, but among the eminent men who have held the prebend of Brownswood may be mentioned Bishop Fox, the founder of Corpus Christi College at Oxford.*

In the neighbourhood of the park we still have Brownswood Road and Manor-House Tavern to remind us of the Manor of Brownswood and of its manor-house, which has now disappeared.

Finsbury Park occupies a part of the site of Hornsey Wood House and grounds. This Hornsey Wood was itself part of a larger forest or park called Hornsey Park, the property of the Bishops of London. This, again, it is hardly necessary to add, was a part of the extensive forest of Middlesex. Hornsey Park has played much the same part in history in the north of London as Blackheath in the south, and has furnished a meeting-place for princes and a field for the display of that pageantry which delighted our forefathers.

The earliest of these took place in 1386, in the eventful reign of Richard II., when the Duke of Gloucester, together with the Earls of Arundel and Warwick, and several other noblemen, resorted to arms in order to oppose Robert de Vere, Earl of Oxford, whom the King had created Duke of

* Lysons, ' Environs of London,' vol. ii., part ii., p. 423.

Ireland. They assembled in this park and were sufficiently
strong to alarm the King, who requested a meeting at Westminster. He there ' gave them fair words, took them into
his chamber, and made them drink together.' This apparently had the same effect as it would have at the present
day, for the drinking was the pledge of friendship, and the
insurgent nobles then disbanded their followers.*

Another historical personage associated with Hornsey
Park is the boy-King, Edward V. His model uncle, Richard
of Gloucester, was escorting him to London, there to have

The Boating Lake, Finsbury Park.

him murdered in the Tower with his younger brother. An
old chronicler tells us ' when the kynge approached nere
the cytee, Edmonde Shawe, goldsmythe, then Mayre of
the cytie, with the aldermenne and shreves in skarlet, and
five hundreth commoners in murraye (*i.e.*, violet), receyved
his Grace reverently at Harnesay Parke, and so conveighed
him to the cytie, where he entered the fourth day of May,
in the fyrst and last yere of his reigne.'† The Duke, with
his numerous followers, all attired in mourning, must have

* Brewer, ' London and Middlesex,' vol. iv., p. 212.
† Hall's Chronicle, p. 351, quoted in Thorne's ' Environs of London.'

formed a sombre contrast to this gay scene. He rode before the King, cap in hand, bowing low to the people and pointing out to them their rightful ruler, who was so soon to be taken from them. The people were too awed and apprehensive of danger to be hearty in their welcome, and time soon showed what just grounds they had for their fear.

The next monarch we have to record at Hornsey Park is Henry VII., who was greeted here by the Mayor and citizens of London after his return from a successful war in Scotland.*

Hornsey Wood must have been a place of particular beauty at the beginning of this century. It was a great place of attraction on Palm Sunday, when it was ravaged for the so-called palms. The popularity of the Hornsey Wood Tavern, which stood just to the south of the present lake, was dependent upon the visitors to the woods. Hone thus apostrophizes the old tavern :

> ' A house of entertainment—in a place
> So rural, that it almost doth deface
> The lovely scene ; for like a beauty-spot
> Upon a charming cheek that needs it not,
> So Hornsey Tavern seems to me.'†

The tavern stood on the summit of some rising ground, and was originally a small roadside public-house, with two or three widespreading oaks before it, which afforded a pleasant shade to its frequenters. The wood itself immediately adjoining the tavern for some time shared with Chalk Farm the honour of affording a rendezvous for the settlement of *affaires d'honneur*.‡ From contemporary accounts we gather that the house was a ' good, plain, brown-brick, respectable, modern, London-looking building.' On the left of the entrance was a light and spacious room of ample accommodation and dimensions, which boasted a fine leather folding screen. The neglect of this screen called forth Hone's indignation, who describes it as bearing some remains of a

* Howitt, ' Northern Heights,' p. 432.
† Hone, ' Everyday Book,' p. 759.
‡ Lloyd, ' Highgate,' p. 297.

spirited painting spread all over its leaves, to represent the
amusements and humours of a fair in the Low Countries.
Hornsey Wood Tavern and its grounds displaced a romantic
portion of the wood, which was a favourite resort of Crabbe.
'On one memorable occasion he had walked further than
usual in the country, and felt himself too much exhausted to
return to town.   He could not afford to give himself any
refreshment at a public-house, and much less pay for a

The Old Hornsey Wood House.   (From an old woodcut.)

lodging, so he sheltered himself upon a haymow, beguiled
the evening with Tibullus, and when he could read no
longer, slept there till morning.'*   Crabbe used to come to
Hornsey Wood to search for plants and insects, and often
he would spend whole afternoons here in his favourite pursuit.
Although called a wood, it was properly little more than a
thicket, for the small trees, shrubs and bushes were so inter-
woven that in some places it was impassable.

* 'Life of Crabbe,' by his son.

There have been two Hornsey Wood Houses. The first, as we have seen, was an unpretentious roadside inn, which well became its situation, and harmonized with its rural surroundings. Hone describes it as being kept by two sisters, Mrs. Lloyd and Mrs. Collier, ' ancient women and large in size.' Their favourite seat was on a bench fixed between two venerable oaks, which furnished a home for swarms of bees, and here the two ancient dames would talk of bygone days as they sipped a friendly glass with their customers. In a ripe old age they both passed to their

The New Hornsey Wood House.

graves within a few months of one another. But Hornsey Wood had to move with the times, and the old place was pulled down and the oaks felled in order to make room for a larger and more fashionable successor. The new proprietor spent some £10,000 in making the alterations, and he formed a lake on the site of the present ornamental waters. This lake, which was well stocked with fish, afforded more hope of sport to the London angler than the New River, and the pleasant walk around it, together with the beauties of the prospect, combined to make the tavern, with its grounds and tea-gardens, as attractive as the present park. Another sport much patronized in the grounds was pigeon-shooting, but in course of time Hurlingham and the Welsh Harp became the head-quarters of this fashionable pastime.* The tea-drinking in the tavern gardens declined in course of time, owing to the great number of places of the same kind being established nearer London. One of the first acts in the laying-out of the park was the pulling down of the Hornsey Wood Tavern, which has given its name to another place of refreshment on the east side of the park.

Skirting the park on the south-eastern boundary is Seven Sisters Road, which was formed in 1831-33, and connects Holloway and Tottenham. The Seven Sisters, from which the road takes its name, is the sign of two public-houses at Tottenham. In front of the one at Page Green, near the entrance to the village, were seven elm-trees in a circle, which are the sisters in question. Tradition says that these trees were planted by seven sisters when about to separate, and it is also said that this is the spot on which many a martyr had been burnt. In 1840 the elms were considered to be nearly 500 years old, but they were then fast going to decay. Thirty years later only their dead trunks were standing, and now all trace of them has disappeared.†
It was intended originally that the portion of the park fronting Seven Sisters Road should be reserved for building sites, in order to recoup a portion of the sum expended in the formation of the park, but in consequence of considerable pressure which was brought to bear upon the late Board, they decided not to reduce the area available for recreation.

Some little difficulty was experienced when the park was first opened owing to a clause in the original Act which gave some adjoining owners a right of way through the park

* Hone, 'Everyday Book,' p. 760.
† Thorne, 'Environs of London,' part ii., p. 622.

between Seven Sisters Road and the Green Lanes. This had to be kept open always, so that the park could not be closed at night-time, and much annoyance and even danger was experienced through the passage of droves of cattle at all hours of the day and night. Eventually, in 1874, these restrictions were modified, and power was obtained in another Act to construct a road skirting the northern and

The Lake, Finsbury Park.

western sections of the park. The three parts of this road are called Endymion Road, Stapleton Hall Road, and Upper Tollington Park. Stapleton Hall was once a celebrated place of entertainment, much after the style of Hornsey Wood Tavern, and the many other places of a similar character in the neighbourhood.

The last point of interest we will call attention to is that underneath the tennis-ground is a huge covered reservoir belonging to the East London Waterworks Company. This reservoir was formed under the powers of the East London Waterworks Act, 1867, in connection with a supply of water from the river Thames, which is brought from Sunbury to the storage-bed in Finsbury Park. The soil above still remains part of the park, as the company were only empowered to purchase an easement and right to construct and maintain the reservoir and pipes.

CLISSOLD PARK.

Clissold Park, perhaps better known locally as Stoke Newington Park, which sufficiently indicates its locality, consists of 53 acres. Although it was opened by the Earl of Rosebery, the first Chairman of the London County Council, on July 24, 1889, the credit for its acquisition is due to their predecessors, the Metropolitan Board of Works, who, however, passed out of office before seeing the fruit of their labours in this connection. It was acquired under the Clissold Park (Stoke Newington) Act, 1887, from the Ecclesiastical Commissioners at a cost of £96,000, made up as follows :

Charity Commissioners	-	-	-	- £47,500
Metropolitan Board of Works	-	-	-	- 25,000
Vestry of Stoke Newington -	-	-	-	- 10,000
South Hornsey Local Board-	-	-	-	- 6,000
Hackney District Board	-	-	-	-	- 5,000
Vestry of Islington -	-	-	-	- 2,500

£96,000

As the park before being opened to the public consisted of the grounds of a large mansion, it had all the benefit of the many fine trees which now form so prominent a feature in its landscape. In this it had a great advantage over its neighbour, Finsbury Park, which was originally nothing but agricultural land, and it must be some years before the trees and shrubs then planted will equal the beauty of those in Clissold Park. The cedars on the lawn are fine specimens, and one of the thorns is said to be the oldest in England.

Another attraction of the park is the New River, no longer new, whose sluggish waters coil round and through it. This is spanned by three miniature bridges, and, alas for the beauty of the landscape! it has to be guarded on both sides by two grim rows of iron fencing, not only for the sake

The Deer-Pen, Clissold Park.

of safety, but to protect the fish with which it abounds from the grasp of would-be anglers. If the site of the park had fallen into the hands of the builders, the river would have continued its course in underground pipes, and so one of the few remaining beauties of Stoke Newington would have been lost for ever. We can well understand Robinson writing: ' Besides the incalculable convenience which the supply of water affords, the New River must ever be considered as one of the greatest ornaments to those places through which it passes;' and he singles out as the

most prominent example the site of our present park.*
The quaint brick towers which are seen from the park
belong to the adjacent pumping station of the New River
Company.

But to those who would not be drawn by the beauties of
the landscape, or the quiet cool shade by the river banks,
other inducements are offered. Clissold Park was one of
the first London municipal parks in which bird and animal
life was specially provided for. Hitherto the various water-
fowl on the lakes were the only attractions of this kind; but
some deer and guinea-pigs were presented by some generous
members of the public, and were placed as a trial in this
park. Every Londoner is familiar with the deer contained
in such large areas as Bushey and Richmond Parks, but to
confine them to the necessarily small space in this park was a
bold experiment. Fortunately, it has proved quite successful,
and the deer enclosure is never without an admiring crowd.
In addition to these, there is also an aviary stocked with British
birds, and the time may not be far distant when the example
of the Zoological Gardens is followed in providing elephant
and camel rides for the juvenile visitors. Among the mis-
cellaneous buildings provided for the comfort and recreation of the public must be mentioned a bandstand, where
weekly performances are given during the season, and
shelters built in a rustic manner to harmonize with the
surroundings.

The mansion in the park, which is situated on the rising
ground at the back of the old church, is at present untenanted, although one or two rooms are used for refreshment and other purposes. Several proposals have been on
foot to convert it into a museum, but this object has not yet
been achieved. It is a plain brick structure, with a classic
colonnade in front. The bricks used in its erection were
made from the clay dug on the portion of the park where
the lakes are now. The date when it was built is n ot known

* Robinson, 'History of Stoke Newington,' p. 6.

exactly, but it is shown in a view of Stoke Newington by Ellis, dated January 1, 1793.* The greater part of this view is taken up with the New River, but the mansion and the old church are seen in the background. The first occupant was Jonathan Hoare, one of the famous race of bankers of that name. The village of Stoke Newington was a favourite resort for bankers at this time—such men as Burnand, Twells, Leicester, and Bevan, having taken up their residence here. It soon afterwards came into the possession of Mr. Thomas Gudgeon, who lived here in 1804. The next occupant was Mr. Crawshay, who was also the owner of about 27 acres of the adjoining land. He obtained a perpetual lease of the property from the Ecclesiastical Commissioners, at a yearly rental of £109 'and a fat turkey,' but provisoes were inserted in the covenant against the cutting down of trees, or the granting of building leases. A romantic story is connected with the mansion while he lived here. It appears he had two daughters, one of whom captured the heart of the Rev. Augustus Clissold, a curate at the neighbouring church. The father, who was gifted with an irascible temper, hated parsons, and would not hear of the match ; the curate was forbidden to visit the house, and the lovers were compelled to communicate with each other through messengers, whom the father threatened to shoot. It is even said that Mr. Crawshay had the walls increased in height, and so the unfortunate pair had to wait and wait, till at last their patience was rewarded, for the irate parent died, the lovers were married, and the curate entered into possession of the mansion and grounds. He altered the name from Crawshay's Farm to Clissold's Place or Park. On the death of the Rev. Augustus Clissold the mansion once more reverted to the Crawshay family, and Mr. George Crawshay became its owner. A memento of his ownership exists in a mural drinking-fountain at the

* A copy of this may be seen hung up (together with other interesting views of Old Stoke Newington) in the adjacent Free Library. They were the gift of E. J. Sage, Esq.

north end of the mansion, which bears the following inscription :

'In memory of three sweet sisters, aged 1, 3, 4, daughters of Wilson Yeates, Esq., interred at Horton, Bucks, 1834. Erected by their sister, Rose Mary Crawshay (widow), 1893.'

His business as the proprietor of large iron mines in Wales did not leave him many opportunities of visiting his Stoke Newington property, so he tried to let it, as he was prevented by a covenant in his lease from cutting the grounds into small plots for the erection of villas. Not

The Mansion, Clissold Park.

being successful in this, the only course left open was to sell the mansion and grounds, and as soon as it was known in 1884 that Mr. Crawshay was desirous of realizing the property, an influential committee was formed to endeavour to make terms for its purchase as a park for the use of the public. These negotiations, however, came to no successful issue, and early in 1886 large notice-boards were displayed advertising the sale of about 20 acres of the freehold estate, over which the owner had absolute control. About the same time Mr. Crawshay wrote to the *Times*, in which he said : ' I have expressly reserved the 5½ acres freehold on which the old house stands because of the extreme beauties of the ground, which, together with the well-timbered leasehold and the shaded and encircling waters of the New River, make up a whole which I trust will never be destroyed.' After this assurance all were surprised to learn that Mr. Crawshay had sold his interest in the whole of the estate for £65,000 to the Ecclesiastical Commissioners, who had already prepared to plot out the land for building purposes. No time was lost in communicating with that body, and the amount fixed as purchase-money for the whole of the park was £95,000. This was raised by various means, as we have already seen; but the Commissioners, who have shown themselves very liberal landlords at other places in London, have a right to be included among the list of contributors. The time occupied in raising the purchase-money extended over two years, and the accumulated interest amounted to nearly £5,000, or a total of £100,000, towards which only £96,000 had been received. The deficit in the interest was most generously foregone by the Commissioners, and Clissold Park thus became open to the public for ever.*

Two members of the committee who had thus secured Clissold Park for the public deserve to be singled out for special mention, viz., Joseph Beck and John Runtz, both of whom have now passed away. A handsome drinking-fountain, on an elevated site near the centre of the park, has been erected as a permanent memorial to their labours. The inscription on the front panel runs:

THIS FOUNTAIN
WAS ERECTED BY SUBSCRIPTION,
A.D. 1890,
IN GRATEFUL RECOGNITION OF THE UNITED EFFORTS OF
JOSEPH BECK,
JOHN RUNTZ,
AS LEADERS OF THE MOVEMENT BY WHICH THE USE OF THE
PARK WAS SECURED TO THE PUBLIC FOR EVER.

---

* For many of the particulars of these negotiations I am indebted to an interesting little book entitled 'The Story of Church Street, Stoke Newington,' by Giltspur.

On one of the side-panels are the simple words :

JOSEPH BECK,

MEMBER OF

COURT OF COMMON COUNCIL

and on the opposite side :

JOHN RUNTZ,

MEMBER OF

METROPOLITAN BOARD OF WORKS.

The adjoining Free Library also contains oil-paintings of both these gentlemen, so that they are not likely to be soon forgotten.

The works of laying out the property for public use were rapidly pushed on, and consisted of the usual formation of paths, drainage, laying on water-supply, and the erection of boundary and other fencing. The lakes which had been filled in were re-excavated, and they have been locally christened Beckmere and Runtzmere.

Stoke Newington (from Saxon *stoc*, a wood) means the new town in the wood, and no doubt this district formed part of the extensive forest of Middlesex, and so late as the year 1649 there were upwards of 77 acres of woodland in demesne.* The little village of Stoke Newington, like many another populous suburb, has only of late years developed into a busy centre of town life. At the commencement of the present century, it consisted of a single straggling street (Church Street) in the midst of fields. Many illustrious names, however, are connected with this former rural retreat. John Howard, the prison reformer ; Defoe, the author of 'Robinson Crusoe'; Dr. Watts, the hymn-writer, and many other celebrities, claimed it as their home, and but few are aware of the classic ground which the modern common-place Stoke Newington covers. Most of the simple beauty of the erstwhile village has also disappeared ; the old church nestling amidst the trees which nearly hide it from view, a

* Robinson, ' History of Stoke Newington,' p. 1.

row of tiled Queen Anne houses, and Clissold Park are all
that remain.

The park is situate in two different parishes—Stoke
Newington and Hornsey—and forms part of the lands of
two distinct manors. The portion around the mansion is in
the prebendal Manor of Neutone (Stoke Newington), whilst
the remainder through which the New River flows is within
the prebendal Manor of Brownswood.* The Manor of Stoke

The New River and Paradise Row, Clissold Park.

Newington, now vested in the Ecclesiastical Commissioners,
has been the property of the Prebendary of Newington from
very early times. In Doomsday Book, Neutone is mentioned
as the property of the Canons of St. Paul's, and their ownership can be traced back to the time of King Athelstan, about
the year 940. It was held by the prebendaries in their own
hands till 1550, when it was first leased to William Patten,
one of the Tellers of the Receipt of the Queen's Exchequer at
Westminster, and Receiver-General of her revenues in the

* For descent of this manor, see p. 312.

county of York. A new lease was granted to Mr. Patten in
1565 for ninety-nine years at £19 per annum, which was
assigned about 1571 to John Duddeleye. He died in 1580,
leaving his property to his wife and daughter. The former
married Thomas Sutton in 1582, and after his death, in 1611,
the manor came into the possession of Sir Francis Popham,
who had married Anne Dudley. It was next the property
of his son John, who died without issue, and was succeeded
by his brother Alexander, who, during the Civil War, rose
to the rank of Colonel in the Parliamentary army, and was
returned member for Minehead, Bath, and Somersetshire
successively. When the prebendal estate was sequestered
and sold, in 1649, under the provisions of the Act for the
abolishing of Deans, Deans and Chapters, Canons, Prebends,
Colonel Popham purchased it of the sequestrators for the
sum of £1,925 4s. 6¼d., so that he became Lord of the
Manor in fee, and continued so until the Restoration, when
the Church recovered its rights, and he returned to his former
state of lessee. The manor remained in the Popham family
till 1699, when the lease was sold to Thomas Gunston. On
his death it came to his sister, who was the wife of Sir
Thomas Abney, Lord Mayor of London, and when he died
she entered into full possession. Lady Abney, by her will,
directed the lease and estate to be sold, and, after the payment of certain legacies, the residue was to be distributed
' to poor Dissenting ministers, to poor Dissenting ministers'
widows, and other objects of charity.' The manor and
estate were put up to auction in 1783, and purchased by
Jonathan Eade for £13,000. He died in 1811, and bequeathed
the manor to his sons William and Joseph Eade. The
Prebendary obtained an Act in 1814 to enable him to grant
a new lease for ninety-nine years to these two gentlemen.
William Eade assigned his interest to his brother in 1815,
who then sold these lands to the late Mr. Crawshay, who
also purchased the reversion from the Prebendary. This
portion of the park, then, within the Manor of Stoke
Newington, was the freehold of Mr. Crawshay.

The Manor of Brownswood, in which the remainder is situate, was vested in the Ecclesiastical Commissioners upon the death of Prebendary Secker, and by them a lease was granted to Mr. Crawshay, who thus possessed the whole property. The park also includes a small strip of the waste of this manor. This strip, containing 24 perches, was enclosed by Mr. Crawshay under the Act 53 George III., entitled 'An Act for enclosing Lands in the Parish of Hornsey,' in satisfaction for his right of common upon the

The Bandstand, Clissold Park.

waste lands in the parish of Hornsey in respect of his copyhold estate. We have already seen how Mr. Crawshay sold his interest in the estate to the Ecclesiastical Commissioners, from whom the land was bought.

The author of the 'Ambulator' (1774) in describing Stoke Newington as 'a pleasant village near Islington, where a great number of the citizens of London have built houses, and rendered it extremely populous, more like a large flourishing town than a village,' goes on to speak of the

church, and afterwards of a celebrated walk of very ancient
origin. He says : ' Behind the church is a pleasant grove
of tall trees, where the inhabitants resort for the benefit of
shade and a wholesome air.'* This ' Mall ' of the Stoke
Newington folk is what is known at the present day as
Queen Elizabeth's Walk.   It is still a pleasing avenue,
although very few of the original trees remain.

   Many are the attempts which have been made to connect
' Good Queen Bess ' with this corner of the world. Tradition
centres round the old manor-house as the place where she
stayed.   This house, which was probably erected in 1500,
was pulled down in consequence of its dilapidated condition
in 1695. About the time of Elizabeth, either before or after,
some relatives of the celebrated Earl of Leicester were living
here.   The Dudley arms were found in one of the houses
built upon its site, and probably came from this mansion.
The parish register also contains an entry of the burial of
a servant of the Countess of Leicester in 1682, and this
possibly confirms the truth of the statement.   While the
Dudleys were here it is said that the Princess Elizabeth came
to this house in hiding during the reign of her half-sister,
Queen Mary. A former Rector of the parish, the Rev.
Thomas Jackson, describes the story as current in the
village in the last century that ' when the Princess was the
hope of the Protestants, exasperated by persecution, she
was brought by her friends to the secluded manor-house,
embosomed in trees, as to a secure asylum, where she
might communicate with her friends, and be ready for any
political emergency. They tell us that an ancient brick
tower stood, in the early part of the last century, near the
mansion, and that a staircase was remembered leading to
the identical spot where the Princess was concealed.' It is
not unlikely, then, that during her seclusion she trod this
very path which now bears her name. When afterwards she
was crowned, it is well authenticated that she visited the
Dudleys, and that on one occasion ' Her Majesty, taking a

   * Quoted in ' Old and New London,' vol. v., p. 530.

jewel of great value from her hair, made a present of it to their daughter, Miss Ann Dudley.'* The Dudleys of Stoke Newington are but little known to history, and no confirmation of the tradition connecting the seclusion of the Princess Elizabeth with this mansion can be found. After the death of Mr. Dudley his widow married Thomas Sutton, who is best remembered as the founder of the Charter-

house School and Hospital. They lived at this manor-house till their removal to Hackney. The Miss Ann Dudley mentioned above married Sir Francis Popham,

Queen Elizabeth's Walk in 1800.

the son of Sir John Popham, Chief Justice of the Queen's Bench from 1592 to 1607, and not only they, but their son, Colonel Alexander Popham, lived in the mansion. The last owner, Timothy Matthews, citizen and grocer, cleared away the buildings and let the site on building lease.† The houses from the churchyard to that called Manor House are on the site of the old manor-house and grounds.

* Bearcroft, ' History of Mr. Sutton,' quoted in Robinson.
† Robinson, ' History of Stoke Newington,' pp. 48-54.

Running almost parallel with Queen Elizabeth's Walk, and forming the western boundary of the park, is another thoroughfare which boasts a rural name, Green Lanes. This is now lined with substantial villas, and contains here and there the remains of what were once, no doubt, continuous avenues of trees. The noise of the tramcars passing

Clissold Park, with Old Stoke Newington Church in the Background.

along it dispels any lingering trace of rusticity which it might possess. In this walk James Mill used to take his daily airing, hand-in-hand, perhaps, with his portentous little son John, if, indeed, two such philosophic minds can be supposed ever to have condescended to so trivial an action.*

The most striking object visible from inside the park is

* *Pall Mall Gazette*, July 26, 1886.

the tall and graceful spire of the new parish church of Stoke Newington. The church itself is a fine cathedral-like structure, built from the designs of Sir George Gilbert Scott, and considered one of his masterpieces.

The church occupies the site of the old rectory-house, which was chiefly built of wood, and had extensive grounds. This was probably built by the Lord of the Manor when the old church was restored in 1563.

The old church, whose churchyard seems to form part of the park, is a very ancient fabric. Its present appearance, peeping from the dense foliage of the trees around it, is quite in accord with its past surroundings. It takes us back in thought to the time when Stoke Newington was an unpretending village with here and there a farm-house dotted among the fields which surrounded it on every side. The date of its erection is lost in the dim past. A small square stone over the south door records the date (1563) of its restoration. It is said before this time to have been a small Gothic structure of stone faced with flint and pebbles. This was then large enough to accommodate the few worshippers of the village. Nothing would better illustrate the growth of the village than a comparison between these two churches, standing side by side. Various attempts were made to keep pace with this growth, and from time to time enlargements had to be made to the parish church. In 1716 it was nearly doubled in size by building on the north side of the churchyard, and still later, in 1723, the chancel was extended.

CHAPTER XVII.

HACKNEY COMMONS.

Clapton Common—Stoke Newington Common—Hackney Downs—Mill Fields.

THE parish of Hackney is particularly rich in open spaces, and though they were recognised as an undoubted boon at the time of acquisition, their value as recreation-grounds is still further increased at the present day, now that the area has become more congested. They comprise Clapton (7½ acres) and Stoke Newington Commons (5¼) in the north, Hackney Downs (41¾), North Mill Field (23¼), and South Mill Field (34¼) in the centre, London Fields (26½) and Well Street Common (20¾) in the south, while the most recent acquisition, Hackney Marsh (337 acres) runs along the greater portion of the eastern boundary. These are quite irrespective of Victoria Park, disused churchyards, squares, and various other small plots of land dotted about in different parts of the parish, which together make a considerable total.

It is difficult to make out the boundaries of these open spaces in ancient maps of the district, for the simple reason that they were then indistinguishable from the surrounding fields of which they formed part. The marsh, which has been protected by its very nature, seems to have undergone least change. The commons themselves lean towards the useful rather than the ornamental. By no stretch of imagination can they be called picturesque or rural; they are simply green islands of turf in the midst of seas of bricks and mortar.

If it were not for extensive fencing, there would hardly be a blade of grass upon them, so great is the use to which they are put.

Of the seven so-called Hackney Commons, only two were open spaces of common really, viz., Clapton Common and Stoke Newington Common. The remainder were lammas lands.* The peculiarity of these latter was that from April 16 to August 11 the exclusive right to the soil and herbage belonged to different proprietors, and for the remaining eight months the herbage belonged to the freehold and copyhold tenants of the manor. The grazing of the lammas lands was managed by a body of men elected annually at the courts of the manor, called marsh drivers. The number of cattle to be turned on to the lands was regulated by the amount of rent paid. A small fee was charged for each head of cattle turned out to graze, and the money so raised was spent in improving the lands. Of late years, owing to the extraordinary development of Hackney, the only lands of any use for grazing purposes have been Hackney Marsh, and to a limited extent the Mill Fields. The number of freeholders of land in the lammas lands was very extensive, and each owner's strip was marked off with posts. Each freeholder had the exclusive ownership of his particular strip, subject to the lammas rights. The peculiar way in which the lammas lands were held once led to a rather amusing incident—amusing, that is, to anyone except the unfortunate freeholder. In the year 1837, the lessee of a certain portion of Hackney Downs ploughed it up for the purpose of growing corn, and owing to the lateness of the season the crop could not be carried off before the middle of August, when the inhabitants entitled to common and lammas rights entered the land and carried away a great portion of the crop.†

* From an Anglo-Saxon word meaning bread mass, it being observed as a festival of thanksgiving for the fruits of the earth on August 1 (old calendar).

† Robinson, ' Hackney,' p. 76.

During this century the lammas lands have been subjected
to some kind of supervision which has prevented extensive
encroachments from being made. A committee was specially
appointed by the inhabitants in 1809 to report upon them,
and they came to the conclusion that there had been many
unlawful enclosures and other violations of the lammas lands.
Their meetings were held at the once famous Mermaid Tavern,
and their report published in 1810 gives some very interesting
details of the lammas lands at that time.   It appears from

Lea Bridge, Mill Fields.

this that the rights of turning cattle on to lammas lands are
manorial privileges, and that they existed six or seven hundred
years before the grant of the Manor of Hackney to a lay
subject.

All the lands comprised in the Hackney Commons were
attached to the Manor of Lordshold, Hackney ; Clapton and
Stoke Newington Commons and the strips in Dalston Lane
and Grove Street (now Lauriston Road) being waste of the
manor, and the remainder, including the marsh, as we have

already stated, being lammas lands. The Manor of Lordshold is co-extensive with the parish of Hackney, and was in
the time of Edward VI. valued at £61 9s. 4d. It is not
mentioned in the Doomsday survey, but was probably
included in the larger manor of Stepney, and as such formed
part of the demesne lands of the Bishops of London. In
the nineteenth year of King Edward I. (1290) Richard de
Gravesend, Bishop of London, and his successors, were
granted by the Crown free warren in Hackney and Stepney
provided that the lands were not included within the boundary
of the great forest of Middlesex. The Bishops of London
continued to enjoy these privileges till Bishop Ridley, in
1550, surrendered the manor to King Edward VI. in consideration of certain other lordships, and in this same year it
was granted to Thomas, Lord Wentworth.* His son and
successor to the estate had great disputes with the copyhold tenants of the manor concerning some of the customs,
benefits and privileges of the tenants, which differences were
appeased by a deed of covenant entered into by them reci-
procally on June 20, 1618, and confirmed by a decree of the
Court of Chancery dated July 22, 1618. These covenants
and agreements were also the subject of an Act of Parliament in 1622-23, which dealt most minutely with all the
customs of the manor.† Among the subjects treated of, the
wastes of the manor are of course included, and by this Act
the copyholders are entitled to ' lop and shred all such trees
as grow before their houses . . . upon the waste ground,
and the same convert to their own use, without any offence,
so the said trees stand for defence of their houses, yards, or
gardens ; and also may dig gravel, sand, clay and loam, upon
the said waste grounds, to build or repair any of their copyhold tenements . . . so always as every of the said copyholders do fill up so much as shall be digged by him or
them.' The Act also provides for the appointment of certain
customary tenants to be drivers and viewers of the wastes

* Robinson, ' Hackney,' p. 303.
† This is printed *in extenso* in Robinson's ' Hackney,' pp. 354-408.

and commons of the manor to be elected annually. The fees they received for marking and the redemption of impounded cattle, after paying travelling expenses, were ' to be employed to the scouring of the common sewers which be upon the said waste ground and commons, and laying of bridges over the said common sewers.'

The Manor of Hackney was separated from the Manor of Stepney by Lord Wentworth in the early part of the seven·teenth century, and remained in that family till the forfeiture of the Earl of Cleveland's estates in 1652. In 1659, William Smith and others, who probably purchased it of the Parliamentary Commissioners, alienated it to William Hobson, who died in 1662, leaving his three daughters co-heiresses. Their husbands were joint Lords of the Manor till 1669, when they sold it to John Foorth, an Alderman of London. Coming now to 1676, we find the manor in the possession of Nicholas Cary and Thomas Cooke, goldsmiths, of London, and later on, in 1694, it belonged to Sir Thomas Cooke. Three years later it was purchased by Francis Tyssen, and remained in that family till 1794, when it passed by marriage to William George Daniel, a Captain in the 91st Regiment, who took the name of Tyssen. The present Lord, Baron Amherst of Hackney, is the grandson of this last owner.

The acquisition of the Hackney commons required no less than three Acts of Parliament. After the passing of the Metropolitan Commons Act, 1866, the local authority thereby constituted was empowered to present a memorial to the Enclosure Commissioners with a view to establishing a scheme for the local management of any Metropolitan common, so that the scheme might be confirmed by Act of Parliament. This was the course adopted with regard to the Hackney commons upon the initiative of the Hackney Board of Works. The question of including the Marsh in the scheme was discussed at the time, but it was decided to leave it out. Eventually the Metropolitan Commons Supplemental Act, 1872, confirmed the scheme of the Commissioners, and the late Metropolitan Board of Works took over

the charge of the commons. The Lord of the Manor con-
tended that this scheme in no way interfered with his rights,
and that he had perfect liberty to carry on any excavation
for gravel and brick-earth on his lammas lands, and, what
was more serious, to enclose any part of the lands with the
consent of the homage. In order to test the matter, he
enclosed a portion of Hackney Downs, a plot on North Mill
Field, and also commenced extensive gravel excavation on the
Downs. These steps led to prolonged litigation, and in the
end the Lord of the Manor was declared in the right. A second
Act of Parliament was therefore required to enable his rights
to be purchased, which was effected by the Metropolitan
Board of Works (Hackney Commons) Act, 1881, the price
being fixed at £33,000. But there still remained the rights
of many freeholders to be acquired in addition to those of the
Lord of the Manor. These were purchased under the third
Act relating to Hackney Commons—viz., the Metropolitan
Board of Works (Various Powers) Act, 1884. Altogether
the total cost of obtaining the commons has been about
£90,000. Having now considered these open spaces generally,
we shall glance at each one separately.

CLAPTON COMMON.

Clapton, formerly Broad, Common is the most northernly
of the Hackney open spaces. Situated on the high ground,
it has an advantage in many respects over the other commons
of the district. For one thing, the air is purer, and it is
even stated that in the early morning, before chimneys and
factories poison the air with their smoke, it is possible to
sniff the ozone of the pure sea-breeze. Of recent years,
before the houses surrounding the common were built, there
must have been an extended view over the surrounding low-
lying country, with the river Lea winding through the green
marshes like a silken thread in some elaborate tapestry work.
It has been said that ' the view over the Lea Valley from the
heights of the mellow, old-fashioned suburb of Clapton is
not inferior in its way to that of classic Richmond Hill itself.

Beyond the quiet meadows are the lines of aspens and poplars, backed by the rounded forms of the elms about Walthamstow, with their masses of deep shadow, and on the sky-line the ridge of Epping Forest, with a spire here and there, sunlit, and standing out against the purples and ambers of the woods.'*

These beauties, which were once visible from the common, are shut out by the row of houses called Buccleuch Terrace, built about a hundred years ago.

On the common is a small pond, much in demand for skating in the winter, which must not be confounded with the better-known Clapton pond some distance down the main road. Behind the pond on the common is Stainforth House, once the residence of Mr. Richard Foster, and more recently of the late Suffragan Bishop of Bedford. Craven House, at the north end of the common (so named after a former owner—Mr. Arthur Craven), was at one time the home of the late Mr. Samuel Morley, M.P. for Bristol.†

On the western side of the common is St. Thomas's Church, not a very ornate structure. In October, 1864, a terrific gunpowder explosion between Plumstead and Erith occurred, which shook the houses in Upper Clapton to their very foundations. St. Thomas's Church came off very badly, for the east wall was split from top to bottom by the force of the explosion.‡

STOKE NEWINGTON COMMON.

This small open space, situated at the west of the parish, had once a claim to rural beauty. The cuckoo and the nightingale were regular visitants in their season, and the whole of the district of Stoke Newington, as we have already seen, was country indeed. But this was long before it became under municipal control. The Great Eastern Railway now passes right through the centre of the common, and any attempt, by planting or other means, to arrive at

* A writer in the *Echo*.
† 'Glimpses of Ancien Hackney,' p. 223.   ‡ *Ibid.*, p. 220.

the picturesque will always be marred by the proximity of the iron road. We need not go back far to find very different surroundings for the common. In the Ordnance Survey map of 1868, the only boundary of the common which is shown at all built over is the west. On the north were large houses in the midst of extensive grounds—Baden Farm and Thornbury Park. On the east was a similar estate— Elm Lodge—and to the south wide-spreading brickfields. Now these have given place to terraces of houses which almost threaten to swamp the few acres of green which remain. It is probable that originally the common, or some part of it at any rate, reached to the highroad on the west. We have a record of one grant of the common land for building.

In 1740, Thomas Cooke, of Stoke Newington, built a large house on the common, and divided it into eight sets of rooms for the accommodation of as many poor families. The land had been let to him by the Lord of the Manor for ninety-nine years, from Midsummer, 1740, at a rent of 2s. 6d. The inmates paid a yearly rent of 2s. per family as an acknowledgment of their tenancy. By his will, the founder of the charity left his houses and land at Eltham in trust, to keep in repair his house, *built on the common in the parish of Hackney*, whilst any surplus was to be divided amongst the inmates of the almshouse. In 1842 the occupants lived rent free, and each family received 4 guineas a year, and two sacks of coals at Christmas.* These almshouses have lately been rebuilt, and will be noticed on the north side of the road. passing out opposite Abney Park Cemetery.

The Great Eastern Railway were empowered by their Metropolitan Station and Railways Act, 1864, to carry their line across Stoke Newington Common in a gallery or covered way, or, if not so covered in, the company were to acquire, and give in exchange to the proprietors of the common, to be held as part and parcel thereof, so much land adjoining thereto as should be equal in quantity to the land taken from

* Robinson, ' Hackney,' p. 394.

the common for the purposes of the railway.   The line was
not covered in, and the company acquired and threw into
the common a plot of land at the south-west corner equal in
area to that taken by them.   This arrangement was completed in 1873 by a formal deed of conveyance to the Lord of
the Manor.   Another exchange took place at the southern
end, in addition to that with the railway company, so that
the present appearance of the common is very different to
what it formerly was.

<h2 style="text-align:center">HACKNEY DOWNS.</h2>

Hackney Downs is a rectangular open space, skirted on
the west by the Great Eastern Railway, which fortunately
does not intersect it.   Its elevation is higher than any
other point within the same radius of the Metropolis.   This
is one of the lammas lands of Hackney, and we have
already mentioned the chief historical fact connected with
it, viz., the removal of the freeholders' crops in 1837.   But
the mention of harvest operations in Hackney seems so
very droll that further particulars of this event, taken from
a contemporary record, may not be out of place.   The
account is as follows :

'A strange scene was witnessed here on the evening of
Monday, August 14.   Owing to the lateness of the season,
a notice, signed by the steward of the manor, had been
issued a few days previously, stating that the Downs would
not be open to the public until Saturday, the 25th, or a
fortnight beyond the time when cattle have hitherto been
admitted.   The crops on the Downs were this year unusually
fine, the greater portion of which remained uncut on Monday
morning, when a few persons made their appearance, and
began to help themselves to the corn, alleging that, as no
person could now legally claim it, they had as good a right
to it as anyone else.   Mr. Adamson, to whom the greater
part of the crops belonged, very naturally disputed their
right, and gave them into custody.   On being examined at
Worship Street, the magistrates had no sooner heard the

facts of the case than they dismissed the men on the ground that, in their opinion, the corn was common property, and could be claimed by no one parishioner more than another. The men soon made known the magistrates' decision, which seemed to justify anyone in helping himself to all he could get. That was the notion which generally prevailed, and which a good many were not at all backward to act upon. From eight o'clock till near midnight, troops of men, women, and children were to be seen coming from the Downs loaded with more wheat than they could carry, strewing the ground with it as they came along, and hurraing and cheering each other at this practical assertion of their rights. Some, who should have known better, even brought horses and carts to aid in the work of plunder. The scene was a most humiliating and disgraceful one. On a small scale was exhibited, we fear, too true a specimen of the temper of an English mob when free, or supposed to be free, from the bridle of the law. The property of one neighbour was at the mercy of hundreds, and his fellow-neighbours seemed to glory in showing how merciless they could be. A number of policemen were on the ground, but, after the decision of the magistrate, they could not effectually interfere ; all they could do was to prevent any breach of the peace between Mr. Adamson's labourers, who were busily engaged in removing their master's property, and those who were as busily engaged in helping themselves.

' In the early part of the day, Mr. Adamson does not appear to have done all that he might to save his crop ; afterwards, when he was threatened with the loss of all of it, more energy was manifested. At ten in the evening several waggons were in motion, and, we presume, continued so during the night, for on the following morning the whole had disappeared. The total loss of the freeholder was estimated at £100.'*

He made one further attempt to imprison the marauders, for he proceeded against several of them in the Court of

* *Hackney Magazine*, October, 1837, p. 175.

Queen's Bench for a riot. On the trial, the Court perceived the error of both parties, and, as a sort of compromise, induced the accused individuals to plead guilty that it might obtain the power of discharging them, and so end the dilemma. Another labourer who was arrested was not so fortunate. It was proved that he was neither a parishioner

View of the Hackney Brook at Hackney Downs about 1838. (From a water-colour drawing in the Tyssen Library, Hackney.)

nor a copyholder, and he was fined 20s., and 2s. 6d. for the value of the five sheaves of wheat he had taken. The account of these proceedings gives us a good glimpse into the Hackney of the past, which consisted of a few large and select houses surrounded by cornfields, which have now been forced back further into the country.

About this time we should have seen another landmark of old Hackney which has now disappeared, viz., the Hackney Brook. This passed along the western boundary of the Downs. It was once a stream of some importance, and, prior to the excavation of the New River in the seventeenth century, appears to have been connected with no other rivulet of any size. It could once be truthfully described as a 'still and rippling spring,' which 'steals its clear waters' home to the river Lea,* but in its later years became 'a ditch of running liquid filth, exceedingly noxious and highly prejudicial to the health of the districts through which it flowed.' The stream had its source in what are now the lakes of Clissold Park; thence, running easterly, it crossed Lordship Road, and continued flowing in that direction till Stamford Bridge was reached. Before the bridge was built, the brook was crossed by means of stepping-stones, which gave the place the name of Stone Ford, corrupted into Stamford.† From here it altered its course from east to south, and ran for some distance parallel to the road leading to Shacklewell. It then flowed past the Downs, and so made its way to the river Lea.

On the western part of the Downs was formerly an ancient spring which had never been known to freeze in the hardest winters. This has had to be filled in, but ample provision for thirsty travellers has been made by the erection of two memorial fountains—one to Mr. G. Gowlland, and another to Mr. Michael Young, both local celebrities.

The playground of the handsome school of the Grocers' Company at the south was formerly part of the Downs, and the enclosure of this portion of the lammas land led to considerable rioting in Hackney. The fence when erected was pulled down, and the playground perambulated by the inhabitants, who resented this encroachment on their rights. But once again the Lord of the Manor gained the day, so

* Author of 'La Bagatelle.'
† John Thomas, MS. 'History of Hackney.'

that the passing of the commons under municipal control is a decided benefit to Hackney.

In making Downs Park Road some years ago, which forms the northern boundary of the Downs, some Roman pottery was discovered, which is interesting in connection with the other relics of these early times brought to light in excavations in the neighbourhood.*

MILL FIELDS.

The Mill Fields take their name from some once famous corn-mills at Lea Bridge, now the property of the East London Waterworks Company. From a notice of sale in 1791, when an undivided moiety of the Hackney Waterworks and Corn-mills was in the market, we gather that the corn-mills were capable of grinding nearly 300 quarters per week.† Some five years later, in 1796, on January 14, there was an immense fire here, which, after burning with amazing rapidity for two hours, entirely consumed the mills, with a quantity of wheat and flour. About 3,000 quarters of this, the property of the Government, were also involved in the common destruction, which is supposed to have been caused by a flour-weigher leaving a lighted candle between two sacks of meal, one of which must have caught fire.

The two commons, north and south, are divided by the Lea Bridge Road, which was formed under an Act (30 George II.) for making a new road from Clapton down to the river Lea. The land for this was taken from the South Mill Field, for which no compensation was paid. It is probable that many enclosures have taken place on these lammas lands. In Rocque's map, 1745, they are shown as one continuous field, and though this cannot be taken as conclusive evidence, it was stated in the report of the committee of inhabitants (1810) that there were then rumours afloat that a considerable portion of these fields had been enclosed, but they had no evidence to prove the accuracy of the statements.

* There is a drawing of this pottery in the Tyssen Library at Hackney.
† From a book of newspaper extracts in the Tyssen Library, Hackney.

The original Lea Bridge was built of wood, with three arches or waterways, the centre of which was 68 feet between the abutments. After standing for seventy-five years, it was deemed insecure, and was rebuilt in 1820, the new iron bridge being 140 feet long. At the corner of North Mill Field, facing the river Lea, is the Jolly Anglers public-house, which appears from its internal vestiges to be upwards of 300 years

Lea Bridge Mills and River Lea about 1830. (From a water-colour drawing in the Tyssen Library, Hackney.)

old. It seems originally to have been built of brick, and must have been very small. The primitive building consisted of the bar, kitchen, cellar, and small bed-chamber over the bar, while the other parts have been subsequently added by different tenants.*

In the brick-field adjoining North Mill Field, which also has a valuable substratum of brick-earth, have been discovered

* John Thomas, MS. 'History of Hackney,' chap. iii., section 4, p. 42.

some remarkable fossils.   Some bones were found here, which
were pronounced by the late Professor Sir Richard Owen to
be those of the woolly-haired rhinoceros.   These remains of
an antediluvian inhabitant of our island had probably been
washed to this spot by some inundation.   Elephants' bones

The River Lea and the Jolly Anglers, Hackney Marsh, in 1850.   (From a
water-colour drawing in the Tyssen Library, Hackney.)

have also been unearthed from the soil, which must have
been trampled upon some thousands of years ago by these
huge monsters.†

North Mill Field is the probable site of a fierce battle
which took place in 527 between Octa, the grandson of

* 'Glimpses of Ancient Hackney,' by F.R.C.S., p. 214.

Hengist, King of Kent, and Erchenwin, the founder of the kingdom of Essex. This latter chief had revolted from the King of Kent, who made a powerful though unsuccessful attempt to win his subjects back to his allegiance. He convened an assembly of the wise men of his kingdom, and placed before them the alternatives of peace and war with the usurper of his power. The vote was unanimous, and upon the advice of one of the sages it was decided that 'the measures for war be immediate in their adoption and prompt in their application, so that the rebels, having no forewarnment of invasion, might be surprised, and the success of the expedition thereby rendered sure.'

The meeting-place was appointed at Hrofeceastre (Rochester), and galleys were ordered to be in readiness by the banks of the Medway; and at sunrise one morning in 527 the Kentish King boarded his galley-ships with 15,000 followers. The ships went straight to the Bay of Halviz (Woolwich), and there a deliberation was held. The King had two proposals for consideration, both having the same object, viz., the surprise of Londinbyrig (London). This he proposed to carry out in one of two ways: either to land his warriors on the west bank of the Ligan (Lea) near the ford (at Temple Mills), and march upon the city in two columns; or to disembark at the upper ford (formerly near Lea Bridge), and marching south to fall upon the city in that direction. He inclined to the latter proposal because he thought the people of London would be less prepared for an attack from the south, and this was the decision eventually arrived at.

So the galleys proceeded upon their way and anchored in the waters of Lochtuna (the lake formed by the Lea overlooked by Leyton). In the meantime the people of London were not idle. The deputy King had obtained information of the proposed expedition, and supposing, as the enemy had sailed to the Lea, that the attack would be from the northeast, he decided to march out to meet the foe. The route taken would be along Bishopsgate Street, Ermin Street (now

Kingsland Road North), and then turning north-east they would make their way to Clapton and thence to North Mill Field. Erchenwin, to prevent the advance of the enemy by any other route than the one he was taking, ordered an advance detachment to post itself upon an ascent from the marsh, so as to command a good view of the surrounding district. It had not long been stationed there before the chief in command observed the approach of a division of the enemy, and a battle at once ensued in which the Londoners were completely victorious.

Erchenwin and the main body then arrived in sight of the Ligan (Lea), and a desperate fight ensued between the full strength of both armies. Octa was conspicuous for his bravery, but when, sorely wounded, he was compelled to retreat, the rest of his followers fled and were slaughtered by the conquering Londoners. On the following day the victorious East Saxons returned to their capital, having thrown off the yoke of the King of Kent ; and so ended the Battle of Hackney.*

* Abridged from the 'History of Hackney' (manuscript), by John Thomas, 1832.

CHAPTER XVIII.

LONDON FIELDS—WELL STREET COMMON—HACKNEY MARSH.

London Fields.

THESE Fields are the nearest open space on this side of London to the city, and as such have been subjected to very rough treatment in their time. The vicissitudes through which they have passed have been more remarkable than those of any other of the Hackney commons. At the time when Hackney boasted of the patronage of many wealthy citizens these Fields seem to have been chiefly devoted to sheep-grazing. In Rocque's map (1745) the wide thoroughfare at the south-west of the fields is shown as 'Mutton Lane.' At the present day two thoroughfares leading off the fields—Sheep Lane and Lamb Lane—preserve the memory of the former frequenters. The sheep seem to have departed to 'fresh fields and pastures new,' but it is not so many years since the marsh drivers were able to let the land during the close-time of lammas lands to a cow-keeper to put on a certain number of cows. With the increase of population, however, the use of the Fields became very much extended, and in course of time the surface was worn so bare that the four months of close-time were not sufficient to enable the grass to grow again and establish itself. As a consequence the Fields 'became in dry weather a hard, unsightly, dusty plain, with a few isolated tufts of turf, and in wet weather a dismal impassable swamp.' In the evidence given before the Select

Committee of the House of Commons on the Metropolitan
Commons Act (1866) Amendment Bill, it was stated that it
was 'the run of the riff-raff and vagabonds at the east of
London during the whole season, night and day.' The
most dissolute practices were carried on, cockshies were put
up, and the scenes were very similar to those at a common
fair. On Sundays the Fields furnished a platform for
itinerant lecturers, who only came here to provoke discussion. Altogether London Fields were not a credit to
Hackney, and the Hackney Vestry prepared a scheme for
dealing with them; but after Parliamentary notices had
been lodged, the idea had to be abandoned owing to the
opposition of the inhabitants. If this scheme had been
carried out, the acquisition for public use would have been
effected without expense to the ratepayers, either by selling
the brick earth, which forms a valuable substratum, or by
selling a belt round the Fields for building purposes so as to
pay for the remainder. But the disgraceful scenes on this
open space are now things of the past, and as we have
mentioned before, by systematic fencing, it is now possible
to see a vestige of green turf in place of the bare surface
once presented.

It is an open question whether any encroachments have
taken place on this common. The Court Rolls record an
attempted pilfering in 1809, which was discovered in time
and the offender punished. The entry runs as follows:
'And the homage aforesaid further present an encroachment made by William Parker Hamond in suffering a part
of London Field called "the nursery" to be enclosed, and
thereby depriving the tenants of this manor and the
parishioners of Hackney from the benefit they usually had,
and of right were and are entitled to have, of the herbage
thereof in common with the rest of London Fields; and
of digging up or causing to be dug up the brick earth
therein and of permitting and suffering horses, carts, and
other carriages going over London Fields to that part
thereof called "the nursery" aforesaid to fetch, take, and

carry away the said brick earth therefrom, and thereby destroying the herbage of London Fields aforesaid, contrary to all justice and reason, and we amerse the said W. Parker Hamond for so doing in the sum of £1,000.' The nursery was on the west of the Fields and extended almost to Queen's Road. Latterly it was known as Grange's Nursery. It seems very probable that the roads between the north of

The Old Cat and Mutton, London Fields, about 1830. (From a sepia sketch in the Tyssen Library, Hackney.)

Lansdowne Road and the common are built upon a former portion of London Fields.

The modern public-house at the south corner of London Fields is on the site of an ancient tavern, dating back at least to 1731, for a newspaper cutting dated June 14 of that year describes how 'yesterday morning a fire broke out near the Shoulder of Mutton alehouse in London Fields near Hackney.' Its present name, the Cat and Mutton, may

23

possibly be a corruption of its older title, the Shoulder of Mutton, which appears to be its original designation.   In 1798 we find the ' Cat ' added as a prefix to the sign, in which year ' the public-house called the Cat and Shoulder of Mutton ' is announced for sale.   As such it gave its name to the lower part of London Fields, or *vice versâ*, for this portion bears some resemblance to a shoulder of mutton.   The following extracts* from the press of the day will introduce us to London or Shoulder of Mutton Fields in a new light, viz., as a resort of highwaymen :

' 16 *Dec.*, 1732.—A few days since a tradesman in the Ward of Farringdon Without was attacked near the Shoulder of Mutton by two fellows, who robbed him of his money and pocket-book.'

About the same time : ' The watch is ordered to begin their patrol at five o'clock in the morning on account of Mr. Baxter being robbed on Wednesday, at that hour, by two fellows, who started out on him from behind the Watch house in the Shoulder of Mutton Fields.'

' *April*, 1751.—William Flora sent by the master of the Rochester Hoy to receive £36 at the Two Blue Posts at Hackney, on his return was robbed by two footpads in the Shoulder of Mutton Fields who made off with the booty.'

But London Fields has afforded considerable sport to others than footpads and highwaymen during its lengthy career.

' On Friday, Sept. 24th, 1802, a cricket-match was played on London Fields for the substantial stake of 500 guineas between eleven gentlemen of the London Fields club, and eleven gentlemen of Clapton.   Although the betting was 5 to 4 on the former at starting, they were defeated by an innings and 49 runs.'

Here is another interesting account of a contest of a different character.   It is headed, ' Extraordinary Pedestrianism ': ' A match which has long been depending, was decided on Thursday afternoon (July, 1813) in London Fields

* From a book of newspaper cuttings in the Tyssen L brary, Hackney.

between a man of the name of Thos. Dudley, aged 74, who has acquired great celebrity in the sporting world by running on stilts, and a sailor who is equally noted, for a short heat. About 3 o'clock they started, for a considerable sum, the wager having been previously made that the sailor was to give the veteran 50 yards at starting, and that the distance which they should run should be 100 yards. The old man came in, leaving his antagonist at the distance of 30 yards, to the no small amusement of a great concourse assembled on the occasion. The old man performed the distance in 10 seconds.'

London Fields has also provided a drilling-ground for soldiers. In June, 1798, there was a brilliant gathering here ' when the 1st and 2nd regiments of the Tower Hamlets Militia were reviewed, the former in London Fields, and the 2nd at Bethnal Green, by the Duke of York, the Duke of Gloucester, Marquis Cornwallis, and the Earl of Harrington. They went through their manœuvres with great credit.'

Later on, in September, 1804, the loyal Hackney Volunteers paraded here with every requisite for marching at a moment's warning.

At the corner of Tower Street, which leads away to the north-east from London Fields, formerly stood a white house with a tower-shaped wing overlooking the Fields. Its site occupied that of the schools at the back of St. Michael's Vicarage. A most eccentric man once lived here, who led a very retired and secluded life, and had a particular antipathy to doctors. A child of his died, and as no medical man was in attendance, a coroner's inquest was demanded. For several days he resisted all intrusion, barricaded his house, and was seen at night-time walking up and down the flat roof with a loaded gun. This was before the police were established. It was some time before an entry was effected, and the man secured, when the inquest was held in due course on the decomposed body of the child.*

* 'Glimpses of Ancient Hackney,' p. 15.

It was from Tower House, at the corner of London Lane, which gave its name to Tower Street, that Milton, in 1656, took his second wife, Catherine Woodcock, the daughter of Captain Woodcock. She died in child-bed about ten months after their marriage, and Milton wrote his twenty-third sonnet, so much criticised by Johnson, to her memory.

Milton had gone quite blind three years before he married his second wife, so that he could not fully appreciate the full beauties of the Hackney of his age. But Milton was not the only celebrity that came a-courting to Tower Street, for it is said that Daniel Defoe, who was a resident at Stoke Newington, walked across the intervening fields, down the country lanes, to see his future wife here. Many of his children were baptized and buried in Hackney Church.*

To the north of London Fields, where Navarino Road now runs, were formerly Pigwell Fields, called variously Pig's-well (Robinson), Pyke-well, or Pit-well. In these fields were certain land-springs (of which there are many in this district) which were collected into a well-head or conduit. This water was then carried by conduit pipes to another conduit at Aldgate, and formed the only water-supply for some hundreds of years for this side of London.†

Well Street (or Hackney) Common.

This common is situated at the south-east corner of Hackney, and adjoins Victoria Park. Well Street, from which it takes its name, does not actually form one of its boundaries, although it is probable that at one time it did, and that the intervening land has disappeared, as common lands have an unfortunate habit of doing. Well Street naturally suggests a well, and the difficulty of locating it has puzzled more than one topographer. This well was situated by Cottage Place, and is believed to have been co-eval with the palace of the Priors of St. John of Jerusalem. Possibly it may have been partially a mineral spring, or, at any rate,

* 'Glimpses of Ancient Hackney,' p. 15.   † *Ibid.*, p. 240.

from its contiguity to a monastic establishment, have had a special holy reputation, and hence the road to it would naturally be named after it.*

The estate which partially surrounds Well Street Common, known as the Cassland estate, belongs to a very important charity, which has done a great deal towards the improvement of the neighbourhood. The trustees have swept away a wretched village of houses—or, more properly, hovels—which had received the name of Botany Bay. It is said that this peculiar appellation was given to it because so many of its inhabitants were sent to the *real place*, not because of their good deeds, it is feared. The charity was founded by Sir John Cass, a worthy Alderman of London, who died in 1718. His father, Thomas Cass, was carpenter to the Royal Ordnance, and the large fortune which he had acquired descended to Sir John, who built two schools near the Church of St. Botolph, Aldgate, in addition to other buildings near them. He devised the whole of his estate, after the death of his wife, for the purpose of providing a free dinner daily for the charity children attending these schools. He died with the pen in his hand, without having fully completed the statement. This led to very lengthy litigation, which was not settled for thirty years after his death.† From investigations which have recently been made, it appears that a carpenter—Mr. Cass—lived where Lauriston Road now runs. He had an only son, who lived and died in the same house. In the minutes of the Select Vestry, the father, Thomas Cass, is first mentioned as being present on April 6, 1686. Subsequently, in 1699, his son appears as a vestryman. No doubt the land at Hackney was very cheap at this time, so that father and son, being thrifty people, they were able to acquire an extensive property.‡

At the southern corner of the common stands the French Hospice, a large building of dark-red brick with stone dress-

---

* 'Glimpses of Ancient Hackney,' p. 179.

† John Thomas, MS. 'History of Hackney.'

‡ 'Glimpses of Ancient Hackney,' p. 168.

ings, standing in extensive and well-timbered grounds, building and gardens alike being an ornament to the neighbourhood.   The hospital stands upon the site of the garden of a
former Rector of South Hackney*—the H. H. Norris, whose
memory is still treasured.   But even Rectors are not above
suspicion, for the committee entrusted with the inspection of

French Hospice, Victoria Park.

the lammas lands reported in 1810, ' There is a rumour in
circulation that some of the land withinside of the Rev. Mr.
Norris' fence is part of Well Street common '; but no
evidence of the fact had been before them.   Perhaps the
hospital, then, stands on what was formerly part of the
common.   Its history is a very interesting one.   Its institution is an outcome of the persecution of the French

* ' Glimpses of Ancient Hackney,' by F.R.C.S., p. 171.

Protestants, many of whom fled to England after the Edict of Nantes was revoked. By dint of carrying on their native industries in England, many of these refugees managed to amass considerable fortunes, and were able to lend a helping hand to their less successful compatriots. One of the results of this charity was the French Hospital, originally founded in Old Street, St. Luke's, by Monsieur de Gastigny, who was Master of the Buckhounds to the Prince of Orange, whilst in Holland, and accompanied him to England on his corona- tion as William III. He bequeathed in 1708 £1,000 towards founding a hospital for distressed French Protestants, and by means of other benefactions the trustees were enabled in 1716 to purchase some land in the parish of St. Luke, upon which a building was erected capable of accommodating eighty persons. A royal charter of incorporation was granted by George I., under the title of 'The Hospital for Poor French Protestants and their Descendants residing in Great Britain.' Its early days were times of prosperity, for owing to increased support the buildings were enlarged, and in 1760 had 234 inmates. But as time went on, the directors, owing to the death of many benefactors, were forced to go somewhere where land was cheaper, and the present site was determined upon. The new building was designed by a descendant of a Huguenot family—Mr. Roumien—and provides for forty men and twenty women. No one is admitted under sixty, and the gates are closed against married couples. The ranks of the inmates are chiefly recruited from the weavers of Bethnal Green and Spitalfields.*

At the western corner of the common are the buildings of another ancient charity—viz., Monger's Almshouses. These were founded under the will of Henry Monger (dated April 17, 1669), a former inhabitant of Hackney, who gave 'a piece of land in Well Street, for six almshouses to be built upon it with brick, and £400 towards the said buildings.' The buildings were intended for six poor men of the parish of Hackney, who could have their wives with them if married,

* *Windsor Magazine*, October, 1895.

but if the man died, the widow had to leave, and so lost her husband and home at one stroke. Attached to the almshouses is an annuity of £12, arising out of land in Hackney Marsh, £9 of which is given in quarterly instalments to the

Monger's Almshouses, erected under the will of Henry Monger, dated 1669.
(From a sepia sketch in the Tyssen Library, Hackney.)

inmates, and the remainder goes for repairs. The election of the almsmen is in the hands of the trustees of Sir John Cass's charity, on the recommendation of the Rector and churchwardens of South Hackney.*

* Robinson, ' Hackney,' vol. ii., p. 374.

The small strips in Dalston Lane and Lauriston Road (late Grove Street) present no special features. They were enclosed and laid out at a cost of £400, and handed over to the Hackney Board of Works in 1884, who have maintained them since that date.

HACKNEY MARSH.

We now come to the last and the largest of the open spaces of Hackney—viz., Hackney Marsh. The title is not euphemious, but it has the merit of antiquity, and so certainly ought to be preserved. It is a large area of flat meadow-land lying on the eastern boundary of Hackney, and intersected and skirted by the river Lea and its tributaries. It is 337 acres in extent, and is at a distance of 3½ miles from the Royal Exchange. The land, like the majority of the Hackney commons, was formerly subject to lammas rights, and so long as these lammas rights were maintained the land could not have been built upon; but at any time an arrangement could have been made between the Lord of the Manor and the severalty owners, and the owners of the lammas rights, to convert the marsh into freehold building land. Forming as the marsh did a splendid air space between the portions of Hackney which were built upon and the rapidly increasing outlying districts between Stratford and Leyton, it became evident that the marsh must be secured for the health and recreation of the people of London, and the Hackney District Board, by resolution in May, 1889, asked the London County Council to purchase or rent the marsh. Meanwhile, in November, 1889, a somewhat trifling incident led to more decisive steps being taken. The Rev. E. K. Douglas, of the Eton Mission, Hackney Wick, brought to the notice of the Metropolitan Public Gardens Association the fact that the lads of a football club connected with the mission had been ordered off the marshes by the Drivers, who had proceeded to carry off their goal-posts. Mr. Douglas was invited to attend the meeting of the association, held on December 4, 1889, when he asked that steps might be taken to get permission for his boys to play football on the marshes. The association, however, decided that the right course would be to take up a larger field of enterprise altogether, and to make application to the Board of Agriculture to grant a regulation scheme under the Metropolitan Commons Act, 1866, by making use of the powers conferred in the little-known and little-used Metropolitan Commons Amendment Act, 1869 (32 and 33 Vict., c. 107), whereby twelve or more ratepayers of the parish in which a Metropolitan common lies can present a memorial to the Board of Agriculture asking for a scheme.

On March 5, 1890, the Association decided to incur expenditure on the prosecution of a scheme for the regulation of the marsh, which would place it under the control of the London County Council, but which would not necessarily entail the purchase by the Council of any existing beneficial interests unless they were proved to be detrimentally affected by the putting in force of the powers which the regulation scheme conferred.

In September, 1890, the Board of Agriculture issued a draft scheme on the lines mentioned for the regulation of the marsh, and on October 31, 1890, signified to the Association its intention to hold an inquiry, which was opened at the Hackney Town Hall on December 1, 1890, by Mr. George Pemberton Leach, Assistant-Commissioner to the Board of Agriculture.

The acquisition in connection with this scheme, however, was adjourned by the Commissioner in order to afford the London County Council an opportunity of buying the marsh. This opportunity the Council took, offering £50,000 for the property, £10,000 of which was to be found by the Hackney District Board. This offer was refused, and the matter dropped for the time; but negotiations were speedily renewed, with the result that the Lord, the commoners, and other owners of rights, combined for the purpose of selling the marsh, and agreed to take £75,000, which, finally, was the amount paid. Of this the London County Council contributed £50,000, the Hackney District Board £15,000, the Lord of the Manor £5,000, and private subscriptions £5,000. The land was finally transferred, free of all its previous existing rights, to the Council under the London Open Spaces Act, 1893, and a formal ceremony to dedicate it for ever to the use and enjoyment of the public took place on Saturday, July 21, 1894. As already stated, the area of the marsh is very considerable, and, owing to its flatness, it has proved a most valuable acquisition to the playgrounds of London, being equally suited for cricket in the summer and football in the winter.

The only drawback to the full enjoyment of the marsh was the periodical flooding to which it was subject. To remedy this, four new cuts were formed to take off the severe bends of the Lea, and so enable the more rapid discharge of flood-water. The old channels were retained, thus forming islands, which by suitable planting have been made pleasing features of the river. In connection with one of these cuts a bathing-pool has been formed, which cannot, however, be used till the Lea is purified. Further, a low flood-bank and gravelled promenade parallel to the Lea were made, and also a small bank alongside the waterworks drain between the Temple Mills and Homerton roads to prevent flood-water from backing up from the south.

During the carrying out of these works, the marsh was visited by a severe flood, which it is to be hoped will be the last. Besides retarding the progress of the laying-out, the rising waters did considerable damage, and the floating plant had to be rescued by means of boats. We may fairly prophesy, however, that the floods of Hackney Marsh are now things of the past.

The buildings around the marsh and the river Lea have some very interesting historical associations. One of the branches of the latter, known as the Mill River, or the Lead Mill River, supplied the water-power to the Temple Mills, so called because they were originally erected and owned by the Knights Templars. After the dissolution of that Order, they became the property of the Knights of St. John of Jerusalem. There were three water-wheels, and by the various adaptation of the machinery thus set in motion, corn was ground, and trunks of trees were bored to form water-main pipes, some of which are still found in digging along the main thoroughfares. Points were also ground to pins and needles, as many as 120,000 needles being pointed in a day. The rough needles were then sent down to Worcester-shire, where the eyes were made and the steel tempered,* and they were returned here to receive the finishing polish. The mills also received royal patronage, for Prince Rupert, grandson of James I., after his retirement from military duties, spent the greater portion of his time here in chemical experiments. As one of the fruits of these labours, he invented a composition for making guns, called Prince's metal, and the guns were bored at these mills after they had been cast. When, however, he died in 1682, the secret of its manufacture died with him.†

The Temple or Rochott Mills were purchased by the East London Waterworks Company in 1834.

This company also possesses extensive works at the north of the marsh, which are partly erected on what was formerly common ground. Under the provisions of their Act of 10 George IV., the company were empowered to take 12 acres from the marsh for extending their works, but by a general clause they were entitled to take as much more as they wanted. When the Bill was before Parliament, it naturally excited some alarm among the tenants of the manor, as will be gathered by the following extract from the Court Rolls :‡

'The homage . . . further present that a Bill is now before Parliament to empower the East London Water Company to take for the purpose of making reservoirs, etc., part of the common lands belonging to this manor called

---

* John Thomas (MS. 'History of Hackney') says these works were carried out at Lea Bridge Mills, but this is not the general view.

† Robinson, 'Hackney,' p. 67.

‡ Book xxii., date April 28, 1829.

Hackney Marshes, that the said company state that they shall only require about 12 acres of the said common, but . . . according to the plans and book of reference . . . the several lands amount to upwards of 77 acres, all of which the homage conceive the said company would be empowered to take should they think fit.'

Another clause in the Bill to the effect that the money to be paid by way of compensation should be applied in aid of the poors rate of the parish of Hackney also raised a storm. From the records of the Water Company, it appears that the total amount of land thus taken from the marsh was a little over 20 acres, for which a sum of £750 was awarded as compensation. This amount was paid into the Bank of England to be laid out in the purchase of lands to be added to the' lammas lands.*

The White Hart, near the Temple Mills, is a very ancient hostelry. It is said to have been built in 1513 (*temp.* Henry VIII.).† A 'toll-bar,' one of the few survivals of that kind in London, still levies a tax of twopence on every horse that passes. In the gardens of the White Hart has stood' for many years a large pollard poplar, the spreading branches of which used to support a capacious platform, approached by a flight of steps, which was capable of seating some twelve to sixteen persons. This has now been broken down.

But the most interesting fact in connection with this part of the marshes is that which occurred in the ninth century, when the Danish Vikings sailed up the Thames and ascended the Lea, penetrating as far as Ware, where a fortified camp was built, and the adjacent villages were sacked. It must be borne in mind that the ships of the Danes were small and nearly flat-bottomed, without much keel, so that the narrow and shallow Lea was quite sufficient for their navigation. The citizens of London, who had turned out to dislodge the foe, were repulsed with heavy loss after a fierce battle, where-

* Robinson, 'Hackney,' p. 66.
† 'Glimpses of Ancient Hackney,' by F.R.C.S., p. 166.

upon King Alfred ordered channels to be cut by which the current of the Lea was diverted and the depth reduced, according to Stowe's Annals, ' soe that where shippes before had sayled, now a smal boate could scantily rowe.' The result was that the Danish ships were left so much aground that their return to the Lea mouth and so into the Thames was rendered impracticable. In the vicinity of the White Hart traces of the channels cut are still in existence, and an ancient boat, supposed to be one of the Danish canoes, excavated on the marsh, is preserved in the British Museum. This stratagem on the part of King Alfred not only accomplished his purpose, but was the means of conferring lasting benefit upon the marshes. What was formerly moist and spongy ground was converted into dry fields of fertile meadow.* This is not the only relic of ancient times, as in 1757 a part of a Roman stone causeway was found on the marsh, together with some Roman coins, which go to prove that there was a Roman highway across the marshes, probably the great thoroughfare from London to Essex. These discoveries were made in the course of widening and deepening the channel of the mill tail of the Temple Mills. In addition to the causeway and coins there was also found a stone coffin, which lay from east to west, some 4 or 5 feet below the bed of the channel. Being firmly sunk in the bed, it was left there, as the best foundation for the new superstructure about to be erected.† In an account of this discovery‡ it was stated that there were found ' an urn full of Roman coins, some in high preservation, from Julius Cæsar to Constantine the Great, with several medals, a stone coffin with a skeleton therein, measuring 9 feet 7 inches long, the inscription on it unintelligible.' It was added, that in removing the old foundation a vault was discovered, in which were several urns but quite imperfect; and what is very remarkable, the vaults for centuries past are supposed to

* Robinson, ' Hackney,' p. 27.
† John Thomas, ' History of Hackney, chap. ii., sec. 2, p. 24.
‡ Gentleman's Magazine, November, 1783.

have been 16 feet under water. From these discoveries it may fairly be surmised that the site of the Temple Mills had anciently been a place of burial on the roadside, or a little distance from the great road which ran across the marsh.

To the left of Sydney Road, formerly known as Wick Lane, was once a beautiful upland field, known as the Hilly Field, a gradually rising ground until it abruptly sloped down to the marshes. This is now occupied by a Board school, also extensive coal-sidings and the railway, while what was still left of it to the south-west is covered by small property. On the right were first the extensive gardens of the last houses in High Street, then a meadow and private gardens and grounds, in the centre of which stood Wick House, latterly the residence of T. Ballance, but some while before that of Levy Smith, whose grounds led down to the silk-mills belonging to him, situated on Hackney Brook, having a mill dam there and worked by water-power. Silk-Mill Row, which by an inscription on a stone tablet was refronted (perhaps rebuilt) in 1820, was a row of cottages for the workpeople. Two branches of the trade were carried on at these mills, first throwing the silk, *i.e.*, preparing it from the raw state, and thus fitting it for weaving. There were latterly two steam-engines employed in place of the water-power, by which upwards of 30,000 spindles were set in motion, and between six and seven hundred men, women, and children were employed. When the manufacture of silk was removed from here, horsehair and flock were dressed and manufactured. This Wick House was for some years the residence of Colonel Mark Beaufoy, F.R.S., who was an authority on nautical and hydraulic matters, and the ancestor of the Beaufoys of South Lambeth. From the manuscript history of Hackney by John Thomas it appears that a nautical clock was stolen from Colonel Beaufoy's observatory in 1806. This clock had four hands, by means of which the distance a ship had sailed could be told from 150 miles down to single yards. This curious machine was put in motion by a log-line, and was considered a great discovery in navigation. The closing career of Wick House was that of a gentleman's school till it was pulled down in course of time. We also gather from a newspaper cutting of 1805 that near this spot a serious coach accident took place. The coach was passing down Hackney Wick, laden with passengers, when a heavy thunderstorm came on. The coachman dismounted to lead the horses for safety, as it was pitch dark, but he missed the usual track, and the

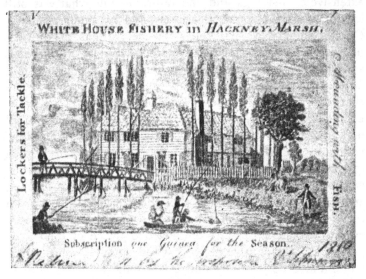

Season Ticket for the White House Fishery, Hackney Marsh, 1810. (From the original in the Tyssen Library, Hackney.)

vehicle, coming on the edge of a precipice, was overturned and lodged in the wash. Many of the passengers were hurt, but the coach was almost broken to pieces, although the horses were saved.

The White House public-house, which is situated in an isolated position in the centre of the marshes, possesses a museum, in which may be seen some interesting specimens of rare birds, which were formerly observed along the banks of the Lea. The rarest of all is a fine specimen of the cream-coloured courser, a native of Barbary and Abyssinia, of which only three or four have been taken in this country during the last century. The specimens of fish include a jack of 25 lb., trout 11½ lb., barbel 13½ lb., chub 7½ lb., carp 11 lb., bream 5¾ lb., and an eel of 4 lb. This latter, however, is altogether eclipsed by a monster fresh-water eel weighing 22 lb. 7 oz., which was caught in 1766. At the period when these were principally taken the White House Fishery was at its height, and had no less than 150 annual subscribers. In the Tyssen Library at Hackney may be seen some of the elaborate cards of membership, as well as a copy of 'The Angler's Companion and Guide to the White House Fishery,' containing a view of the White House and map.

Dick Turpin frequently made this house his home, and was from time to time in concealment here after some of his predatory excursions. In fact, this part of the world seems to have been a favourite haunt of highwaymen, the marshes especially, owing to their solitude, being very often the scenes of daring robberies.

Some law-breakers of a different kind are associated with the marsh. In 1682 the Rye House conspirators had prepared blunderbusses, muskets, and pistols, which were to be brought by the river Lea from the marsh almost to the gate. These arms they designated 'swan-quills, goose-quills, and crow-quills.' They also had ordered powder and shot by the appellation of 'ink and sand.'*

In the rainy season the footpaths across the marshes were often impassable, the Lea overflowing its banks considerably. One of the most serious of these floods took place in January, 1841, when, after a rapid thaw accompanied with heavy rain-showers, the marshes and the low-lying lands on both sides of the river Lea presented a large sheet of water from Stratford to Tottenham Mills. The accumulation of the water on the marsh caused great injury to the railway, the banks being undermined in several places, so that the running of

* John Thomas, MS. 'History of Hackney,' p. 29.

24

the trains was stopped for some time. Communication was kept up between Homerton and the White House in its isolated position on the marsh by means of boats, and many houses in the lower part of Hackney were. flooded. A still older record in 1775 tells us that the waters were so much out on Hackney Marsh that the inhabitants were obliged to live in their upper apartments and have their provisions brought to them in a boat.

In 1766 an Act was passed for the further improvement of the river Lea by means of canals or cuts in different parts of its course from Hertford to the river Thames. Among these it was provided that a cut should be formed, leading from the proper channel of the river, between Lea Bridge and the buildings belonging to the Hackney—now the East London—Water Works, close by the Pudding Mill stream. This Hackney cut, which passes through the marsh, was completed in 1770, and was opened on September 17, when many barges and boats immediately passed up to try if it were navigable, and it proved to answer extremely well.*

By a later Act, passed in 1850, for the further improvement of the navigation of the river Lea, power was given to the river trustees to make certain new cuts, which involved the taking of some more of the lammas lands, for which the sum of £449 1s. 6d. was awarded as compensation.

In 1641 a divine named John Thomas, in an appendix to a small pamphlet entitled 'The Booke of Common Prayer Vindicated,'† called attention to the discovery of a ' base sect of people called Re-baptists in Hackney-marsh neere London.' The account of the proceedings must be given in the author's own words : ' About a fortnight since, a great multitude of people were met going towards the river in Hackney Marsh, and were followed to the water-side, where they all were baptized againe, themselves doing it to one another, some of which persons were so feeble and aged that they were fayne

* *Gentleman's Magazine*, September, 1770.

† There is a copy of this curious publication in the Tyssen Library at Hackney.

to ride on horsebacke thithere. This was wel observed by many of the inhabitants living there abouts and afterwards one of them christened his owne child and another tooke upon him to church his owne wife, an abominable act, and full of grosse impiety.'

The grazing rights have always been of considerable value. All parishioners that paid parish rates in house or lands to the amount of £10 had the right to depasture cattle upon the marshes upon payment of fixed annual charges. These charges have of late years been applied to improving the approach road and the marsh generally, which not many years ago were impassable at wet seasons.

The marshes, with their great extent of level surface, afford exceptional facilities for cricket and football. The Earl of Meath, at the dedication ceremony, declared them to be the most magnificent playground in the world. Before their acquisition for public use, they were the rendezvous on Sunday mornings of a peculiar crowd, made up of gunners, rabbit-coursers, mouchers, and ' broken sports,' all of whom were particularly welcomed by the local publicans, whose receipts were considerably swelled by their presence. On these meadows was established a rival Hurlingham, sparrows at a penny apiece taking the place of the aristocrat's pigeon. Occasionally the victims would be larks at 2d. or starlings at 3d., and on special days even a rabbit. Judging from the performances, the sportsmen of Hackney Marsh were hardly likely to clear the country of game. A writer in the *Pall Mall Gazette** witnessed a match between two rival shots : ' They bought a rabbit and gave it ten yards law, and both fired and missed it. A black retriever dog brought the wretched creature back unhurt, and the performance was repeated, the dog once more carrying it back. They were actually going to try the same thing again, when the writer picked up the rabbit and broke its neck. Rather to his surprise, they good-naturedly agreed to cry their bet off, as the beast had had enough ; he had been netted from ferrets, been coursed twice

* December 15, 1893.

24—2

that morning, and had eight shots fired at him, and so was
held to have yielded a fair share of sport.

The White House, Hackney Marsh.

' It would be a libel on the marsh gunners to adduce this
as an average example of their prowess, but it was an actual

occurrence, and illustrates the drawn-out cruelty and mutilation incidental to this kind of Sunday morning recreation.'

But rabbit-coursing was perhaps even more popular, and was extensively carried on here in all its cruelty and brutality. Such scenes remind us of an event which happened in 1791, when one ' Friday afternoon a bull was baited near Temple Mills, upon Hackney Marsh, on which occasion it is judged that, on a moderate computation, upwards of 3,000 people had assembled by four o'clock. The bull was brought to the stake soon after that hour, and after twelve dogs had run at him, he broke loose. A strange scene of uproar and confusion ensued, hackney coaches and jockey-carts driving furiously in every direction, horsemen riding against each other, many hundreds of people tumbling one upon another, and the rest running different ways to avoid the fury of the enraged animal, which tossed a girl about nine years old, who fortunately, however, received no material hurt. The bull was again brought to the stake, and worried by eight more dogs, one of which attacked him at a time. The bull was now a third time brought to the stake, and after being again baited, was led from the ring. . . . The bull being again brought to the stake, was baited till the approach of evening, when he was wickedly let loose among the crowd, which by this time had greatly increased, by a concourse of people of all descriptions, not only from London, but the adjacent villages. While at liberty, the bull tossed an elderly man, but he received no injury.'*

On the same afternoon a desperate prize fight was fought between a chimney-sweep and a butcher, as a rival attraction to the bull-baiting.

A more legitimate contest took place in 1737, when a famous race was run, or rather swam, from Tyler's Ferry to the bridge on the marsh by two horses. There was pretty good sport, the winning horse coming in first by two lengths.†

* From a book of newspaper extracts in the Tyssen Library, Hackney.
† Robinson, ' Hackney.'

The marsh has now settled down to a more peaeeful and uneventful existence, but one incident has happened which deserves to be recorded. On the day of the dedication ceremony, two of the constables employed here were presented with certificates for saving life. They had jumped into the deep waters of the Lea with their uniforms on to rescue some persons from drowning who had accidentally fallen in. It was fitting that they should receive the rewards of their gallantry on a day which will be always memorable in Hackney as the occasion when this magnificent playground was dedicated to public use for ever.

CHAPTER XIX.

HAMPSTEAD HEATH.

THIS fine open space of 240 acres is situated on the summit of one of the highest hills round London, and the lights of London, as seen from its broad expanse, have occupied the attention of poet and artist alike. Perhaps there is no spot around the Metropolis which is more identified with the holiday life of the Londoner than the heath. To a Cockney ''Ampstead 'Eath' is *par excellence* the place to spend a happy day. He seeks recreation near at home, and he finds here more liberty than in the trim elegance of the parks. It is irksome for him to be ordered to keep off the grass, or to be told that his dog must be led with a string or some other suitable fastening; and so Bank Holiday sees even this huge recreation-ground of the northern heights uncomfortably crowded. At these times the by-laws are relaxed, and some idea of the scene the heath presents can be gathered by the illustrations given. As many as 100,000 have been known to come to the heath on a Bank Holiday, and on one occasion this popularity was the indirect cause of a serious accident. On Easter Monday, 1892, two women and six boys were suffocated by the dense crowd descending the stairs at the railway-station; but, fortunately, there has been no repetition of this, and this regrettable contretemps has not caused any diminution in the number of visitors.

The views from Hampstead Heath have often been compared with those of Richmond Hill, and certainly no other

place so near London can boast of so varied or extensive a
prospect. Goldsmith has described the view from the top
of the hill as finer than anything he had seen in his wander-
ings abroad, and yet he wrote the 'Traveller.' Apart from
the beauties of the heath, many of the approaches to it are
well worth a visit, especially from the south, in the neighbourhood of Christ Church. The many fine avenues of trees
have led Hone to call Hampstead the 'place of groves.'*
One of the finest of these is that called Judges' Walk or

Side-Shows, Hampstead Heath, on Bank Holiday.

King's Bench Avenue. The story is that when the plague
was raging in London, the sittings of the Courts of Law
were transferred to Hampstead, and that the heath was
tenanted by gentlemen of the wig and gown, who were forced
to sleep under canvas, like so many rifle volunteers, because
there was no accommodation to be had in the village for love
or money.† Mr. Baines, in his recent work, quotes some
interesting correspondence on this point. It appears that

* Hone, 'Table-book,' p. 810.
† Cassell's 'Old and New London,' vol. v., p. 459.

Sir Francis Palgrave found by accident in the Record Office the formal account of the assize which was really held under these old trees in 1665, but unfortunately the reference has been lost.*

The perambulation of the heath is best commenced from Hampstead Heath Station, where immediately upon our entrance we find ourselves in a well-laid-out garden, with a background of bright flowers. This garden occupies the site of one of the New River Company's ponds, which has been

Judges' or King's Bench Walk, Hampstead Heath.

filled up. The adjoining round house belongs to the same company. No one can help admiring the ruggedness of the scene before him. In front is the wide-stretching heath, formed by the sand-digging into a series of hills and dales. Here and there a golden sand-bank, which has not yet been covered with a green carpet of turf, forms a pleasant relief in the landscape, whilst all around there are belts and groups of trees to form a leafy background to the whole. An avenue of willows at the commencement of East Heath Road leads

* Baines, 'Records of the Manor, Parish, and Borough of Hampstead,' 1890, p. 116.

us past the donkey-stand, which is one of the institutions of Hampstead Heath, to a higher spot where the scene is again varied. In the immediate distance can be seen across the valley the noble mansion of Kenwood Towers, whilst behind the hill are Highgate Church and the classic dome of St. Joseph's Retreat. On the left, stretching past the viaduct, is a perfect forest of timber, and, as a contrast, on the other side the unbroken lines of bricks and mortar, from which many a tower and steeple rises to the sky. Still ascending, we come now to the high ground by the Vale of Health, and look down on what might still be a pleasant place if we could only shut our eyes to the hideous taverns which force themselves into notice. We have risen now above the houses, and can see right across London; but the culminating point is reached by Jack Straw's Castle, where there is nothing but open country stretching forth at our feet. Advantage was taken of the high ground and the openness of the country to erect a semaphore telegraph on the ground to the west of this point, which still bears the name of Telegraph Hill. This was the first in the line of communication between Chelsea Hospital and Yarmouth.* Up to this point Hampstead Heath has been indeed charming, but as nothing in comparison with what we now see. The heath here is much wilder, and covered with a wealth of gorse and bracken, while the trees are more varied, stately firs and graceful birch being for the first time seen. Much has been written about the beauties of the prospect from the elevated ridge leading from Jack Straw's Castle to the Spaniards, and every word of praise is well deserved. There is an unbroken view extending to Finchley, Hendon, Harrow, and even Windsor Castle may be seen on a fine day. In a hollow close by the Whitestone Pond is the ancient pound, enclosed with a brick wall, dated 1787. Crossing now to North End, and exploring the West Heath, we have finished our perambulations, and are able to consider the steps which eventually led to the preservation of this fine open space.

* Park, 'Hampstead,' p. 259.

The growing popularity of Hampstead as a residential district was perhaps one of the first reasons to make the past Lords of the Manor anxious to turn the broad acres of the heath to account as a huge building estate. It was on the ground adjoining, called the East Park Estate, that Sir Thomas Wilson, Lord of the Manor, first attempted to commence building operations in 1831, and from that time down till 1871 there was a continual agitation for the preservation of the heath. The Lord of the Manor (who as tenant of an entailed estate could only grant leases for his life, or for twenty-one years) was equally active in his exertions to obtain powers from Parliament to grant long building leases. It was urged by the supporters of Sir Thomas Wilson that he had no intention to build on the heath, and that the agitation against him was promoted for private reasons. But the evidence given by him before a Committee of the House of Commons appointed to inquire into the question of open spaces in the Metropolis does not bear out this view, as will be seen by the following :

'Q. (*Committee*). Do you consider Hampstead Heath private property ?

A. (*Sir Thomas Wilson*). Yes.

Q. To be paid for at the same rate as private land adjoining ?

A.. Yes.

Q. Do you consider that the inhabitants of the neighbourhood have rights on the heath ?

A. There are presentments in the Court Rolls to show that they have none.'

In 1857 the appointment of the late Metropolitan Board of Works gave a fresh impetus to the proceedings, and one of the first acts of the newly-formed body was to appoint a committee to consider the necessity for providing more parks and open spaces for the Metropolis. This committee reported ' that considering the advantages which Hampstead Heath presents for promoting the health of the Metropolis, and its value from the beauty of its site as an ornament to the

capital, and considering also that the acquisition of the site
of the heath and of such adjoining lands as it may be
desirable to connect therewith will, if the purchase thereof
be delayed, involve a very much larger expenditure than
would be required at the present time, it is important that
the heath and the adjoining lands above referred to should
be purchased for the public use at as early a period as

Swings on Hampstead Heath on Bank Holiday.

possible.' But at this time the late Board's attention was
taken up with the proposal to form Finsbury Park, and the
question had therefore to remain in abeyance. In January,
1858, the attention of the late Board was called to the
application to Parliament by the Hampstead Vestry for
power to purchase Hampstead Heath and certain lands
adjoining to form a park, and to impose upon the Board the

duty of acquiring the funds required for that object. This Bill was not opposed by Sir Thomas Wilson, because it is said that the purchase-money for the heath would have been at least £400,000; but the Board, although agreeing with the proposed purchase, were not quite satisfied with the terms of the Bill, and they successfully opposed it in Parliament. The Select Committee of the House of Commons, however, reported that, although they had decided to negative the preamble, they were 'strongly impressed with the public utility of the proposed purchase of Hampstead Heath for the purpose of the recreation and health of the labouring classes of the Metropolis, and they wished to impress upon the Metropolitan Board of Works the urgent necessity of taking the matter into their serious consideration, with a view to secure Hampstead Heath for the public without any unnecessary delay, as, owing to the peculiar circumstances of the case, the Committee fear that the selling price of the property will be largely increased if deferred much longer.' Such a decided expression of opinion could hardly fail to carry weight with the Board; but although steps were taken to ascertain the several interests in the land, no practical result followed, and the question stood in abeyance for several years.

In 1866 a most important measure was passed, as far as the interests of commons are concerned, namely, the Metropolitan Commons Act, which prevented the enclosure of such lands within a radius of fourteen miles from Charing Cross. This fact, together with local representation, again revived the agitation, and the Chairman of the late Metropolitan Board of Works approached Sir Thomas M. Wilson with a view to ascertain whether he was prepared to negotiate for the dedication of Hampstead Heath to the public use, and, if so, upon what terms. This interview was altogether unfruitful, and the price mentioned by the Lord of the Manor for the heath, regarding it as building-land, was from £5;000 to £10,000 per acre, which was of course out of the question, and it appeared useless to continue the negotiations

at that time.   In the month of December of this same year
(1866) Sir Thomas Wilson began to build simultaneously on
two different parts of the heath.   On the highest and most
prominent part, near the flagstaff, the foundations of a house
were laid.   At another part, near Squire's Mount, only the
sods were removed and building materials brought upon
the ground.   A protest was at once made by some of the
copyholders, headed by Mr. J. Gurney Hoare, but the steward
refused to enter this upon the Court Rolls, because he said it
contained a statement prejudicial to the rights of the Lord of
the Manor and the only entry made was to the effect that the
acts complained of had been done, without specifying by whom.
As the building operations still went on, a lengthy action was
commenced by Mr. Hoare on behalf of the copyholders
against the Lord of the Manor, with the object of their
being declared entitled to the rights of common, to dig
gravel, sand, and loam for their necessary use, and to use
the heath for recreation ; and at the same time to restrain
the Lord from selling the sand, loam, and gravel, otherwise
than for the proper use of the demesne lands, from destroying the trees or pasture, and from building on or enclosing
the heath.   Before this suit was brought to a termination,
the Lord of the Manor died, and in January, 1870, it was
ascertained that his successor, Sir John Wilson, was willing
to negotiate for the sale of the heath.   The result of these
negotiations was that Sir John and all others concerned
decided to co-operate with the late Board in obtaining the
necessary Parliamentary powers, the purchase-money for
the whole of the interest of the Lord of the Manor being
fixed at £45,000, with an additional sum of £2,000 for
expenses.   A Bill was at once promoted in the next session
of Parliament, which received the royal assent on June 29,
1871.   The late Board of Works took formal possession of
the heath on January 13, 1872, and dedicated it to the use
of the public for ever.

   It will be seen from this account that the earlier proposals
were to form the heath into a park, but the peculiar beauty

arising from its wildness would have been lost if this had been done. Subsequently some important additions of land, such as Judges' Walk, the lovely Wildwood Avenue, and other plots not included in the original contract for the heath, were purchased in order to preserve its picturesqueness, and various planting works have been undertaken, but no extensive laying-out has been done, and so the natural features which have endeared it to so many artists have been preserved. At the time when the heath was purchased its surface had been much spoilt by the extensive sand-digging carried on; but the hand of Nature is very kind, and these depressions have now been transformed into grassy dells.

North End, Hampstead.

This sand is confined to the heath and is not found in neighbouring fields, a fact which has puzzled many geologists. It is admitted that this deposit of sand has been caused by some operation of Nature, probably by the action of some former sea or lake. The waters having subsided, or having been turned into some other channel immediately after making this deposit, the slight exterior coating of mud or slime which they left behind was not sufficient by any natural process to form a soil capable of agricultural cultivation. But this deposit of sand was not made equally over the previously deposited mound of clay, with a consequence that in some places the clay remained as the outward stratum,

or was covered, not with sand, but with loam.   Where this
was the case the soil was productive, but where the sand
remained it was barren and useless for cultivation, and
consequently remained as common land, unenclosed and un-
cultivated.   The remaining rich soil was at an early time
sufficiently valuable to form a fitting gift from a King to a
prominent Minister, and eventually resolved itself as the
Manor of Hampstead, to which the sand-covered tract was
attached as waste land, so that it would be more correct
to speak of the heath being confined to the sand than
*vice versâ.**

The word Hampstead (originally Hamestede) is a corruption
of ' homestead,' and the name has probably been given to the
district from the fact that, in its early history, all that was
habitable was a small farm of some 500 acres, with no other
habitations near except a few hovels occupied by farm
labourers.   This homestead formed, no doubt, an occasional
residence for the monks of Westminster Abbey.   Long
before the early times when Hampstead was the possession
of the monks, the district formed part of the ancient forest
of Middlesex, through which the Romans constructed their
highway—Watling Street—on the west of Hampstead.
Both Camden and Norden make Watling Street to have
crossed the heath, but the evidence is against this theory.
Roman remains have, however, been discovered close to the
heath, just at the commencement of Well Walk.   In the
summer of 1774 a Roman sepulchral urn, large enough to
hold 10 or 12 gallons was dug up, but it was broken to
pieces before it was got out.†

The first grant of property here, the date of which is very
uncertain, was by King Edgar to one of his noblemen, called
Mangoda, who is mentioned in the King's charters as *nobilis
minister*.   By a second charter, dated 986, the Manor of
Hampstead, of which the heath was the waste land, was
given to the monks of Westminster by Ethelred.   The gift

* Park, ' Hampstead,' pp. 47, 48.
† *Gentleman's Magazine*, vol. xlvi., p. 169.

was confirmed by William the Conqueror. It remained the property of the Abbey till the dissolution, when it formed part of the endowment of the new bishopric of Westminster, founded in 1540. Dr. Thirlby was the first and the only Bishop of Westminster, and during the nine years of his tenure he alienated nearly all the property of the see, and it must have been a good thing for Westminster when he was translated to Norwich, whereupon his former see was reduced to a deanery, and the Manor of Hampstead reverted to the Crown. It was granted in 1551 by Edward VI. to Sir Thomas Wroth, in whose family it remained till 1620, when it was sold to Sir Baptist Hickes, afterwards Lord Campden. His son-in-law, Sir Robert Noel, next obtained the manor by marriage, and so brought it into the Gainsborough family. The third Earl of Gainsborough sold the manor to Sir William Langhorne, Bart., who also purchased the Manor of Charlton, in Kent, both of which descended, as we have already seen,* to Sir Thomas Maryon Wilson.

One of the peculiar customs of the Manor of Hampstead was recently revived, when a court baron was held to effect a seizure in the case of the copyhold hereditaments of one of the tenants, who had been deceased over seven years, but whose heirs, although they had been 'proclaimed' three times, had not 'been admitted' to the 'copyhold hereditaments of which he died possessed.' Every search had been made to find the heirs but without avail. It therefore became the steward's duty, according to the manorial customs, to order 'a seizure in the name of the lord,' and the property was forfeited for want of an heir.

Hampstead at the present day is looked upon with decided favour as a fashionable suburb, but such has not always been the case. In the reign of the bluff King Hal, its chief inhabitants were washerwomen ; and here the clothes of the nobility, gentry, and chief citizens of London used to be brought to receive that whiteness which only country air can

* See p. 167.

25

give.* Coming on now to the seventeenth century, it was
chosen by a few of the more venturesome citizens as an
occasional residence; but it was not till the beginning of
the next century that it came into popular favour. The
event which brought Hampstead into prominence was one
which has exerted such a mysterious influence on so many
parts of the Metropolis—viz., the discovery of medicinal
springs. In the course of our travels to the various parks
and open spaces, we have met with these on many occasions
at Dulwich, Streatham, and other places. The story of
these fashionable wells is almost identical. At first, after
an accidental discovery, some fortunate invalid was reported
to have received a miraculous cure from the use of the waters.
Then they began to be recommended by certain physicians,
aware perhaps of the efficacy of novelty with the fanciful.
Immediately they became the height of fashion, and some
enterprising proprietor introduced music and entertainment
to attract the robust as well as the invalids. Finally, for no
apparent cause, except that some rival spring had been
discovered, their mushroom popularity was over, and their
decline rapid, and it is only by consulting past records that
we are reminded of their existence. This is the story of the
Hampstead wells.

It is not certain when they were discovered, but they were
held in some public esteem in 1698. Dr. Gibbons (the
Mirmillo of Garth's 'Dispensary') was the first physician to
recommend them for medicinal purposes, and his example
being followed by many others, they sprang into popularity
at once. The wells were furnished with a tavern, situated
near the East Heath in Well Walk, and the newspapers of
the time are full of advertisements of the advantages offered,
which comprised a dancing-room, raffling shops, and a bowling green. The following is a typical advertisement: 'The
wells are about to be opened with very good music for
dancing all day long, and to continue every Monday during

* From a MS. 'History of Middlesex,' quoted in Brewer's 'London
and Middlesex,' vol. iv., p. 190.

the season; there is all needful accommodation for water-drinkers of both sexes; and all other entertainment for good eating and drinking; very good stables for fine horses; and a further accommodation of a stage coach and chariot from the wells at any time in the evening or morning.'*

It was about this time that a comedy by Baker, called 'Hampstead Heath,' was produced at Drury Lane, which is chiefly interesting because it contains a satirical description

Whitestone Pond, Hampstead Heath.

of the frequenters and the amusements of Hampstead. The opening scene commences:

'*Smart.* Hampstead for awhile assumes the day; the lovely season of the year, the shining crowd assembled at this time, and the noble situation of the place, gives us the nearest show of Paradise.

'*Bloom.* London now, indeed, has but a melancholy aspect, and a sweet rural spot seems an adjournment o' the nation, where business is laid fast asleep, variety of diversions feast our fickle fancies, and every man wears a face of pleasure. The cards fly, the bowl runs, the dice rattle. . . .

* *Postboy*, May 10, 1707.

25—2

'*Smart*. Assemblies so near the town give us a sample of each degree. We have court ladies that are all air and no dress; city ladies that are overdressed and no air; and country dames with brown faces like a Stepney bun; besides an endless number of Fleet Street sempstresses that dance minuets in their furbeloe scarfs.'*

But in addition to the dancing-room and the bowling-green, there was another very accommodating institution in connection with the wells, called Sion Chapel, where any couples could be married who brought a license and five shillings; and an advertisement in *Read's Weekly Journal* of September 8, 1716, informs us that 'Sion Chapel, being a private and pleasure place, many persons of the best fashion were married there. Now, a minister is obliged constantly to attend, and therefore notice is given that all persons, on bringing a license, and who shall have their wedding dinners in the gardens, may be married in that said chapel without giving any fee or reward whatever.'†

The popularity of the Hampstead wells lasted for less than fifty years, and the patronage of its former frequenters was transferred to the New Tunbridge wells, the site of which is, curiously enough, also used as a place of recreation, under the name of Spa Green. But the springs had done their work; they had caused a great increase in the residents, who could not fail to be charmed by the beauty of the district, and ever since this time Hampstead has held its own among the outlying suburbs.

The great assembly-room of the tavern seems to have been put to better uses, and was converted into a chapel somewhere about 1733, under the title of Well Walk Chapel,‡ and continued to be used as such for over a century. The spring, which now flows very slowly, is covered with a massive stone fountain adorned with coats-of-arms, and

* Quoted in Park's 'Hampstead.' The original was lent to Mr. Park by John Kemble for his history.
† Quoted in Howitt's 'Northern Heights,' p. 27.
‡ Howitt, 'Northern Heights,' p. 29.

bears the inscription ' Chalybeate Well,' so that any visitor may try its healing virtues without paying the 1s. for admission which was formerly the charge.   It also bears on its face a granite tablet :

' To the memory of the Honble. Susanna Noel, who, with her son Baptist, third Earl of Gainsborough, gave this well together with six acres of land to the use and benefit of the poor of Hampstead. 20th Dec., 1698.

> ' Drink, traveller, and with strength renewed
> Let a kind thought be given
> To her who has thy thirst subdued,
> Then render thanks to Heaven.'

Before leaving Well Walk with its shady elm-trees, we must notice the seat at the end nearest to the heath, which has taken the place of a wooden bench, the favourite resting-place of the poet Keats.*

From Well Walk it is but a short distance to another house which was associated with the drinking of the waters. Before Hampstead had quite lost its popularity as a watering-place, most of the aristocratic patrons had deserted the Wells Tavern for the Upper Flask, which also boasted its card-rooms and its bowling-green.   This house situated in Heath Street, on the right-hand corner of East Heath Road, has long been a private residence, but its past historical associations are very interesting.   During the time it was a public tavern, it was the summer resort of the Kit-Cat Club, which boasted among its members such names as Steele, Pope, and Dr. Arbuthnot.   These litterati used to sip their ale under the venerable old mulberry-tree which flourished till 1876, when the weight of the snow in a heavy storm at Christmas time broke it down and destroyed it. Sir Richard Blackmore in his poem, ' The Kit-Cats,' has alluded to their visits to Hampstead :

> ' Or when, Apollo-like, thou'rt pleased to lead
> Thy sons to feast on Hampstead's airy head—
> Hampstead, that, towering in superior sky,
> Now with Parnassus does in honour vie.'

---

* Hone, ' Table-book.'

The Upper Flask has also been made famous by Richardson in his novel of ' Clarissa Harlowe,' who makes his heroine escape here for a short time from the pursuit of Lovelace. Mrs. Barbauld, long a resident at Hampstead, says 'she well remembers a Frenchman who paid a visit to Hampstead for the sole purpose of finding out the house where Clarissa lodged, and was surprised at the ignorance or indifference of the inhabitants on that subject. The Flask Walk was to him as much classic ground as the rocks of Mallerie to the admirers of Rousseau.*

Richardson is not the only author to bring Hampstead Heath into prominence. Macaulay in describing the beacon-flames that warned England of the approach of the Armada, describes how 'High on bleak Hampstead's swarthy moor they started for the north'; and, again, Charles Dickens, who frequently used to ride to Jack Straw's Castle, tells us of Bill Sikes in his flight after the murder coming to Hampstead Heath. 'Traversing the hollow of the Vale of Health, he mounted the opposite bank, and, crossing the road which joins the villages of Hampstead and Highgate, made along the remaining portion of the heath to the fields at North End, in one of which he laid himself down under a hedge and slept.'

The last innkeeper who had possession of the Upper Flask was Samuel Stanton, whose nephew and successor of the same name was styled ' gentleman ' in 1737. He left it to his niece, Lady Charlotte Rich, and in 1771 it was purchased by George Steevens, the well-known commentator of Shakespeare, who resided here till his death in 1800.† He possessed an ample fortune, and spent a large amount in the improvement of the house and grounds.

Following now the main road from the Upper Flask across the heath, we arrive at another historical inn, Jack Straw's Castle, now being rebuilt. The mantel-tree over

* ' Life of Richardson,' quoted in Thorne's 'Environs of London,' part i., p. 282.

† Howitt, ' Northern Heights,' p. 127.

the kitchen fireplace is said to have been made from the gibbet-post on which was suspended the corpse of Jackson, a notorious highwayman.

Jack Straw, it will be remembered, came into prominence in Wat Tyler's rebellion. He was in charge of the insurgents who burnt the Priory of St. John of Jerusalem, thence striking off to Highbury, where they destroyed the house of Sir Robert Hales, and afterwards encamped on Hampstead Heights. The original 'castle' of Jack Straw consisted of a mere hovel, or a hole in the hill-side. If the rebels had been successful in their ambitious projects, Jack Straw was to have been king of one of the eastern counties, probably Middlesex.

Jack Straw's Castle, Hampstead Heath, 1891.

It was on the slope behind the 'castle' that the corpse of John Sadleir, the fraudulent M.P. for Sligo, was found on the morning of Sunday, February 17, 1856. Beside it was a small phial which had contained essential oil of almonds, and also a silver cream-jug from which he had taken the fatal draught. He was connected with several enterprises, and by means of forgeries and misrepresentations had duped many. He continued to deceive till the very last, and it was not till after his suicide that the extent of his infamy was brought to light. He was led to take his life by the action of Messrs. Glyn, the London agents of the Tipperary Bank, of which he was the principal manager. They returned its drafts as 'not provided for,' a step which was

followed a day or two after by the Bank of Ireland.*
Hampstead is an awkward place for a suicide to select. In
the event of a jury returning a verdict of *felo de se*, the Lord
of the Manor is entitled to the whole of the goods and
chattels of the deceased of every kind, with the exception of
his estate of inheritance. Sadleir's goods and chattels were
already forfeit, but the cream-jug was claimed and received
by the Lord of the Manor as an acknowledgment of his
right, and then returned.† John Sadleir figures in Dickens'
'Little Dorritt' as Mr. Merdle. 'I shaped Mr. Merdle
himself,' he writes, 'out of that glorious rascality.'

Leaving now Jack Straw's Castle and continuing our walk
along the breezy Spaniards Road, we arrive at the other
end of the heath at the Spaniards Inn. This is built
upon the site of the toll-gate (marked in Rocque's map as
Spaniard Gate) erected by the Bishops of London at the
Hampstead end of the road, made in the fourteenth century
through their land to the North of England, when the
Roman highway, Watling Street, had become neglected and
ruinous.‡ It derives its name from the fact that it was
taken originally by a Spaniard as a place of entertainment.
Subsequently a Mr. Staples 'improved and beautifully
ornamented' its gardens, and made 'pleasant grass and
gravel walks, with a mount' commanding extensive views
into seven counties. The walks and plats were embellished
'with a great many curious figures, depicted with pebble
stones of various colours, viz., a rainbow and star; the sun
in its glory; the seven stars; the Star and Garter; motto and
crown; half moon; a coat-of-arms; the twelve signs of the
zodiac; Tower of London; Hercules' pillars; the blazing star;
a dial on the grand mount; Adam and Eve; Salisbury spire;
the Roman eagle,' and a host of others equally curious.§

* *Gentleman's Magazine.*
† Thorne, 'Environs of London,' vol. i., p. 284.
‡ Park, 'Hampstead,' pp. 15, 252.
§ From a manuscript description of Middlesex, quoted by Park,
pp. 252, 253.

This extensive list of subjects has long ago disappeared, together with the mount and a greater portion of its views, but the garden is still a pleasant one.

The Spaniards played an important part in the Gordon riots of 1780. Dickens describes in ' Barnaby Rudge ' how the rioters, after sacking Lord Mansfield's house in Bloomsbury Square, 'marched away to his country seat at Caen Wood, between Hampstead and Highgate,' bent upon destroying that house likewise, and lighting up a great fire

South View of the Spaniards, Hampstead Heath. (From a print by Chastelaine, in the King's Library, British Museum.)

there, which from that height should be seen all over London. But in this they were disappointed, for a party of horse, having arrived before them, they retreated faster than they went, and came straight back to town.' This does not tell us anything about the generous act of the landlord of the Spaniards, which led to the decamping of the besiegers. Fresh from their destructive work in London, on their way to Caen Wood they had to pass this inn, when the landlord, learning their object, stood at his door and invited them to drink. He threw open his cellars to this hot and thirsty crew, already broiling from the fire in Bloomsbury, and whilst they caroused he had a messenger speeding his way to the barracks for a detachment of the Horse Guards. By the time that the rioters had exhausted the barrels of the Spaniards, they found this troop drawn up across their way to Caen Wood. The steward and Mr. Wetherall, the medical man of the family, had also sent out ale in abundance from the cellars of Lord Mansfield ; and thus, at once tottering under the fumes of beer and confronted by the soldiery, the mob fled, and left Caen Wood to stand peacefully through more peaceful times.'*

To turn from rioters to a loyal demonstration we must pass on to 1803, when the Defence Act was just carried. In the summer there was a great gathering on the heath, which all Hampstead turned out to see. The loyal parish was literally up in arms to give effect to the measure, and no fewer than 700 good men and true took the oath of allegiance as volunteers.† Subsequently the heath was resorted to for rifle practice, and in 1808 a target-bank was formed. The same fate befell this as the rifle ranges at Wimbledon, for the homage upon the representation of Lord Erskine and others complained ' that in consequence of several corps of volunteers, militia, and other military, having of late resorted to the target-ground which was formed on Hampstead Heath for the use of the corps of Hampstead volunteers exclusively, and by reason of the frequent firing with ball at the targets set up against a mound or bank thrown up by the permission of the lady of this manor adjoining her freehold land, next the said heath, the peace and tranquillity of the manor and parish of Hampstead is very much broken and disturbed, and such firing is not only extremely prejudicial to the comfort of the inhabitants of Hampstead, but

* Howitt, ' Northern Heights,' p. 354.
† Baines, ' Hampstead,' pp. 452, 453.

from the unskilfulness and irregular conduct of some of the
military who come from places out of the parish to practise
firing on the said ground and heath, has already been productive . . . of very serious injuries and accidents to persons
. . . traversing the said heath.' The target-bank had
therefore to be dug down, and the site was levelled in the
following June.*

* From the Court Rolls, May 30, 1808.

CHAPTER XX.

HAMPSTEAD HEATH (continued).

THE large but plain white house adjoining the Spaniards is celebrated for having been the residence of Lord Erskine, who has been pronounced by other distinguished lawyers the greatest forensic orator that England has ever produced. It is surrounded by high walls that shut out the view of its grounds from the sight of the curious, and the chief characteristic about its appearance is the long portico leading into it from the road. When Lord Erskine came to live here the house was not of much importance, but it had extensive grounds and commanded a fine view of the picturesque surroundings. He at once set about improving it, and having planted it with evergreens of different descriptions, he gave it the name of Evergreen Hill.* He is also said to have planted with his own hand the extraordinarily broad holly hedge separating his kitchen-garden from the heath, opposite to the Fir-Tree Avenue.† The present name of the mansion is Erskine House, after its famous occupant. The greater part of the leisure time of this legal light was spent in his garden, which was on the opposite side of the road, and connected with the house by a subterranean passage. At a time when he was at the very pinnacle of his profession he describes his private life thus in a letter to a friend: 'I am now very

* Park, 'Hampstead,' p. 319.
† Howitt, 'Northern Heights,' p. 57.

busy flying my boy's kite, shooting with a bow and arrow, and talking to an old Scotch gardener six hours a day about the same things, which, taken altogether, are not of the value or importance of a Birmingham halfpenny, and scarcely up to the exertion of reading the daily papers.' *

Many of the famous men of the day came to Hampstead as the guests of Lord Erskine, and here occurred his last meeting with Burke, from whom he had been estranged for

Erskine House, Hampstead Heath, in 1869.

some time owing to a difference in politics. Their parting, as described by Erskine, is very affecting: ' What a prodigy Burke was! He came to see me not long before he died. I then lived on Hampstead Hill. "Come, Erskine," said he, holding out his hand, "let us forget all! I shall soon quit this stage, and wish to die in peace with everybody, especially you." I reciprocated the statement, and we took a turn round the grounds. Suddenly he stopped. An extensive prospect over Caen Wood broke upon him. He stood

* Quoted in Howitt's ' Northern Heights,' p. 73.

wrapped in thought, gazing on the sky as the sun was setting. "Ah, Erskine!" he said, pointing towards it, "this is just the place for a reformer ; all the beauties are beyond your reach—you cannot destroy them."'*

But even lawyers are sometimes found napping, and Lord Erskine found out to his cost that he had made a great mistake in selling Evergreen Hill, and buying a barren estate in Sussex, where he is said to have set up a manufactory of brooms, which was the only valuable product of his property. His Hampstead house remains much the same as he left it, but the subterraneous tunnel has been filled in, and the grounds connected by it with the rest of the property are now in the possession of Lord Mansfield. Chief Justice Tindal afterwards lived in this house.† It has been pointed out that the contemporary residence of three great legal lords at Hampstead in the persons of Lords Erskine, Mansfield, and Loughborough is one of the most remarkable associations of the place, and the residence of Erskine there will ever remain as one of its greatest glories.‡

The middle house of the three near the Spaniards was once the residence of Sir W. E. Parry, the Arctic explorer. From his garden at the back, looking due north over the low range of the Middlesex hills, Sir Edward must have seen the streamers of the Aurora Borealis flaming into the sky, reminding him of his ice-bound Arctic home of former years.§

It was in April, 1842, that Parry went to Hampstead for the benefit of his health. 'I cannot express,' he wrote, 'how I continue to enjoy, and, I am sure, to profit by, the lovely views from Hampstead and its charming air.'‖

The detached house known as The Firs was built by a tobacconist, Mr. Turner, of Fleet Street, who planted the grove of fir-trees in front, and made the road from here to North End.

* Quoted in Howitt's 'Northern Heights,' p. 80.
† Baines, 'Hampstead,' p. 429.
‡ Howitt, 'Northern Heights,' p. 81.
§ Baines, 'Hampstead,' p. 472.
‖ 'Memoirs of Parry,' by his son, p. 267.

The chain of ponds on the heath forms a prominent feature. They are fed by the numerous springs for which Hampstead has long been noted. In the reign of Henry VIII. the question of an increased water-supply for the Metropolis was being considered because that 'eyther for fayntness of the springes, or for the drinesse of the earth, the accustomed course of the waters comminge from the olde springes and auncient heades, are sore decayed, diminished, and abated,' as the Act passed in the thirty-fifth year of his reign, 'concernynge the repayring, makyng, and amendynge of the Cundytes in London,' quaintly puts it. 'For remedy

The Fir-trees, Hampstead Heath.

whereof, Sir William Bowyer, knight, nowe mayre of the saide citie . . . not onely by diligent searche and exploracion, hath founde out dyvers great and plentyfull sprynges at *Hampsteade-heath*, Marybone, Hackney, Muswelle-hylle . . . but also hath laboured, studied, and devised the conveyaunce thereof, by cundytes, vautes, and pipes to the saide citie.' The Mayor and citizens were empowered by this Act to lay pipes, dig pits, and erect conduits in the grounds of any proprietors wherever required, 'Provided always . . . that if the sayd mayre and comminaltie of the citie of London . . . do fetch and convey any water from

any springe or springes within the saide heath called
Hampsted Heath unto the sayde citie . . . then they . . .
shall for ever pay unto the Bysshop of Westminster for the
tyme being, at the feast of Saint Michaell the archaungel,
one pounde of pepper; in and for the acknowledgement of
hym and them for the lordes and very owners of the saide
heath.' These works were carried out by a later Mayor,

Sluice-House on Hampstead Heath.

Sir John Hart, in 1589-90, but the springs were afterwards
leased out by the City. In 1692 the lessees were incorporated under the name of the Hampstead Water Company,*
whose works were afterwards transferred to the New River
Company.

In course of time the supply was not equal to the demand,
and the company then proceeded (1835) to sink a well in

* Park, 'Hampstead,' pp. 71-74.

the vicinity of the ponds at the south of the heath. The work was long and difficult, but at the depth of nearly 400 feet an excellent spring of water was discovered. The great depth, however, necessitated the use of a steam-engine to raise the water to the surface, for the accommodation of which the 'Round House,' which still stands, was built.* The wells have now been superseded by other sources of supply, but the ponds are available for different sports at various seasons of the year: in winter, skating, in the summer, model yachting and fishing, whilst bathing is carried on all the year round. The water is very deep and dangerous, so that it is necessary to have a boatman on duty in case of any sudden emergency.

In addition to the chain of ponds there are other isolated pieces of water—the Leg of Mutton Pond on the extreme west, the Whitestone Pond near Jack Straw's Castle (the soil of which is not part of the heath), and another large pond also belonging to the New River Company, near the Vale of Health, which was added in 1777. The Whitestone Pond was originally only a small one, but it was enlarged and otherwise improved by the vestry in 1875. It takes its name from the white milestone which stands just inside the shrubbery near the pond. The Leg of Mutton Pond was formed, and part of the road from Child's Hill to North End was raised and improved, during a severe winter as the result of works instituted for the relief of the unemployed poor by Mr. Hankin, an overseer of the parish. About 1825 the road was known as Hankin's Folly.†

The works carried out by Sir John Hart were undertaken at the time when the course of the river Fleet was much choked up. This ancient river, which is now nothing more than a sewer, had its source in a spring which rose at the foot of Hampstead Hill, and fell into the Thames at Blackfriars. It was once large enough to admit of ten or twelve ships laden with merchandise coming up to Fleet Bridge. Even after the Fire of London, it was cleared out

* Baines, 'Hampstead,' p. 212.   † *Ibid.*, p. 208.

26

so as to admit barges of considerable burden as far as Holborn Bridge.*

The fanciful name of the Vale of Health is a relic of the time when Hampstead was a watering-place. Its former name was Hatches Bottom. It would be a picturesque spot but for the huge unsightly tavern, with its towers and

Well Walk, Hampstead, showing Keats' Favourite Seat.

battlements, which spoils all the beauty of the rustic cottages under the shade of the willows. But we have another grudge against this Vale of Health tavern. To make room for it a cottage, which was the home of Leigh Hunt, had to be pulled down. It was 'the first one that fronts the valley,' and Shelley and Keats were often there. In fact, it was through his many visits to his friend's house that Keats

* Park, 'Hampstead,' p. 73.

obtained that liking for Hampstead which led him to make it his residence from 1817 till he left England in 1820. Whilst at Hampstead he wrote 'Ode to a Nightingale,' 'Eve of St. Agnes,' 'Isabella,' 'Lamia,' and 'Hyperion,' and commenced 'Endymion.' It is well known how sad were the last years of this poet. Leigh Hunt, as well as Hone, gives us a melancholy picture of him : ' It was on the same day, sitting on the bench in Well Walk (the one against the wall), that he told me, with unaccustomed tears in his eyes, that his heart was breaking.'* Both Keats and Shelley died abroad, as if to emphasize by their death in foreign lands that they were outcasts from England. But though rejected by their contemporaries, their memory is treasured by their many admirers of to-day.

The distinguished foreigner Prince Esterhazy occupied, it is said, about 1840, a house in the Vale of Health, which has long since been pulled down. It is strange that he who was reputed to be so wealthy should have chosen so modest a home. It may perhaps have been to be near the medicinal spring.†

Halfway down the road leading from Jack Straw's Castle to North End is Hill House, which was formerly the seat of Mr. Samuel Hoare, the banker, whose descendant, Mr. J. Gurney Hoare, was one of the most prominent of those who resisted the encroachments of the Lord of the Manor upon the heath. Hill House has acquired fame through the visits of the poet Crabbe to its genial host. His son writes : ' During his first and second visits to London, my father spent a good deal of his time beneath the hospitable roof of the late Samuel Hoare, Esq., on Hampstead Heath. He owed his introduction to this respectable family to his friend Mr. Bowles . . . and though Mr. Hoare was an invalid, and little disposed to form new connections, he was so much gratified with Mr. Crabbe's manners and conversation, that

---

* Leigh Hunt, ' Byron and his Contemporaries,' vol. i., p. 440, quoted in Thorne's ' Environs of London.'

† Baines, ' Hampstead,' p. 477.

their acquaintance grew into an affectionate and lasting intimacy. Mr. Crabbe in subsequent years made Hampstead his head-quarters on his spring visits, and only repaired thence occasionally to the brilliant circles of the Metropolis.'*
The place was evidently congenial to his writing, for the poet himself writes : ' My time passes here I cannot tell how pleasantly. To-day I read one of my long stories to my friends. . . . I rhyme at Hampstead with a great deal of facility, for nothing interrupts me but kind calls, or some-thing pleasant.' Crabbe was not the only poet to honour Mr. Hoare, for Coleridge, Wordsworth, Rogers, and Camp-bell, were frequent visitors when he was in residence. Campbell writes of his friend : ' The last time I saw Crabbe was when I dined with him at the house of Mr. Hoare at Hampstead. He very kindly came to the coach to see me off, and I never pass that spot on the top of Hampstead Heath without thinking of him.'† Wordsworth, too, refers to his connection with Crabbe at Hampstead :

' Our haughty life is crowned with darkness,
Like London with its own black wreath,
On which with thee, O Crabbe ! forth-looking,
I gazed from Hampstead's breezy heath.'

Following now the road to North End, which becomes more beautiful at every step, we come into Wildwood Avenue, a spot which must charm every true lover of Nature. From thence the road, or rather lane, descends very rapidly into the quaint little village of North End. It is hard to believe that London is only five miles away, so quiet and so rural is the scene, and it is likely to remain so, too, since the village is surrounded on three sides by the heath. The chief historical associations of North End are with, the fine mansion known as Wildwoods. There is a Wildwood marked upon the map in Park's ' Hampstead,' close by the Spaniards, and there is another house of the same name

* ' Life of the Rev. George Crabbe,' by his son, quoted in Thorne's ' Environs of London,' vol. i., p. 289.
† Quoted in ' Old and New London,' vol. v., p. 454.

at North End, neither of which is to be confounded with Wildwoods, the large mansion at the foot of the hill on the left-hand side. The house, formerly known as North End House, must be at least 200 years old, that is, the original part of it, for it has been enlarged and altered considerably at different times. The place it occupies is named Wildwood Corner in Doomsday Book.*

The great Lord Chatham lived in this house for a while, at that time when the strange and mysterious malady attacked him which made him as helpless as a child. He was here in 1766 and 1767, at that period when State matters at home and abroad demanded the most rigorous attention. And yet the Prime Minister remained at North End, 'inaccessible and invisible,' and the country was as a ship without a rudder. During this melancholy time he used to be driven about the heath in his carriage, with the blinds drawn up, and shunning the frequented parts as much as possible. Mr. Howitt, writing in 1869, says: 'The small room, or rather closet, in which Chatham shut himself up during his singular affliction—on the third story—still remains in the same condition. Its position from the outside may be known by an oriel window looking towards Finchley. The opening in the wall from the staircase to the room still remains, through which the unhappy man received his meals or anything else conveyed to him. It is an opening of perhaps 18 inches square, having a door on each side of the wall. The door within had a padlock, which still hangs upon it. When anything was conveyed to him a knock was made on the outer door and the articles placed in the recess. When he heard the outer door again closed, the invalid opened the inner door, took what was there, again closed and locked it. When the dishes or other articles were returned, the same process was observed, so that no one could possibly catch a glimpse of him, nor need there be any exchange of words.'† Since Lord Chatham was at

* Baines, 'Hampstead,' p. 50.
† Howitt, 'Northern Heights,' p. 90.

Hampstead the house has been transformed by a later owner and has had another story added, but the statesman's room is still retained.

Near the summer-house at the top of the grounds a murder was committed, more than a hundred years ago—at least, so says tradition.   One of the female servants of the owner then living there is said to have been killed by the butler.   No record of this has at present been found, but there are those who assert that the ghost of the murdered woman still walks in the garden.*   What more confirmation could be required ?

Close by this thatched summer-house is an elm which still goes by the name of the Gibbet-tree.   Between this tree and another there formerly stood a gibbet, on which was suspended the body of Jackson, a knight of the road, for murdering Henry Miller at this spot in 1673.   His victim was buried on March 20, 1673, and in the following year was published ' Jackson's Recantation ; or, The Life and Death of the Notorious Highwayman, now hanging in Chains at Hampstead, etc. ; wherein is truly discovered the whole mystery of that wicked and fatal profession of padding on the road.'†

One of the houses at North End, viz., Golder's Hill, was the residence, till his death, of the celebrated surgeon, Sir Spencer Wells, Bart.   Parts of the mansion are very old, but it has been altered and enlarged at various times.   The present imposing and modern appearance of the house is due to some works carried out in 1875, when a new front was added.   The extensive grounds, of some 36 acres, have a certain wildness of their own which make the estate one of the most picturesque in the neighbourhood of London, and in this respect they compare favourably with anything on the Heath itself.   The eminent landscape authority Mr. Robinson, in describing the property, says :

' Places where the simple and essential conditions for beauty

* Howitt, ' Northern Heights,' p. 91.
† Park, ' Hampstead,' p. 305.

in planting and design are understood or illustrated are far too rare, and it is all the more pleasing to meet with an example of artistic treatment of a garden almost in London, on the western border of Hampstead Heath.

'As regards design and views it is the prettiest of town gardens, and the conditions of its beauty are so simple that there is little to be said about them. An open lawn there is, rolling up to the house; groups of fine trees and wide and distant views over the country.

'A sunken fence separates the lawn from some park-like meadows; and beyond, the country north of London opens up, without any building visible on either side or in the foreground. From almost every other point of view these trees form a picturesque group and afford a welcome shade in summer. The whole of the front of the house, it must be understood by those who have not the opportunity of seeing the place, is an open lawn without any of the impedimenta usual in such places.'*

The estate is plentifully supplied with water. Near the mansion is an ornamental lake spanned by a rustic bridge. At the farther end is another lake surrounded with sedge-grass, furnishing a quiet haunt for the many moor-hens whose shrieking notes are the only sounds heard. Close by is a delightful valley, through which a trickling stream lazily meanders. The wild beauty and picturesqueness of the scene must appeal to every lover of Nature, and it is hard to realize that the centre of smoky and noisy London lies only five miles away.

On the death of Sir Spencer Wells the property was put up for sale as a building estate, and was at once purchased by Mr. Barrett on behalf of a committee who wished to secure it as an addition to the Heath. This object was attained with the aid of the municipal and local authorities, and the possession of such an estate would be a matter for boasting on the part of any city.

It is said by some that Golder's Hill was once the resi-

---

* W. Robinson, 'The English Flower-Garden,' 1896.

dence of the poet-physician Mark Akenside, the author of
'Pleasures of the Imagination.' His friend Dyson bought
a house for him at North End in the hopes that he might
create a medical practice among the many invalids who came
to Hampstead to drink the waters. But Akenside's manner
was not such as to inspire confidence, and after a short stay
of under two years he returned to town. If Akenside lived
here it must have been in the old and comparatively small
part of the modernized mansion.

This part of the heath also has its ancient inn, the Bull
and Bush, famous now chiefly for its good dinners and its
tea-gardens, which command extensive views of the surrounding counties. It is an old-fashioned tavern, and we can well
believe the tradition that it was once a farmhouse. It is
also said to have been the country seat of Hogarth, who
planted the yew bower of the garden. Among other celebrities
who have visited it may be mentioned Addison, Gainsborough,
Sir Joshua Reynolds, Garrick, Sterne, Foote the comedian,
and Hone the antiquary.* It can be but little altered since
their time, and there is every prospect of its remaining a
sylvan retreat for some time to come.

A neighbouring farmhouse was once the home of William
Blake, the imaginative artist and poet, and also of John
Linnell, another celebrated landscape artist. It is only
natural that the beauties of Hampstead should have
attracted many an artist. Clarkson Stanfield, the eminent
sea-painter, occupied a venerable house here; but his presence
at Hampstead was not due to his desire to paint the surrounding landscape, but rather to secure a quiet retreat.
Constable, who lived in Well Walk, and Linnell were both
essentially landscape painters, and they have treasured up
for us on canvas the scenes which were so dear to them.
The list of artists is a long one, and we can do no more
than mention such names as Romney, Morland, Haydon,
and Herbert, who either lived at or frequented Hampstead
in the pursuit of their art.

* Baines, 'Hampstead,' pp. 234, 235.

Whitefield, the prince of preachers, found on Hampstead Heath an auditorium large enough to accommodate the vast audiences he was accustomed to draw together. He thus records his visit in his diary :

'*May* 17.—Preached, after several invitations thither, at Hampstead Heath, about five miles from London. The audience was of the politer sort, and I preached very near the horse course, which gave me occasion to speak home to their souls concerning our spiritual race. Most were attentive, but some mocked.'*

The heath, like most places of the kind near London, was once the resort of highwaymen. Dick Turpin is said to have had a house at Hampstead; and Claude Duval is credited with a particular liking for the roads in the neighbourhood of the heath. But one of the most curious facts in connection with its history is that so late as the thirteenth century it was infested with wolves, and was consequently as dangerous to cross then as it was in comparatively recent times because of the highwaymen.†

The elections for the county of Middlesex were held on Hampstead Heath till the year 1700-1, when the first announcement appears of their taking place at Brentford. The following items of news, extracted from papers now extinct, bear upon the subject :

'Yesterday was the election for the county of Middlesex held at Hampstead Heath; the candidates being Sir William Roberts and Esquire Ranton, against Sir Francis Gerard and Mr. Middleton.'—*True Protestant Mercury*, March 2-5, 1681.

'We hear now that Admiral Russell and Sir John Worsnam stand candidates for Middlesex, against Sir Charles Gerrard and Ralph Hanton, Esq.; the election to commence on Thursday sevennight upon Hampstead Heath.' —*Flying Post*, October 19-22, 1695.

Besides being the scene of excitement at election times,

* Continuation of the Rev. Mr. Whitefield's ' Journal,' 1739, quoted in Park's ' Hampstead,' p. 239.

† Howitt, ' Northern Heights,' p. 45.

the heath was at one time famous for its horse-racing. The course was on the west side of the heath, but in consequence of the bad company attracted by the races, it was found necessary to suppress them altogether.

'On Monday last (September 4, 1732) at the race run at Hampstead Heath for ten guineas, three horses started; one was distanced the first heat, and one drawn; Mr. Bullock's Merry Gentleman won, but was obliged to go the course the second heat alone.'—*Daily Courant*, September 6, 1732.

But though the races of Hampstead are things of the past, the energy of the inhabitants has found a vent in another institution, viz., the annual bonfire on November 5. Under the auspices of the Hampstead Bonfire Club, the memory of Guy Fawkes is preserved with becoming dignity, and the procession which takes place on this occasion is worthy of a Lord Mayor's Show. Some cynic has stated that this is the result of the 'preposterous health of the inhabitants,' but as a matter of fact, the proportion of residents who attend the fire carnival is probably less than one in twenty, and it is always a heavy day for the railway companies. In addition to the *pièce de résistance*, in the shape of the huge bonfire lit near the Vale of Health, the programme includes a mimic bombardment of Jack Straw's Castle and a procession of cars illustrative of contemporary topics, which parades the principal streets of Hampstead.

The trees on the heath will always remain as one of its greatest attractions. Park gives an account of a remarkable hollow elm at Hampstead, the position of which he was unable to define. It had an entrance door cut out in the bottom of the trunk, from which ascended a winding staircase to the top of the tree. At the top was an octagonal turret to enable visitors to see the views across the surrounding country. There is a print of it (now exceedingly rare) in the British Museum, dated 1653, which contains the following description of the tree:

'1. The bottom above ground in compass is 28 foote.

2. The breadth of the doore is 2 foote.

3. The compass of the turret on the top is 34 foote.

4. The doore in height to goe in is 6 foote 2 inches.

8. The height to the turret is 33 foote.

11. The lights into the tree is (are) 16.

18. The stepps to goe up is 42.

19. The seat above the stepps six may sitt on, and round about roome for foureteene more.

All the way you goe up within the hollow tree.'

Some copies of this print are embellished with verses by Robert Codrington, who describes the tree with the aid of a little poetic exaggeration. He also speaks of the prospect from the top, which included ' smooth Richmond's streams . . . Acton's mill . . . Windsor's castle . . . Shooter's Hill ' and groves and ' plains, which further off do stand.'*

Another poetic genius, Edward Coxe, of Hampstead Heath, has also left some lines 'to commemorate the preservation of the nine elms on Hampstead Heath.' The poetry seems to point to Lord Erskine as the one who would have cut down these trees with 'impious strokes' and ' sacrilegious hand.'†

Before leaving Hampstead Heath it is only right to say a word or two about the many so-called encroachments. It must be borne in mind that the Lord of the Manor has a perfect right with the consent of the homage to grant parcels of the waste land of his manor. These tenements, which were originally copyhold, *i.e.*, held by copy of the court roll, could then be enfranchised, and become the absolute freehold of the parties who effected the enfranchisement. It is in this way that the majority of the houses have been built upon what was formerly part of the heath. From 1608 to 1866 there were 450 such grants of the waste, making a total in all of 83 acres 1 rood 21 poles.‡ The usual fine paid for this was five shillings per rod. The Court Rolls were burnt in 1684, and there was doubtless at this time some uncertainty and confusion in the manor as to the

* Park, ' Hampstead,' pp. 33-37.   † *Ibid.*, pp. 40, 41.
‡ From the Court Rolls.

copyhold titles.   Probably some advantage was taken of this
to encroach on the waste.   At all events, in 1686 several
unauthorized encroachments on the waste appear to have
existed, which the homage were ordered to survey, and as
a result several tenants were admitted as copyholders in the
following form: 'The homage present, upon their oaths,
that [A. B.] has incroached [   ] rods of land of the waste
ground of the lord of 'the manor; but what estate he has or
by what rent due to the lord of the manor, whose land he
holds, the same is not stated, because nothing thereof remains
on the court rolls that they can find.   The lord of the manor
nevertheless, at the request and humble petition of the aforesaid [A. B.], out of his generosity and benevolence towards
his tenants, then, by his steward aforesaid, did give licence
and liberty to the said [A. B.] to enclose the aforesaid [   ]
rods of the lord's waste ground and use the same for himself, etc.'

The tenants who were thus admitted were debarred from
rights of common, although the resolution is considered as
being *ultra vires*.   'No tenants of this manor who hold any
tenements or cottages by copy of court roll which have been
built upon the waste of this manor or who hold any parcels
of land . . . enclosed from off the waste of this manor have
any right of common in respect of such tenements . . . for
any cattle on Hampstead Heath.' *

* Court Rolls, May 28, 1759.

CHAPTER XXI.

HAMPSTEAD HEATH EXTENSION, OR PARLIAMENT HILL.

THE lands known by this title comprise 267¼ acres in the parish of St. Pancras, and form, with Hampstead Heath, an unbroken recreation-ground larger than Regent's Park. We have dwelt at some length upon the beauties of the surroundings, and ·the range of views from Hampstead Heath, but to a very great extent their enjoyment is dependent upon the openness of the adjacent fields. Just for one moment imagine Parliament Hill to be in the hands of the builders. Those who are acquainted with the site will remember that although much of the ground is very high, yet a great portion of the heath falls below the level of the adjacent land. This is especially noticeable on the south-east side, which forms a long narrow valley. What would be the outlook then from these low-lying portions? Instead of green fields, ornamented with hedgerow timber, the same bank of bricks and mortar would raise its head as meets the eye looking Londonwards. It is true that from the elevated Spaniards Road the view towards the north could not be much marred, but although it would still be open to the breeze, that breeze would carry with it the smoke of a thousand chimneys. The result, then, of building upon Parliament Hill would have been to shut in Hampstead Heath to a very great extent, and even that large area would appear comparatively insignificant.

All this was foreseen by those who advocated the

acquisition of Hampstead Heath, and so, ever since 1857, there has been a constant agitation to preserve, not only the heath, but also the lands adjoining, which so much contribute to its charm. Hampstead Heath had been used and enjoyed for twelve years before any active steps were taken to secure Parliament Hill fields as an addition to the heath. Then, at a public meeting held in January, 1884, at the Holly Bush Tavern, it was decided to appoint a committee for the purpose of extending Hampstead Heath and preserving Parliament Hill. This Hampstead Heath Extension Committee was a most influential one; among the lengthy list of noblemen and others appeared such names as the Archbishop of Canterbury, Cardinal Manning, and the Marquis of Salisbury. The name of the chairman, the Duke of Westminster, has been associated with more than one movement of this kind, and too much cannot be said in praise of the indefatigable vice-chairman, the Right Hon. George Shaw-Lefevre, M.P., to whose individual efforts much of the success of the undertaking was due. The object of the committee was to acquire the land belonging to Sir Spencer Wilson, adjoining Hampstead Heath, known as the East Park Estate, and so much of the adjoining property of the Earl of Mansfield as his lordship was willing to sell to the public. The scheme was warmly supported by the press, but great difficulty was experienced in obtaining any definite price for the land to be acquired, so that when the question of purchase was brought before the late Metropolitan Board of Works, that body decided, in November, 1885, to take no further steps, because of the very large sum which was asked for the land in question. Such a decision would have crushed a less enthusiastic body than the Hampstead Heath Extension Committee, but they pertinaciously stuck to their guns, and set to work to obtain definite offers from the owners of the lands, and also promises of pecuniary assistance from large public bodies. When they appeared before the late Board, in November, 1886, they had the satisfaction of stating that they had succeeded in inducing

the owners of the Mansfield estate and Sir Spencer Wilson to give an option of purchase of all those portions of their estates which were necessary for carrying out the scheme, the sums amounting in the aggregate to £294,000, to which had to be added various legal and other expenses, making the total required about £305,000. But this was not all, for they had obtained an Act of Parliament (Hampstead Heath Enlargement Act, 1886) empowering the late Metropolitan Board of Works to purchase the land, and authorizing a contribution of £50,900 from the City Parochial Charity Funds. This, together with promises from the vestries of Hampstead and St. Pancras of £50,000, reduced the burden to be thrown upon the ratepayers to about £200,000. These facts considerably altered the case, and the question was re-opened, and eventually, on October 14, 1887, the Board resolved to contribute one half of the cost of acquiring the land, such sum not to exceed £152,500. This still left a sum of nearly £50,000 to be raised by public subscription, and the committee were equal even to this emergency, and the battle of Parliament Hill was won.

The actual cost of acquisition was £301,702, towards which the Hampstead Vestry contributed £20,000, St. Pancras £30,000, Marylebone £5,000, Charity Commissioners £50,000, and public subscriptions £46,000, leaving a balance of £150,702 to be provided from the rates. A small addition of $2\frac{1}{4}$ acres was made in 1890 by the purchase from the New River Company at a cost of £6,500 of the disused reservoir in Highgate Road, which was about to be built upon.

The viaduct which spans one of the ponds on Parliament Hill is a standing reminder of the narrow escape which Hampstead Heath had from being built over. It was intended as part of a carriage-drive towards a house which the Lord of the Manor contemplated erecting on the heath. It is by no means an unpicturesque feature, as will be seen by the illustration.

Parliament Hill, which had thus been rescued from the

grasp of the builder, forms a charming addition to the open spaces of London. On entering from Gospel Oak Station, we find ourselves in an extensive level area, which is devoted to cricket and lawn tennis. On the left the ground gradually rises to the summit of Parliament or Traitors' Hill, from which commanding views of Hampstead Heath and Highgate are obtained. Turning now in the direction of the ponds, we have one of the finest views in the direction of Highgate.

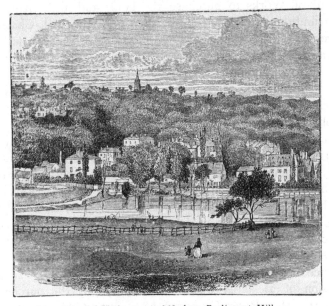

View of Highgate in 1868 from Parliament Hill.

Standing on the high ground close to the southernmost pond, we can look across a valley thickly studded with trees to the graceful church of Highgate crowning the summit of the hill. The eastern boundary, Millfield Lane, is still rural enough to claim the designation of ' lane,' although much of its charm has vanished. The next point to be visited is the Tumulus, railed in and planted with firs, from which elevated spot we can survey all the surrounding landscape—the chain of ponds at our feet, to the north the dense Ken Wood, and all around the undulating meadows, the very sight of which fills us with fresh vigour. The land lying between the Tumulus and the heath was used for brick-fields before being acquired for public use, and it is needless to say how great an improvement to the general surroundings the abolition of these works has been. The rest of the land undulates gently towards the elevated Spaniards Road, when we are on the heath once more.

The East Park estate, purchased from Sir Spencer Wilson, formed part of the demesne lands of the Manor of Hampstead, the descent of which we have already traced.

The remainder, a part of the Ken (or Caen) Wood estates, is in the parish of St. Pancras, and is part of the Manor of Cantalowes. There are four manors described in the Doomsday Book as being in the parish of St. Pancras—Tottenhall, Pancras (including land near the old church and round about Somers Town), Ruggemere and Cantalowes. This last was then the property of St. Paul's, the entry running as follows : 'The canons of St. Paul's hold four hides. There is land to two ploughs. The villanes have one plough, and another plough may be made. Wood for the hedges, pasture for the cattle, and twenty pence rent. There are four villanes who hold this land under the canons, and seven cottages. Its whole value is forty shillings ; the same when received in King Edward's time, sixty shillings. This land laid and lies in the demesne of St. Paul's.'

The first change of owners occurs in 1108, when we find that William Blemund gave to the monastic church of Holy Trinity (now called Christ Church, Aldgate) 'his wood with the heath-ground . . . in the parish of St. Pancras.' This gift was confirmed by a charter of the King in 1227, where the property is described as being 'close to the park of the Lord Bishop of London on the south side.'

At the time of the dissolution of the priory in 1531, the land reverted to the King.

It remained as Crown property till 1544, when it was granted, together with the 'Millfields,' to two private gentle-

27

men, who two years later sold their interest to Lord
Wrothesley. In 1588 it was in the possession of a Mr.
Woodruffe, who disposed of it to a Mr. Gardiner.

From 1590 to 1640 there is a break in the records. In
this latter year we find that Sir James Harrington, of *Caen
Wood*, had a child baptized in Hornsey Church. In 1660
Sir James sold the estate to a Mr. Bill, who held the patent
office of King's printer. Pepys records how that he and
Lady Bill were sponsors at a christening, and that good
dame, not liking her name, called herself Lady Pelham.

Caenwood or Kenwood House, Highgate.

John Bill died in 1680, and directed that at the death of
his wife the estate should be sold.

In 1689 George Withers was in residence; and some time
prior to 1698, William Bridges, Surveyor-General of the
Ordnance, who was buried in the Tower. Coming now to
the eighteenth century, we find that in 1718 the property
belonged to William Dale, an upholsterer, of St. Paul's,
Covent Garden, who bought it out of his gains in the South
Sea Bubble. He mortgaged it for £1,575 to the Earl of
Hay; but as he paid neither principal nor interest, the

estate was sold, and presumably the Duke of Argyle was the purchaser, for he was in residence at Caen Wood in 1725. The Duke died in September, 1743, and bequeathed Caen Wood to his nephew John, third Earl of Bute, 'as a small consideration of the high esteem' in which he held him. Unfortunately the general public did not endorse this opinion, for Lord Bute is only remembered now by the extreme degree of public detestation in which he was held. In 1755 Lord Bute sold Caen Wood to its most illustrious possessor, the great Lord Mansfield, in whose family it has remained ever since.*

Lord Mansfield, when plain Mr. Murray, before his elevation to the peerage, was a frequent visitor to Hampstead and Caen Wood.

> 'The Muses, since the birth of Time,
> Have ever dwelt on heights sublime;
> On Pindus now they gathered flowers,
> Now sported in Parnassian bowers;
> And late, when Murray deigned to rove
> Beneath Ken Wood's sequestered grove,
> They wander'd oft, when all was still,
> With him and Pope on Hampstead Hill.'

Apart from these names, there are many other celebrities to whom the spot has been familiar. Sir Walter Besant, in making an eloquent appeal to the public for the preservation of Parliament Hill, concluded: 'As for the modern associations of these fields, they are many and . . . well known. They are shared with the recollections of Hampstead and Highgate. Here wandered Keats, Shelley, Leigh Hunt and Coleridge. Here in an earlier generation walked Addison, Steele and Pope. Here lived Akenside and Johnson. There is no end to the literary interest of Highgate and Hampstead. But sacred associations will not save the fields. Nothing will save them but money.'† Fortunately the

---

* The details of the descent of Caen Wood are an abstract of J. H. Lloyd's 'Caen Wood and its Associations.'

† Sir Walter Besant, 'Traitors' Hill,' *Cornhill*, vol. vi., p. 638.

fields have been saved as well as the associations, and the spots which have been honoured in the past by the presence of these famous men will remain as open and free for future generations.

Parliament Hill is made up of a series of hills and valleys, and two of these eminences call for special remark, for around them cluster the chief historical associations. The southern one nearest to Gospel Oak is Parliament Hill proper, whilst the northern, nearest to Ken Wood, is called the Tumulus. It was once distinguishable by a fine clump of firs, which have now gone, with the exception of two bare trunks, which appear in the landscape like the shattered masts of a ship-wrecked vessel.

We will deal with Parliament Hill first. Many are the attempts which have been made to account for the singular name it bears. The usually accepted theory is that this is the spot where the cannon of the Parliamentary forces were planted to defend London from the Royalists in that mighty upheaval which brought Oliver Cromwell to the front. Parliament Hill seems a very unlikely position to have been chosen for such defence works, seeing that close by are more commanding situations on Hampstead Heath and Highgate Hill. But, apart from this, it would have been too far from the Metropolis, and, so far as is known, the extreme posts northward of the Parliamentary fortifications were at Islington and Pentonville.*

Another more probable explanation is that the spot was connected in some way with the Parliamentary elections for the county, or possibly with some older form of Parliament, such as the Hundred-moot or Folk-moot. These latter were held in May and October. Professor Hales, in supporting this explanation, says : ' The fact of there being a barrow on the hill does not render the " moot " theory less probable, but rather the opposite. Hills with barrows upon them, and barrows themselves, were, in fact, often used as moots. The hill assemblies seem to have been glad to avail themselves of

* Lloyd, ' Highgate.'

the reverence attached to such situations. The place where the dead lay (even the dead of another race) was not likely to be rudely disturbed.'*

But the eminence also bears another name, viz., that of Traitors' Hill. This appellation may have been invented to account for a tradition that lingers round the spot, but for which there is no confirmation. It is to the effect that the conspirators in the Gunpowder Plot took up their stand here on November 5, 1605, to watch the blowing up of the Houses of Parliament to be carried out by Guy Fawkes. But in all probability, as soon as the plot was discovered, Catesby and the rest of the conspirators were galloping away from London, so that the theory of their calmly standing on Parliament Hill on this eventful 5th of November must be dismissed at once. Professor Hales suggests another and a much more probable association of this hill with traitors, less remembered, it is true, than the Gunpowder Plot conspirators, but none the less traitors. They went by the name of the ' Fifth Monarchy men,' and were a Puritan sect who supported Cromwell's Government in the expectation that it was a preparation for the ' Fifth Monarchy,' *i.e.*, the monarchy which should succeed the Assyrian, the Persian, the Grecian, and the Roman, and during which Christ should reign for a thousand years.† The leader of the sect was Thomas Venner, a wine-cooper, who also preached at a conventicle in Coleman Street.

On January 6, 1661, Venner and his crew issued forth on their errand of taking London. ' They marched up and down several streets, and killed one or two persons, then " hastened to *Cane Wood*, between Highgate and Hampstead, where they reposed themselves for the night." In fact, they reposed three nights. On Wednesday the unhappy bigots ventured into London again, and were in no long time finally suppressed. A few days afterwards Venner and another (one Hodgkins) were hanged, drawn, and quartered over against

---

* Professor Hales, Lecture on ' Parliament Hill.'

† ' Encyclopædia Britannica,' article ' Fifth Monarchy Men.'

the meeting-house from which they had marched forth in their frenzy less than a fortnight before.'*

It should be mentioned that there is another 'Traitors' Hill' in the grounds of Lady Burdett-Coutts close by, which forms a striking object from the western side of Highgate Cemetery. But there is still another class of traitors who may have given their name to either of these eminences. It is said that they were the followers of Jack Cade, and some authority is given to the tradition by the statement made by Stow that Thomas Thorpe, Baron of the Exchequer, was beheaded by the insurgents at Highgate.†

Shadowy as are the legends connected with Parliament Hill, those associated with the northern eminence, the Tumulus, or mound, are still more so. But tradition can lead us out of most difficulties, and the reason handed down for the *raison d'être* of the Tumulus is as follows : At some very remote time, so it is said, the inhabitants of the old Roman town of Verulamium (St. Albans) were anxious to make it the capital of this part of England. Finding London a dangerous and growing rival, they set out to attack and destroy it. But in this they were disappointed, for the Londoners met and defeated their enemies at this spot, and buried their bodies in the mound which we now see so jealously guarded to-day.‡ Sir Walter Besant asks : 'Was such a thing ever possible? It was once possible, within certain limits of time—say during the first century before Christ and the first half-century after. When Cæsar invaded Britain, internal war was prevailing through the aggressive policy of Cassivelaunus, King of the Catuvelauni, and especially between that tribe and the Trinobantes. Now, the capital of King Cassivelaunus was the city of Verulam, and one of the principal towns of the Trinobantes was London. As the former folk held Western Middlesex and a part of Hertfordshire, and the latter the rest of Middlesex

---

* Professor Hales, 'Parliament Hill.'

† Lloyd, 'Highgate.'

‡ Howitt, 'Northern Heights,' p. 330.

with Essex and part of Hertfordshire, the common frontier
was of great length.   In the year 55 B.C., or shortly before
it, the Catuvelauni fought with and slew Imanuentius, the
Trinobantine King, and drove his son Mandubratius into
exile, and so far reduced and humbled the Trinobantes that
they threw themselves under Cæsar's protection.'*   With
Cæsar's departure, the King of the St. Albans tribe became

The Tumulus, Parliament Hill, 1870.

as aggressive as before, and his action was imitated by his
successor.   'The memory of some battle in this long-raging
warfare may probably enough be preserved in the tradition
attached to the barrow still to be seen near Hampstead
Heath.   One may well suppose that it was a battle of special
note and importance since it made so lasting an impression

* Sir Walter Besant, 'Traitors' Hill,' *Cornhill* (new series), vol. vi.,
p. 635.

on the popular mind, and we may very plausibly conjecture that it was the very battle in which fell King Imanuentius himself. Looking at the lie of the country from the southern hill, we might suppose that the invaders had advanced from the north through the dip between the Hampstead and Highgate hills, and so entered the Valley of the Fleet, and were making for London, when the Londoners, marching up that valley, met them at this spot, and dyed the stream with their own and their enemies' blood. Standing on the barrow, and looking north, one may picture very well the rush of those fiery Britons down the slopes, and the hand-to-hand encounter in the valley.'* But the matter-of-fact nineteenth century demands something more than these romantic stories, and is always anxious to exchange fancy for fact. Learned societies and scientific men generally had long been speculating as to what the tumulus consisted of, and what it contained, and pressure was brought to bear upon the London County Council with a view to the mound being opened in the interests of antiquarian research. It must be confessed, however, that the neighbourhood of London is not a favourable one for conducting archæological examinations of this sort. The presence of an obtrusive public somewhat hampers any operations, and the probability is that any prominent tumuli would have been already rifled in the hope of obtaining buried treasure. A popular belief was very much current to the effect that the Tumulus was the burial-place of Boadicea, Queen of the Iceni.

It was hoped that the opening of the Tumulus would set at rest all the many rumours, and, nothing daunted by the many difficulties in the way, the London County Council decided to undertake the work. Mr. C. H. Read, F.S.A., a British Museum expert, most generously offered to superintend the opening, and he received great help from Mr. George Payne, F.S.A., of Rochester, and one bleak day in October, 1894, the work was commenced. As the operations

* Professor Hales in *Athenæum*, November 17, 1883, and January 26, 1884.

evoked such general interest at the time, we cannot do better than quote extensively from Mr. Read's report. He says:

'This barrow lies on the northern slope of the hill, immediately between the Vale of Health on the west and Parliament Hill on the east. Its appearance before excavation was that of a circular mound sloping gradually on the north and south sides to a nearly level base, and entirely surrounded by a ditch varying from 16 to 20 feet in width. On the E.N.E. and W.S.W. sides a bank of earth was thrown up, making a broad rib towards these two points, extending to the ditch on either side. Upon the top of the mound are standing the bare trunks of two fir-trees, all that remain of a group that is said to have been planted about a century ago, and was finally destroyed by lightning within the last five-and-twenty years. An old hedge remains upon the inner side of the ditch.

'The mound is not a true circle, the diameter being about 135 feet to the outside of the ditch from east to west, while from north to south it is about 10 feet wider. The height of the centre of the mound above the ground-level would be about 10 feet.

'It is hard to say how the tradition connecting this mound with Queen Boadicea came into being, but I have not been able to find any other than modern mention of it. Traditions of the kind are frequent enough, but they more commonly attribute the erection of ancient mounds or encampments to a race than associate them with any individual, though instances of the latter are known also, such as Cæsar's Camp in the same parish as the barrow. But all over England are to be found Danes' camps and Danes' dykes; when the latter are examined they are usually found to be of pre-Roman origin, and Danesbury Camp, near Northampton, which has been recently explored, was proved, by the numerous remains of weapons and implements, to be without question an ancient British cemetery of perhaps the first century B.C. It is scarcely necessary, however, to bring forward evidence to prove that popular nomenclature is seldom supported by historical facts. In the present instance there is an obvious improbability in the popular attribution of the barrow as the burial-place of the Queen of the Iceni. After her overwhelming defeat by the Romans, Boadicea is said by Tacitus to have put an end to her life by poison, and Dion Cassius states that the Britons gave her a sumptuous funeral. It is unfortunate that there were no British writers to hand down their side of the question. Whether Boadicea really poisoned herself, or whether it only suited the Roman policy to say so, we shall probably never know, nor does it much affect the present matter. The statement of Dion Cassius is more important. Though he wrote more than a century after Boadicea's death, it seems unlikely that he would have expatiated upon the splendour of her funeral rites without some kind of authority for the statement. And the importance of his account lies in the fact that, if any such ceremony took place, it would scarcely have been in the immediate proxmity of London. It seems obvious not only that the Romans would never have permitted such a gathering of the Britons, but also that if the Iceni wished to bury their Queen in a fitting manner, they would do so in their own country, and therefore if the tomb of Queen Boadicea still exists, it must be looked for in Essex or Suffolk, not on Hampstead Heath.'

The conclusions that this expert came to after a thorough examination were :

' 1. That it is without question an artificial mound, raised at a spot where there was originally a slight rise in the ground.

' 2. That a great quantity of additional material was added to it, chiefly on the northern and eastern sides, and probably within the last two centuries.

' 3. That the tumulus had not been opened before.

' 4. That it is very probably an ancient British burial mound, of the early Bronze period, and therefore centuries before the Christian era. The burial was probably by inhumation, and the bones have entirely disappeared, a circumstance by no means uncommon. In this interpretation of the evidence my opinion is supported by that of Canon Greenwell, whose lengthened experience of these burials enables him to speak with an authority beyond question upon this point.'

The examination of the Tumulus was followed with some superciliousness by Celtic scholars, who are agreed that Queen Boadicea was buried in North Wales, and not near London. One of their. number, the Archdruid Morien, the author of ' The Light of Britannia ' (a work dealing exhaustively with the religious philosophy of the ancient British Druid bards and their symbols) contributed the following letter to the press at the time the Tumulus was opened:

' SIR,

' I have just visited the alleged grave mound of Queen Boadicea on the summit of Parliament Hill, and, after a careful inspection, I have come to the conclusion that it is not a tumulus at all, but one of those structures which the ancient Druids called a " Gwyddva " (*dd* as *th* in " then ").

' A " Gwyddva " signifies literally the Presence Place, meaning in ancient British a tribunal or pulpit, from the summit of which the officiating Druidic priest offered up prayer, and on which he also performed certain ritual practices " in the face of the sun and in the focus of light."

' The mound on Parliament Hill is one of the Llans or High Places of the British Druids.

' The Druids, like all ancient peoples, believed the earth resembled an island in shape, and standing out of the sea, and that its verge, or border, was where earth and the rational horizon were supposed to meet. Bees were sacred in the eyes of the Druids, and for that reason the beehives of the old straw pattern were constructed after the pattern of each of the Druidic Holy Hills. Homer, in his description of the shield of Achilles, which is a symbol of the round half of the world of the ancients, states that when the shield was completed " He poured the ocean round." Re-

ferring to the surface of shields as a mirror symbolical of
the reflection of the wisdom of the Creator, as shown on
the material earth, Homer states, " There shone the image
of the Master Mind!"

' In the eyes of the Druids every island of the sea was a
world. Thus Britain itself was a world.

' Now, each of the mounds like that on the summit of
Parliament Hill was a model of the shape of the whole
earth as understood by the Druids. Running around the
base of the mound on Parliament Hill are traces of a deep
trench. The trench must have been deep in ancient times,
for it is still a striking feature, though now carpeted with
verdure. That trench was formerly full of water, like the
similar one around Avebury, Wilts, a mile in circumference.
The mound symbolized the whole earth, and the trench full
of water around it the sea encircling the earth. By the
officiating Druid standing on the apex of the mound and
engaging in prayer it was implied that he stood on the top
of the whole earth, and that therefore he literally was nigh
unto God.

' The mound bore several characters :

' 1. As the whole earth, it was the Church, and to this
day the enclosure of a church is called Close, which is
obviously derived from the ancient British " Clas " (island).

' 2. The mound was also the symbol of the earth as the
garden of the sun (Adonidis Hortus).

' 3. The earth as a cemetery. It was Mynydd-y-Marw,
otherwise Mount Meru—that is to say, Mound of the Dead.
There is some mysterious connection between this name and
the name Mount Moriah, which I am inclined to believe was
anciently spelt Morsjah.

' 4. Each of those sacred British mounds being the
sanctuary—the Llan and Holy Hill—it was the spot where
each Act of the British Legislature, called " Rhaith " by our
British ancestors, was ratified and sanctioned in the presence,
as it were, of the Almighty Himself, " in the face of the Sun
and in the focus of light " of the Holy Hill.

'We thus see why the hill of the London mound is still called "Parliament Hill."

'It is profoundly interesting to recollect that in one spot within the British dominions, viz., the Isle of Man, the old custom of assembling around a mound to give sanction to legislative work is still duly observed on each July 5.

'The Sacred Mound in the centre of the Isle of Man is called "Tynwald," which is a corruption of the old British Twyn-y-Wlad, meaning "the Holy Hill of the Country." On July 5 the Governor of the Isle of Man and the members of the House of Keys proceed from Douglas, and partake of the Sacrament of the Lord's Supper in the pretty church of St. John, near the "Tynwald." Meanwhile the Manx people encircle the mound. Then the Governor ascends to the summit of the Holy Hill of the Country, and the M.P.'s of the island sit on the slopes, and the Governor reads the various Acts newly passed by the local Parliament, called the House of Keys; and the people express approval and willingness to obey the said Acts of the Legislature. That approval of the people has the force of the sign manual of the Sovereign, Queen Victoria.

'There is not the least doubt similar scenes were often witnessed in distant ages on the summit and around the mound on "Parliament Hill."

'I am, etc.,

'MORIEN, THE DRUID.'

'ASHGROVE, TREFOREST, GLAM.'

Although this systematic search in opening the Tumulus did not result in bringing to light any hidden treasure, as it was thought by many that it might, a most remarkable discovery of treasure-trove was made on July 21, 1892. A little boy, aged three years, was amusing himself in turning over the mole-heaps in the neighbourhood of the Tumulus with a wooden spade and pail, when he came across a bright article which aroused the attention of those who were with him. The digging was continued for a depth of some 7 or 8 inches, with the result that several gilt articles of solid silver and beautiful workmanship were unearthed and taken home. The father of the little fellow, Mr. Haynes— himself an éxperienced traveller and explorer, and the donor of some valuable foreign curiosities to the South Kensington Museum—acting upon legal advice, gave information to the district coroner, for, in accordance with a law passed in the reign of Edward I., an inquiry has to be held upon all articles thus discovered. The formal coroner's inquiry having been held, it was proved that the objects had been found as

Sheep on Parliament Hill.

stated, and constituted treasure-trove, and they were con- sequently handed over to the Treasury, who have deposited them in the South Kensington Museum, where they are now on view. They consist of two spirit or scent flasks with screw tops, a small flat cup with handles, two sockets and nozzles of candlesticks, and one small portion, probably the handle of a cup or a portion of a candelabrum, the weight of the whole being 59 ounces.*

* *Chambers' Journal*, 1893.

The district at the extreme south-east of Parliament Hill is called Gospel Oak, and takes its name from an oak which is shown on the plan in Park's 'Hampstead.' This was situated at the boundary-line of Hampstead and St. Pancras parishes, and its name serves as a relic of the times when it was usual to read a portion of the Gospels under certain trees in the parish perambulations, equivalent to 'beating the bounds.'* This was done every Ascension Day, and Herrick alludes to the custom in connection with a 'gospel-tree' in the following lines :

> ' Dearest, bury me
> Under that holy oak, or gospel-tree,
> Where, though thou see'st not, thou mayst think upon
> Me when thou yearly go'st in procession.'†

An ancient mill is once said to have crowned the summit of Parliament Hill, and so gave rise to the name of the adjoining Millfield Lane. The mill is mentioned in a description of the Caen Wood estate in 1660 : ' 280 acres of land well covered with large timber, and is set out as a capital messuage of brick, wood, and plaster, eight cottages, a farmhouse and *windmill*, fishponds, etc.'‡ Millfield Lane was formerly a delightful country retreat which abounded with hedgerow timber, but it is losing some of its rustic charm. Leigh Hunt, who knew it well, says: ' It was in the beautiful lane running from the road between Hampstead and Highgate to the foot of Highgate Hill that, meeting me one day, he (Keats) first gave me the volume (of his poems) If the admirer of Mr. Keats' poetry does not know the lane in question, he ought to become acquainted with it, both on his author's account and its own. It has been also paced by Mr. Lamb and Mr. Hazlitt, and frequented like the rest of the beautiful neighbourhood by Mr. Coleridge, so that instead of Millfield Lane, which is the name it is known by on earth, it has sometimes been called *Poets' Lane*, which is an appellation it richly deserves. It divides the grounds of Lords

---

* Larwood, ' History of Signboards. † Herrick, ' Hesperides.'
‡ J. H. Lloyd, ' Caen Wood and its Associations.'

Mansfield and Southampton, running through trees and sloping meadows, and being rich in the botany for which this part of the neighbourhood of London has always been celebrated.'*

Charles Mathews the elder lived in Millfield Lane. His residence, which he called Ivy Cottage, has been enlarged by succeeding proprietors, and is now called Brookfield House. It is very easily distinguished as the many-gabled house opposite the southernmost of the ponds. Mathews was born in 1776 at No. 18, in the Strand, where his father

Charles Mathews's House, adjoining Parliament Hill.

was a theological bookseller. He commenced acting at an early age, although he acquired no special reputation till 1803, but his name is inseparably linked with his 'At Home' entertainment, which he inaugurated in 1818. His mimicry and general versatility made him a great favourite. It is said that the bleak situation of his house, and the consequent force of the wind, which used to beat upon it very violently, much alarmed Mrs. Mathews. One night, after they had retired to rest, she was awakened by one of these sudden

* Leigh Hunt, 'Byron and his Contemporaries.' Quoted by Thorne.

gales, which she bore for some time in silence; at last, dreadfully frightened, she awoke her husband, saying: 'Don't you hear the wind, Charley? Oh dear! what shall I do?' 'Do?' said the only partially-awakened humorist. 'Open the window and give it a peppermint lozenge; that is the best thing for the wind.' His humour did not desert him on his death-bed. His medical attendant had given him some ink from a phial which stood in the place of the medicine bottle, and on discovering his error he cried out: 'Good heavens, Mathews! I have given you ink!' 'Never —ne-ver mind, my boy, ne-ver mind,' said the mimic; 'I'll —I'll swallow—bit—bit—of blotting paper.'* Ivy Cottage contained a set of apartments devoted to the fine collection of theatrical portraits, autographs, and engravings, which are now in the possession of the Garrick Club. But the house also contained other interesting treasures; among them was the casket, made from the wood of Shakespeare's mulberry-tree at Stratford-on-Avon, in which the freedom of that town was presented to Garrick in 1769, on the occasion of his jubilee.†

The next house to Ivy Cottage on the same side, called Millfield Cottage, is said to have been occupied for a short time by John Ruskin. He was consulted on the point, but he does not recollect anything further than that he lived with his father and mother when a child either at Hampstead or Highgate.‡

The estate on the right-hand side of Millfield Lane, known as Fitzroy Park, was the seat of Lord Southampton. Fitzroy House, a large square brick building, with capacious and finely-proportioned rooms, was erected about the year 1780. The grounds were tastefully laid out with gravel walks and carriage-drives, shaded by well-grown trees. The Earl of Buckingham resided here in 1811, but in 1828 the mansion-house was taken down, and the estate sub-

* Palmer, 'History of St. Pancras.'
† 'Old and New London,' vol. v., p. 411.
‡ Howitt, 'Northern Heights,' p. 418.

divided into several plots upon which villas were built.*   In
one of these villas lived Dr. Southwood Smith, the popular
physician, author of ' The Philosophy of Health.'†   In 1837
Dr. Smith was appointed by the Government to inquire into
the state of the poor, with a view to see how far disease and
misery were produced by unhealthy dwellings and habits. His
inquiries led to the passing of the Act for procuring improved
drainage, and ultimately to the establishment of the Public
Board of Health, of which he became a leading member.‡
He was also Physician to the London Fever Hospital,
and compiled a treatise on fever, which became a standard
medical work.   He died in Florence in December, 1861.

    Parliament Hill has its chain of ponds as well as Hamp-
stead Heath.   They are five in number, two of which are on
Lord Mansfield's property.   William Paterson, the founder
of the Bank of England, and the originator of the ill-fated
Darien scheme for the formation of a canal across the
isthmus of Panama, conceived the plan of collecting the
springs of Caen Wood into ponds and reservoirs.   His
company was established in 1690 for the supply of water to
Hampstead and Kentish Town, and was a great success till
the competition of the New River Company drove it from
the field.§   The New River, it will be remembered, had
been brought to London in 1614, but it was not a popular
undertaking, so that there was room for some time for this
younger company which has now gone.   The ponds still
remain, and a great delight they are to many, especially in
a severe winter, when they afford excellent skating.   This
is kept up till a late hour with the aid of torches and
Japanese lanterns, and the ponds present a very active and
picturesque scene with their thousands of skaters whirling
round and round in the crisp night air.   But all the year
round they are available for fishing, model yacht sailing,

    * Prickett, ' Highgate,' p. 79.
    † Thorne, ' Environs of London,' vol. i., p. 354.
    ‡ Quoted in Howitt's ' Northern Heights,' p. 325.
    § Howitt, ' Northern Heights,' pp. 330, 331.

and bathing unless they are frozen over, and even this does not prevent some enthusiasts from breaking the ice to have their morning dip. At one time, it is said, long before Izaak Walton breathed, the saintly monks who lived on the Ken Wood estate formed the fresh running waters of the Fleet into reservoirs for the breeding of fish, and thus originated the ponds. The fasts of the Church were very numerous, and the supply of salt fish was very limited owing to the difficulties of transit. These ponds well stocked with fresh fish would thus form a valuable possession to any monastery. *

Both Parliament Hill and Hampstead Heath were brought considerably nearer to many Londoners by the erection of a footbridge over the line of the North London Railway at Gospel Oak in December, 1895. The bridge is of steel, with blue Staffordshire brick, and cost £2,400, while the approach road from Gospel Oak was formed for half that amount. By means of the bridge a rapidly-growing district has now been directly connected with Parliament Hill, to reach which it was formerly necessary to walk a distance of nearly a mile.

* Lloyd, ' Caen Wood and its Associations,' p. 27.

CHAPTER XXII.

HIGHBURY FIELDS.

HIGHBURY FIELDS, 27½ acres in extent, are situated at the junction of Holloway Road and Upper Street, Islington. They were acquired in 1885 at a cost of £60,000, half of which was contributed by the Vestry of Islington. The area of the original fields was 25½ acres, but a subsequent addition of 2 acres at the extreme north brought the acreage up to its present extent. Under the Act of Parliament by which the purchase was authorized, the playing of music and public meetings were prohibited on the original ground, but no restrictions are attached to the small extension, where band performances are given in the summer. Lawn tennis is extensively played here, and cricket in the early morning is allowed on the lower field. Although the fields are enclosed at night time, they are not laid out as a park, for apart from a shrubbery at the margin, and gravelled walks, the area is left as a grass surface. Before the fields were purchased by the late Metropolitan Board of Works, the Great Northern Railway bought the land through which the tunnel passes, which carries their line from Finsbury Park to Canonbury. The surface was then leased by them to the late owner for 999 years from 1876, at an annual rent of £30. This liability was taken over when the fields were acquired for public use, but with this exception the land is freehold.

Highbury Fields are situated in the parish of Islington, which till almost a recent period was a district of open fields and fertile meadows, where cows were grazed to afford the milk-supply for the Metropolis. The fields of Islington were the favourite resort of Londoners who came here to drink milk warm from the cow, and to eat cakes dipped in cream and other dairy delicacies. Lord Macaulay, speaking of the state of London towards the close of the reign of Charles II., remarks that ' on the north, cattle and sportsmen wandered with dogs and guns over the site of the borough of Marylebone, and over far the greater part of the space now covered by the boroughs of Finsbury and of the Tower Hamlets. Islington was almost a solitude, and poets loved to contrast its silence and repose with the din and turmoil of the monster London.'* At the beginning of the same reign these fields for a short time presented a very different aspect, when the poor of London at the time of the Great Fire were flocking here in thousands from the burning city. Evelyn describes† very graphically how ' the poore inhabitants were dispersed about St. George's Fields and Moor Fields, as far as Highgate, and several miles in circle, some under tents, some under miserable huts and hovels, many without a rag, or any necessary utensils, bed or board, who from delicatenesse, riches, and early accommodations in stately and well-furnished houses, were now reduced to extremest misery and poverty.' There were some ' 200,000 people of all ranks and degrees, dispersed and lying along by their heapes of what they could save from the fire, deploring their losses, and though ready to perish for hunger and destitution, yet not asking one penny for relief.'

The district now known as Highbury was originally a part of Newington. The first mention of Highbury is in the year 1444 in a book which contains the names of the donors to the Hospital of St. John of Jerusalem. In it is the following entry : ' Domina *Alicia* de *Barowe* dedit dominium totum de *Highbury* et *Newton*, cum pertinentiis.'‡ This name

* ' History of England,' 8vo., 1849, pp. 349, 350.
† Evelyn's ' Diary,' September 5, 1666.
‡ Quoted in Tomlins' ' Perambulations of Islington,' p. 197.

of Highbury was at first confined to the immediate vicinity
of Highbury (Manor) House, so that the title Highbury
Fields is strictly correct.   Afterwards the district of Highbury comprised places further distant, which before were
simply called land in Islington.

The probable meaning of the word Highbury is 'high
barrow,' and points to the fact that in early times the
eminence upon which Highbury Fields stand was used as a
place of defence.   The ancient moat which formerly surrounded Highbury House may have been the remains of an
earlier means of defence—in fact, a part of a Roman encampment, for the word 'barrow' suggests some earthwork
thrown up either for defence, or for the burial of the slain.
From a mezzotint engraving of Highbury Place published in
1787, Highbury Hill seems to have been abrupt and steep on
the north and north-west, and to have been artificially
rounded or shaped, which work may consistently be attributed to the Romans.*

The Roman occupation of Highbury Fields must mainly
rest upon conjecture, but another encampment here is well
authenticated.   During Wat Tyler's insurrection (1381) a
detachment of the rebels under Jack Straw, after burning
and destroying the magnificent priory in St. John's Street,
proceeded for a similar purpose to the Prior's country-house
at Highbury.   According to Holinshed, the band of insurgents 'who tooke in hand to ruinate that house' was
estimated at 20,000.   They carried their plan of devastation
into complete effect, pulling down by main force those parts
of the building which withstood the attacks of the devouring
element.   This destructive mob, then, must have occupied
the site of our present Highbury Fields, and very unwelcome
visitors they were.†   This incident accounts for the old name
of Highbury House, viz., Jack Straw's Castle—identical with
that of the well-known tavern at Hampstead Heath.   The
moat surrounding the house, over which the insurgents

* Tomlins, 'Perambulations of Islington,' p. 176.
† Brewer, 'History of London and Middlesex,' vol. iv., p. 235.

passed, was filled in in 1855 ; and, popular as the fields now are, it is extremely unlikely that they will ever see such a crowd as this again.

The site of Highbury Fields, which is shown in old maps as Mother Field, was alienated from the Manor of Highbury about 1780, when Sir George Colebrooke, the Lord of the Manor, sold the old mansion called Highbury House or Castle, together with these adjoining lands, to John Dawes, from whose descendant they were purchased as a recreation-ground. The earliest owner of the manor that can be traced is Bertram of the Barrow, the ancestor of the Lady Alice who gave the manor to the Hospital or Priory of St. John of Jerusalem. Upon the dissolution of the monasteries the manor was given, or intended to be given, to Thomas Cromwell, Earl of Essex, but before he could enter into possession he was attainted, and his estates were forfeited to the Crown. Without going into details of the various leases granted by the Crown, it may be stated that the manor was settled on the Lady Mary (afterwards Queen) before the death of Edward VI., and remained as Crown property till the reign of James I., who bestowed the manor on his eldest son. He died in 1612, and the manor once more reverted to the Crown till James I. granted it to his son Charles, who, after he came to the throne, bestowed it, in 1629, on Sir Allan Apsley. By subsequent stages it was sold, in 1630, to Thomas Austen, from whom it descended to Sir John Austen, Bart., and then, in 1723, it was sold to James Colebrooke, the ancestor of Sir George Colebrooke, Bart., the banker who sold the portion of the manor in which we are interested to John Dawes, as we have before mentioned.*

This owner, who was a wealthy stockbroker, proceeded, in 1781, to erect a handsome house on the moated site where the Prior's mansion had formerly stood. When the workmen were preparing to lay the foundations of the house, they discovered a number of pipes made of baked red earth, resembling those used for the conveyance of water about the

* Nelson, ' Islington,' pp. 133, 134.

time of Queen Elizabeth.   There were also some tiles, said
to be Roman, but which were more probably of Norman
manufacture.*   The house, which had cost Mr. Dawes
nearly £10,000 to build, was sold upon his death in 1788, for
£5,400, to William Devaynes, M.P., and a director of the
East India Company.   He in turn sold it to a celebrated
Highbury gentleman, Mr. Alexander Aubert, F.R.S., who
spent a considerable sum in altering and improving the
estate.   He also erected a large observatory near the house,
in the arrangement of which he was assisted by his friend
John Smeaton, the eminent engineer whose name will always
be remembered in connection with the Eddystone lighthouse.
Aubert was always a good friend to Smeaton, and did a great
deal towards advancing him in his profession.   He also
revised and corrected for publication the account of the
building of the Eddystone.   When the 'Loyal Islington
Volunteers' were established, in 1792, mainly through the
efforts of Mr. Aubert, he was appointed their chief officer, so
that he was distinguished alike in the ranks of peace and
war.   Upon his death, in 1805, the house and grounds
were put up to auction in 1806, and were purchased by a
Mr. Bentley.†

Highbury Fields are particularly fortunate in being surrounded with substantial and well-built houses.   At the
beginning of this century Highbury Place was described as
being one of the finest rows of houses in the environs of the
Metropolis, inhabited by eminent merchants and other
persons of opulence.‡   The thirty-nine houses comprising
this row are built on land the property of Mr. Dawes, from
whom Highbury Fields were bought.   Leases were granted
by him during the years 1774 to 1779 to Mr. John Spiller,
by whom the present houses were erected.   In addition to
the large gardens behind, they had allotments in the
meadow ground in front, now Highbury Fields.   Before

* Ellis, 'Campagna of London : Islington,' p. 89.
† Nelson, ' Islington,' pp. 139, 140.
‡ Nelson, 'History of Islington,' 1811, p. 175.

Highbury Place was built there existed in the lower field, opposite to what is now No. 14, a conduit for supplying the City with water. Hence, in ancient maps of the district we find this field called the 'Conduit field'; and Camden, in 1695, speaks of 'an old stone conduit' situated between 'Islington and Jack Straw's Castle.' It must have been more than 200 years old when he wrote, for it was

Highbury Terrace, Islington, 1835.

made as part of a scheme of Sir William Eastfield, Lord Mayor in 1438, who 'caused water to be conveyed from Highberry, in pipes of lead, to the parish of St. Giles without Cripplegate, where the inhabitants of those parts incastellated the same in sufficient cisternes.'*

The question of London's water-supply is one that engrosses a considerable amount of attention at the present day, and fresh sources are being eagerly sought after. We can imagine,

* Stow, 'Survey,' quoted in Nelson.

then, the importance attached to these springs in the north of London, especially when it is remembered that the New River had not at this time been brought to the Metropolis. It was customary for the citizens to visit the conduit-heads, a duty which was made very pleasant by reason of the feasting, which was paid for out of the City purse. There were also other attractions, described in an account of one of these visits by Strype. On September 18, 1562, 'the Lord Mayor, Aldermen, and many worshipful persons, rode to the conduit-heads to see them, according to the old custom: then they went and hunted a hare before dinner, and killed her; and thence went to dinner at the head of the conduit, where a great number were handsomely entertained by the Chamberlain. After dinner they went to hunt the fox. There was a great cry for a mile, and at length the hounds killed him at the end of St. Giles's, with great hollowing and blowing of horns at his death; and thence the Lord Mayor, with all his company, rode through London to his place in Lombard Street.' The supply from these conduits was at best scanty, and the water had either to be fetched from them, or else water-carriers had to be paid to bring it. The vessels they carried the water in were called tankards, and held about 3 gallons. The last instance that is recorded of their actual use is connected with Highbury, by a servant of James Colebrooke, the Lord of the Manor, whose business in town was carried on at a house behind the Royal Exchange.*

When the conduit-house was removed, the spring was arched over with brick, and its site marked by an upright stone. Before Highbury Place was built the conduit remained open as a watering-place for cattle, and afterwards it supplied these houses with water by means of pipes, which were connected with wells or reservoirs behind the houses. In 1692 an official report was made to a special committee of the Corporation, which fully describes the route by which the water was conveyed from the conduit in Highbury Fields to Cripplegate. It runs as follows: 'And we have

* Nelson, 'Islington,' p. 149.

also . . . viewed the springs and water belonging to the Citty neare Islington; and find the same in two heads, *one covered over with stone, in a field neare Jack Straw's Castle,* which is fed by sundry springs in an adjacent field, and is usually called the White Conduit, the water whereof is conveyed from thence, in a pipe of lead, through Chambery* Park, to the other conduit in Chambery-field; and from thence the water of both the said heads so united is conveyed, in a pipe of lead, cross the New River, in a cant, into the Green Man fields, and entering from thence a garden . . . at about forty foot distance from Frogg-lane, into a field on the east side thereof; and from thence, cross the North-east corner of a garden at the hither end of Frogg-lane, into a field belonging to the company of Clothworkers; and from thence, through the field next to, and west of the footway from Islington, unto the stile by the Pest-house, where it crosseth the said way, and so along the east side thereof, cross the road at Old-street, and under the bridge there, into Bunnhill-fields; and from thence, on the west side of the said field, by the Artillery garden, crossing Chiswell-street, into and down the middle of Grubb-street, into Fore-street, and so on the south side thereof to the conduit at Cripplegate : and we cannot find that the said waters are employed to any other use than to the service of the said conduit.'†

All traces of the conduit have now been removed, and there is nothing to denote the site of this ancient landmark, dating back to the fifteenth century.

When the houses in Highbury Place were first built, there was considerable difficulty in letting them, and the first tenants of Nos. 2 to 8 had leases granted at from £34 to £36 per annum; their present value is quite three times this sum. At No. 38 lived for a good many years Abraham Newland, chief cashier of the Bank of England, whose signature was as well known as that of F. May in recent years. The son of a baker of Southwark, he was born in 1730, and

* Canonbury.   † Ellis, 'History of Shoreditch.'

at the age of eighteen was appointed a clerk in the Bank
of England, from which position he rose by successive
gradations to be chief cashier after thirty-four years' service.
He was a most conscientious official, and for twenty-five
years he never slept a single night out of the Bank. It was
his custom to come to Highbury Place in his carriage after
dinner, and take tea with his housekeeper; then to go for
a short walk in these fields, and afterwards return to the
Bank to sleep. Mr. Newland resigned in 1807, owing to
the infirmities of old age, and died in the same year. He
was buried at the Church of St. Saviour, Southwark.* His
name was popularized in a successful song of the day, from
the pen of Charles Dibdin junior, manager of Sadler's Wells
Theatre:

> ' There ne'er was a name so bandied by fame,
>   Thro' air, thro' ocean, and thro' land,
>   As one that is wrote upon every bank-note,
>   You all must know Abraham Newland.
>   Oh ! Abraham Newland,
>   Notified Abraham Newland !
>   I've heard people say, " sham Abraham " you may,
>   But you mustn't sham Abraham Newland.'

Perhaps the best of the other verses is this:

> ' The world is inclin'd to think Justice is blind,
>   But lawyers know well she can view land ;
>   But, Lord, what of that—she'll blink like a bat,
>   At the sight of an Abraham Newland !
>   Oh ! Abraham Newland,
>   Magical Abraham Newland !
>   Tho' Justice, 'tis known, can see through a mill-stone,
>   She can't see through Abraham Newland.'

Another celebrated resident at Highbury Place was John
Nichols, the historian of Canonbury, and for nearly fifty
years editor of the *Gentleman's Magazine*. He was a partner
of William Bowyer, the celebrated printer. He was a friend
of Dr. Johnson, and seems to have been an amiable and

* Nelson, ' Islington,' pp. 176-180.

industrious man, very popular with his friends. He died suddenly in 1826, while going upstairs to bed.

On the opposite side of what is now Highbury Place, and just to the north of Highbury Fields, was the large barn or farm attached to the manor-house. This gave way to a small ale-house, which in course of time developed into a tavern with tea-gardens attached. The Court Baron for the manor used to be held here, and when the business outgrew the limits of the original building, a large barn belonging to the adjoining farm was added to the premises, and so it legitimately received the name of Highbury Barn, which hitherto it could only take as being on the site of an ancient barn.* Under the proprietorship of Mr. Willoughby, who died in 1785, the tavern prospered exceedingly, and his son made extensive additions to the grounds in order to accommodate the numbers attracted to Highbury. A dinner has been served here for 800 persons, on which occasion upwards of seventy geese might have been seen roasting at one fire. The tavern was afterwards provided with a theatre and a dancing-room, and all the attractions of a modern Vauxhall Gardens.

A remarkable society, known as the Highbury Society, used to meet here in years past. It was a friendly association of Protestant Dissenters, who combined together at a time when the privileges of that body were greatly endangered by a Schism Bill, which was directly levelled against all Nonconformists. Queen Anne died on the day on which this Act was to have received the royal sanction, in celebration of which event this society was formed. The meetings were first held at Copenhagen House, but from 1740 onwards Highbury was the place of rendezvous. From a short account of the society, published in 1808, we gather the following particulars regarding their meetings. It appears that the party, who walked from London, after a short stop at Moorfields, proceeded to Highbury, and, to beguile the way, it was their custom to bowl a ball of ivory in turn at objects

* Nelson, ' Islington,' p. 155.

in their path. After a slight refreshment they repaired to
the field for exercise ; but in those days of greater economy
and simplicity, neither wine, punch, nor tea were introduced, and eightpence was generally the whole individual
expense incurred. A particular game called hop-ball formed
the recreation of the members of this society at their meetings. On the board (dated 1734) used for the purpose of
marking the game, the following motto was engraved: ' Play
justly, play moderately, play cheerfully; so shall ye play to
a rational purpose.' The principal toast at their annual
dinner in August was: ' The glorious first of August, with
the immortal memory of King William and his good Queen
Mary, not forgetting Corporal John ; and a fig for the
Bishop of Cork, that bottle-stopper.' How this toast
originated is not known, but it probably arose out of some
of the events which led to the formation of the society.
John, Duke of Marlborough, the great friend of the Protestant and Whig interest, was in all probability the ' Corporal
John ' of the toast. In the winter time the members used
to dine together weekly on Saturday, from November to
March.* The Highbury Society, with its dinners and other
oddities, is no more, as it was dissolved about the year 1833.
Highbury Barn was finally closed in 1871, in consequence
of the repeated refusal of the license, owing to the riotous
behaviour of many of the night visitors. In 1883 the greater
part of the site was covered with buildings, and a large
public-house, the Highbury Tavern (No. 26, Highbury
Park), alone commemorates this once popular place of
amusement.†

* From a ' Report of the Committee on the Rise and Progress of the
Highbury Society,' printed 1808.
† W. Wroth, ' London Pleasure-Gardens of the Eighteenth Century,'
1896, p. 165.

CHAPTER XXIII.

ISLAND GARDENS, POPLAR—ROYAL VICTORIA GARDENS.

THERE is a marked similarity between these riverside gardens. They are both reclaimed marsh ground, situated in comparatively new but rapidly developing districts. There is a railway-station adjoining each, and if the scheme for establishing a free ferry between Greenwich and Poplar had been carried out, the likeness would be still more marked. The principal feature of both places is the long river terrace, which affords excellent views of the shipping passing up and down the Thames. Another similarity in their past history is the periodical flooding to which they have been subject, owing to breaches in their embanking walls.

ISLAND GARDENS, POPLAR.

Island Gardens, those nearest to town, are situated at the extreme south of the Isle of Dogs, opposite Greenwich Hospital. This is almost the only portion of the river-front of the Isle of Dogs which is not used for wharfage or commercial premises. Prior to 1830 this district was practically uninhabited, but the outer fringe is now wholly taken up by the various shipbuilding yards and the many other industries connected with the docks, which are bringing a large resident population here. In fact, it is only the depression in trade, in consequence of the removal of the greater portion of the shipbuilding from the Thames, that has prevented the land being swallowed up for trade purposes some years ago.

The gardens, which are about 2½ acres in extent, supplied a long-felt want in this populous district. Although the surroundings are far from picturesque, from the river promenade a fine view of Greenwich Hospital, on the opposite bank, is obtained, while in the background on the east and west rise the wooded heights of Shooter's Hill and Greenwich Park.

The ground when acquired for public purposes was in a very rough and neglected condition, and paths had to be formed, drained and fenced, which, together with other works, cost nearly £2,000. A residence had been built at one end of the ground, part of which is occupied by the foreman, whilst the remainder is used as a free library. Near the centre of the gardens an inexpensive bandstand, surrounded with a rockery, has been erected, where performances are given during the season. In a corner of the ground is a gymnasium; but the principal feature of the laying out has been the formation of a gravelled promenade along the river-front, which is nearly 700 feet in length. This is liberally provided with seats, and affords splendid views of the river and its surroundings.

The question of acquiring this ground as an open space had occupied public attention on more than one occasion before this end was attained. The land had been let by the Admiralty (who held a lease from the freeholders, the trustees of Lady Margaret Charteris) to the Cubitt trustees, with a reservation that no buildings, except certain villa residences, were to be erected without their consent. One of these villas (Osborne House) was built, and the foundations for another prepared, but it was found that there was no demand for residences of this class in the locality.

After the site had narrowly escaped being built upon for wharfage purposes, negotiations for its acquisition as an open space were again opened in 1889, and some four years later these came to a successful issue. Three parties had rights in the land which had to be purchased—the Cubitt trustees, the Admiralty, and the trustees of Lady Charteris.

The sum of £5,000, which represented the value of the Cubitt trustees' interest, was subsequently reduced to £3,000. They stated that this material reduction from the former price was in the nature of a gift for the public benefit, having regard to the use to which the land was to be put. The amount finally arranged with the Admiralty for the original plots was £3,000, but in addition to this they most generously added as a free gift the land occupied by Osborne House, for the rent of which they received £18 per annum. A further sum of £500 was given to the Cubitt trustees for their interest in this additional plot, and the reversion of the freehold of the whole property was purchased from the trustees of Lady Margaret Charteris for £2,200, making a total of £8,700, towards which the Poplar District Board of Works contributed £3,500. After the laying-out had been completed, the gardens were publicly opened amid great enthusiasm on Saturday, August 3, 1895.

Turning now to consider the history of this riverside recreation-ground, the first thing to be explained is the very peculiar title of Isle of Dogs which the district bears. The name does not seem to be older than the time of Elizabeth, before which the place was called Stepney or Stebonheath Marsh. The first use of the present name is found in some unenvious company in a record of the trial of James Naylor, the Quaker, for blasphemy. The debate as to the prisoner's punishment turned on the delightful alternatives of slitting or boring his tongue, cutting off his hair, whipping, or exiling him to Bristol, the Scilly Isles, Jamaica, the Isle of Dogs, or the Marshalsea.*

Among the many theories advanced for the origin of the name may be mentioned that of Maitland, who writes in 1756 that the Isle of Dogs was first so denominated by sailors from the great noise made by the King's hounds that were kept here during the residence of the Royal Family at Greenwich. He probably took his information from Strype's

* J. G. Miall, 'Footsteps of our Forefathers,' 1851.

29

edition of Stow's 'Survey,' 1720, who mentions the 'Marsh . . . usually known by the name of the Isle of Dogs, so called because when our former Princes made Greenwich their country seat, and used it for hunting (they say), the kennels for their dogs were kept on this marsh, which, usually making a great noise, the seamen and others thereupon called the place the Isle of Dogs.' Other versions of the same story add that the various Princes resided during the sporting season at Greenwich Palace, and kept their dogs here as a convenient spot close to Waltham and the other royal forests in Essex.

A second legend gives an entirely different reason for the name. It runs as follows:

'It is called the Isle of Dogs, as is reported, from a waterman's murthering a man in this place, who had a dog with him, which would not leave his dead master, till hunger constrained him to swim over to Greenwich; which being frequently repeated, was observed by the watermen plying there, who, following the dog, by that means discovered the body of the murthered man. Soon after, the dog returning on his accustomed errand to Greenwich, snarled at a waterman who sat there, and would not be beaten off, which encouraged the bystanders who knew of the murder to apprehend him, who thereupon confessed the fact, and, after due prosecution at law, was hanged on this spot.'*

Another variation of this peculiar name must also be mentioned: 'In some ancient writings possessed by the Corporation of the City of London, this marsh is termed the *Isle of Ducks*, a mode of denomination that has not been noticed by any topographer, but which may readily be supposed to allude to the number of wild-fowl which formerly frequented the spot.'† In a map dated 1740, in the possession of the Commissioners of Sewers, it is also called by this name. There was formerly a small spot on the south side of High Street, Poplar, at the west end, known as Duck Island,

* Griffiths, 'River Thames,' 1746, p. 43.
† Brewer, 'Beauties of England and Wales.'

and this appellation may have been misapplied to the whole district.*

The chief characteristics of the Isle of Dogs in the past seem to have been the number of windmills on the shore, and the rich pastures within the marshes. The memory of the windmills still survives in the name of Millwall, which embraces the whole of the western sides of the so-called island. Among the many old views showing these mills we may mention one of London and Westminster, published in 1752, taken from One Tree Hill in Greenwich Park. On this are shown seven windmills upon the river-bank, opposite Deptford, with a small building attached to each mill.† It is natural that the marsh should have been extremely fertile, intersected as it was with creeks, and surrounded with water. Most of the old descriptions of this place lay particular emphasis on this fact. An old historian says: ' Such is the fertility of this marsh, that it produceth sheep and oxen of the largest size, and very fat. They are brought out of other counties and fed here. I have been assured by a grazier of good report (saith the Rev. Dr. Woodward) that he knew eight oxen sold out of this marsh for £34 each. And all our neighbourhood knew that a butcher undertook to furnish the club at Blackwall with a leg of mutton every Saturday throughout the year that should weigh twenty-eight pound, the sheep being fed in this marsh, or he would have nothing for them ; and he did perform it.'‡ Some other writers went so far as to say it had the richest grass in the country, and others that it was a kind of convalescent home for cattle, which were sent here when on their last legs in order to be fattened up for the market. In Norden's map of Middlesex (1593) it is referred to as the Isle of Dogs Ferme.

The historical notes attaching to this little river garden are not very numerous. All the interest clings to the opposite shore around the many fine buildings—the parish

* Cowper, ' History of Millwall,' 1853, p. 16.   † *Ibid.*, p. 19.
‡ Strype's edition of Stow's ' Annals.'

church of St. Alphege, erected on the traditional site of his
martyrdom by the Danes; the classic pile of Greenwich
Hospital; the peculiar domes of the Greenwich Observatory.
All of these would have a long tale to tell. The maps of
some fifty years back are content to mark the site of the
gardens as 'reed ground.' It was very nearly being acquired
by the late Metropolitan Board of Works some years ago
for another purpose. When the question of providing communication between the two shores of the Thames below

Greenwich Hospital from Island Gardens.

London Bridge was under serious consideration, one of the
proposed sites for a free ferry similar to that at Woolwich
was between Greenwich and the Isle of Dogs. The landing-
stage for the boats, if this scheme had been carried out,
would have been on the western end of the present recreation-
ground. There was a clause in the Bill to enable the late
Board to lay out as a recreation-ground so much of the land
as was not required for the purposes of a ferry. At the time
when this proposal was under discussion, the Earl of Meath,

then Lord Brabazon, offered, on behalf of the Metropolitan Public Gardens Association, to lay out this land for public use. This scheme was not carried out, but the question of communication between the two shores has been solved in another way. A tunnel is about to be built under the Thames similar to that at Blackwall, for one of the entrances to which a small portion of the gardens will be required.

There has been a ferry between these points for nearly three hundred years, and perhaps longer. Pepys records his crossing here on July 24, 1665. The extract is so very amusing that it is worth being repeated. It appears that he had gone down to Deptford on business, 'and by-and-by went over to the ferry and took coach and six horses nobly for Dagenham.' After 'spending the day most pleasantly with the young ladies,' they prepared to come home. ' We set out so late that it grew dark, so as we doubted the losing of our way, and a long time it was or seemed before we could get to the water-side, and that about eleven at night, where, when we came, all merry, we found no ferry-boat was there nor no oars to carry us to Deptford. However, afterwards, oars were called from the other side at Greenwich; but, when it came, a frolic, being mighty merry, took us, and there we would sleep all night in the coach in the *Isle of Dogs*. So we did, there being now with us my Lady Scott, and with great pleasure drew up the glasses, and slept till daylight, and then some victuals and wine being brought us, we ate a bit, and so up and took boat, merry as might be ; and when come to Sir G. Carteret's, there all to bed.' A day or two later he had another stay in the Isle of Dogs, which was not quite so pleasant. Owing to the plague, the Admiralty officers had resolved to meet at Deptford, where he arrived at six in the morning, and he continues : ' By water to the Ferry, where, when we came, no coach there, and tide of ebb so far spent as the horse-boat could not get off on the other side of the river to bring away the coach. So we were fain to stay there in the unlucky *Isle of Dogs*, in

a chill place, the morning cool and wind fresh, above two, if
not three hours, to our great discontent.'

The many references he makes to this ferry give us some
idea of its importance as a means of communication in those
days. Frequently mention is made of messengers coming
and going by this route to Hackney and other parts of the
north of London. The walk across the marsh must have
been very dreary, if not dangerous, especially at night time.
In 1812 the ferry was owned by a society called the Potters'
Ferry Society. In that year an Act was passed creating a
statutory ferry for horses and vehicles in favour of the Poplar
and Greenwich Ferry Company, and by a later Act the
company were empowered to levy a toll of a penny for each
passenger landed from the foot-passenger ferry as a return
for the sums expended by them in the formation of the roads
thereto. The undertaking had originally cost some £200,000,
but was unremunerative. In view of the heavy claims for
compensation, the question of purchasing the ferry rights
had to be abandoned.

Discoveries which have been made at different times during
the excavations for the dock basins seem to make it probable
that a great change has come over the physical features of
the Isle of Dogs. In place of the marsh land as it is now
known, there must have been at some very remote period a
forest here. A very extensive list of quotations is given in
Cowper's ' History of Millwall' bearing on the subject, from
which we take the following : ' In the Isle of Dogs a forest
of this description was found at 8 feet from the grass, consisting of elm, oak, and fir trees, some of the former of which
were 3 feet 4 inches in diameter ; accompanied by human
bones and recent shells, but no metals or traces of civiliza-
tion. The trees in this forest were all laid from south-east
to north-west, as if the inundation which had overthrown
them came from that quarter.'* Another authority supposes
that the cause was an earthquake. The writer, in describing
the discovery of the subterranean forest, goes on to say,

* Weale, ' Survey of London,' p. 36.

' Some violent convulsion of Nature, perhaps an earthquake, must have overturned this forest, and buried it many feet below the present high-water mark; but when or how it happened is beyond the tradition of the most remote ages.'*

Lysons mentions 'that a great quantity of fossil nuts and wood were found' in digging for one of the basins of the East India Docks. This was in 1789; but exactly the same things had been found in the century before, as Pepys mentions them under the date September 22, 1665 : ' At Blackwall. . . . Here is observable what Johnson tells us, that in digging the late dock they did, 12 feet under ground, find perfect trees overcovered with earth. Nut-trees, with the branches and the very nuts upon them, some of whose nuts he showed us. Their shells, black with age, and their kernels, upon opening, decayed, but their shell perfectly hard as ever ; and a yew tree, upon which the very ivy was taken up whole about it, which, upon cutting with an adze, we found it to be rather harder than the living tree usually is.'†
As these remains have been found all over the Isle of Dogs, from the water's edge into Essex, the original forest must have been of considerable extent. In the Doomsday survey, under the heading of Stebonheath, mention is made of a wood, long since disappeared, which was perhaps part of this ancient forest.

The land in the Isle of Dogs, being below the level of the Thames, has always been liable to flooding. At various times the embankments have burst with disastrous consequences to the land-owners. One of the most serious of these took place in 1449, when a breach in the embankment was made, some 20 rods in length, by which 1,000 acres were flooded.‡

The banks were never properly repaired, and as a consequence similar calamities occurred in later years. Some

* ' Encyclopædia Londiniensis,' 1812, vol. xi., p. 408.
† Pepys' ' Diary,' Cassell's edition, 1664-5, p. 139.
‡ Cowper, ' History of Millwall,' p. 40.

account of these, taken from the original minute-books of the Poplar Commission of Sewers, may be found interesting.

'1 *Nov.*, 1629.—At this day many of the Commissioners then present, did meet uppon the marshe called Stebbinheath marshe, also Poplar marshe, where the great breach hapened on the 23rd day of October last, and survayed and viewed the said breach and the outer wall there, and being thereby fully satisfyed that the great breach was soe deepe and dangerous and had soe torne the marshe adjoining very farre into the said marshe, they weare fully satisfyed and resolved that the said breach could not be stopped in the place where the mayne wall was before, but that for the safety of the whole levell, it was necessary that another insett and inner wall should be made with a horse-showe (*i.e.*, horseshoe wall) to preserve the whole levell, and therefore did approve and allow the work already begunne, and did order and did decree that they should proceed to make a strong wall fitt for the defence of the said levell in the place where now the said insett is begunne, and that workmen, materialls, and all things necessary together with money to pay for the same bee forthwith provided for the full finishing and accomplishing of the said worke.'

By far the most serious of these breaches took place in 1659, and the minute on this occasion, dated March 28, 1660, states that 'the said unfortunate breach happened on Thursday, the 20th of March, 1659,' and they ordered, 'Whereas it appeareth unto this court by the presentment of the jury for the said marshe, who uppon a veiwe taken this present day of the walls and banks of the said marshe doe present uppon their oathes that the charge to make upp the great breach which happoned on the 20th day of March last will amount to the summe of £12,000 or thereabouts, the Commissioners do therefore order and impose a tax of fortie shillings the acre uppon every acre of ground, to be paid by the severall owners of lands within the said marshe on the 4th day of Aprill next unto Henry Dethicke, gent., at his house in Poplar, in the countie aforesaid, and to be expended by him for and towards the payment of such officers and workmen as shall be employed, and for such materialls as shall be bought and made use of in making upp the said breach.'

This is the breach referred to by Pepys under date March 23, 1660: 'In our way we saw *the great breach* which the late high water had made, to the loss of many thousand pounds to the people about Limehouse.'

In order to prevent a recurrence of these disasters, the Admiralty expended £8,000 in building the present river wall.

ROYAL VICTORIA GARDENS, NORTH WOOLWICH.

These gardens are situated on the Essex shore of the Thames, in what is called the ecclesiastical district of North Woolwich. Nearly half the parish of Woolwich is on this side of the water, from which has arisen the local saying that more wealth passes through Woolwich than any other town in the world, referring, of course, to the rich cargoes of the ships that pass along the Thames between the two halves of the parish. A curious tradition (very similar to one connected with Battersea) is current to explain why Woolwich should thus have extended its boundaries to the opposite shore. It is to the effect that a native of Woolwich was found drowned on the opposite shore, in Essex, and that the parish in which he was thrown refused to bury him; on this he was buried by the parish of Woolwich, which afterwards claimed the land where the body was discovered, and obtained a verdict in a court of law.*

For many centuries Woolwich was nothing more than a small fishing village, and its rise to importance is of comparatively recent date. North Woolwich is of still more modern growth. To go back only a few years, we find that it consisted of a few cottages, and the Old Barge House, which was the landing-place for the Woolwich ferry-boats. It is now fast developing into an important place, and the

* 'Beauties of England and Wales.'

establishment of several large manufactories in the immediate vicinity will cause it to increase still more. It is asserted by several histories, on the authority of an old manuscript, that prior to 1790 North Woolwich contained a number of houses and a chapel-of-ease, but there is probably some mistake about the record.

For this growing population on this side of the Thames a recreation-ground has been provided in these gardens, which, under the name of North Woolwich Gardens, have long been known as a place for dancing and amusement on payment of admission money. They occupy 10 acres of land on the banks of the river Thames, immediately facing Woolwich, including a raised esplanade, which furnishes a pleasant view of the river. The remainder of the gardens is below the level of the water, thickly planted with shady groves of trees. The old tea-gardens were about to be laid out as wharf property, but a committee, of which the Duke of West-minster was chairman, intervened, and raised sufficient sub-scriptions to purchase the ground. The total cost was £19,000, including £10,000 from the Charity Commissioners, £1,000 from the London County Council, and £500 from the East Ham Local Board. By the express permission of the Queen, the present title of Royal Victoria Gardens was given to the recreation-ground, which had thus been secured.

Coming through the principal entrance, a stretch of lawn faces us, dotted with flower-beds, and beyond this is a long avenue stretching the whole length of the gardens. At the commencement of this avenue, by the lodge, are four statues, relics of the former tea-gardens. One pair of these represents the shepherd and shepherdess so common an ornament at these kind of places, and the other two must have formed part of a grotto or cave. Underneath the trees room has been found for some tennis-courts, whilst a clear space on the other side forms a playground for the children. Turning now to the right, several flights of steps give access to the river-front, along which a gravelled promenade has been formed. This is the favourite place in the gardens. The vessels passing by on the Thames—large steamers to and from the port of London from every part of the world, sailing barges, with their picturesque brown sails, and other craft of every description—present an ever-changing scene. Apart from this, the immediate view is commonplace, the only interesting feature to break the monotony of the chimneys and other commercial premises on the Woolwich shore being the square tower of the old parish church. In the background can be seen the wooded hills of Bostall, St. Nicholas' Church, and Plumstead Common, and the huge mass of Shooter's Hill towering above them all. Behind the promenade on the terrace is another walk under the chestnut-trees, which are so thickly planted as to make it nearly dark even in the most brilliant sunshine. In the summer months a band plays upon the terrace.

Although never obtaining the celebrity of Vauxhall or Cremorne, the gardens attached to the Pavilion Hotel, nevertheless, attracted large crowds to North Woolwich. At the time when they were opened, in 1851, these places of resort were in the height of their popularity. The principal amusement was dancing on an extensive outdoor platform, which was kept up till a very late, or rather early, hour. A small menagerie was among the list of permanent attractions, whilst occasionally there were 'barmaid' and 'monster baby' shows. They were under the management of the late well-known amusement caterer, Mr. William Holland, who first came into prominence in connection with these gardens. They were not a profitable speculation, and about the same time he became lessee of the Surrey Theatre, where he produced several successful pantomimes. The 'people's William' took a delight in relating his multitudinous experiences, and he used to tell how the receipts at the Surrey went to pay the losses at his Thames-side gardens. Another amusing incident he used to tell was how he escaped in a balloon at North Woolwich Gardens from the unpleasant attentions of a process-server. When the popularity of tea-gardens declined, North Woolwich shared in the general downfall, and had to be closed. Most of the buildings which were used for the shows were burnt down just after the gardens were acquired for public use, but before they had been laid out and formally opened.

Communication between the gardens and the opposite shore is maintained by a free steam-ferry, the first established in England. Ferries between North and South Woolwich have existed since very ancient times. In 1308 a messuage and a ferry at Woolwich were sold by William de Wicton to William Atte Halle, mason, for £10.

In 1320 Lambert de Trykenham conveyed one messuage, ·50 acres land, 40 wood, 40 heath, and 14s. rent, in Woolwich and elsewhere, and a ferry across the Thames at Woolwich, to John Latymer and Joan his wife.

In 1340 these lands, rent, and the ferry were conveyed by William Filliol and Mary his wife to Thomas Harwold and his heirs for 100 silver marks.

Some ten years before this last sale the people at Woolwich had petitioned the King to suppress two rival ferries at Erith and Greenwich, on the plea that their competition seriously injured their receipts. The Woolwich ferry is there described as a royal ferry, 'farmed of the King,' which may be accounted for by supposing it as an appurtenance to the royal manor of Eltham.*

The old ferry was a little to the east of the present one on the other side of the gardens, the landing-place being the 'Old Barge House.' This is a very modern structure in spite of its venerable name, but was originally nothing more than an old barge with a hut built upon it. It is said at one time to have been a floating residence, but was firmly established afterwards on shore. The owner of the barge built a cottage on the inland side, and in course of time the present tavern came into existence. Travelling from the opposite shore was an expensive luxury by this ferry. The charges were 3s. 6d. for a horse and cart, and 9d. to 1s. per head for cattle according to the number. The proceeds

* Vincent, 'Records of the Woolwich District.'

of the present ferry at the same rates would be enormous. During the first year of working no fewer than two and a half million foot-passengers and over a hundred thousand vehicles were carried across. The three steam ferry-boats which perform the double journey in twenty minutes form conspicuous objects on the Thames at Woolwich. With their two huge funnels a long distance apart, they look like importations from the Mississippi. They can accommodate a thousand foot-passengers each, and the raised platforms can carry ten vehicles with horses. They are named the *Duncan, Gordon,* and *Hutton.* The first two names are in memory of two patriotic and devoted public servants, the late General Gordon and Colonel Francis Duncan, both of whom were, during a considerable portion of their lives, closely connected with Woolwich by reason of their military duties. The third boat is named after Sir John Hutton, a former Chairman of the London County Council.

Not very many years ago, Woolwich and Barking, on the opposite shore, were important fishing villages, the principal trade being in salmon. The growth of the Metropolis, and the fouling of the water by the ever-increasing quantity of sewage turned into it, have both made salmon-fishing a thing of the past; but by means of the extensive filtration and precipitation works, the contamination of the river is reduced to a minimum. To show how the river is becoming more purified, we may mention that a live haddock was caught off the gardens in March, 1895, but it will be some long time yet before the salmon will return to its old haunts.

As North Woolwich had practically no population at all before 1837, it is hardly to be expected that much history would attach to the site of these gardens. In ancient times this district was probably swampy forest ground. One of our oldest historians, Holinshed, in his Chronicles, published in 1577, says of Essex: ' I find also by good record that all Essex hathe in times past wholie been forest ground, save one (Cantred or) Hundred, but how long it is since it lost the said domination, in good sooth, I do not read.'

Between this forest and the river, the lands were subject
to inundation at every flow of the tide. These 'marshes,
bordering on the Thames, in what is now called the parish
of East Ham, were available property at the time of the
Saxon Heptarchy, for it is recorded that King Offa* endowed
the monastery of St. Peter's, Westminster, with 2 hides of
land in Hamme.' This gift was subsequently confirmed by
King Edgar, and afterwards also by King Edward the
Confessor, in a charter dated January, 1066, wherein
amongst other grants made to Westminster Abbey by the
Kings, his predecessors, 2 hides of land in 'Hamme' are
recited. In Doomsday Book the estate owned by St. Peter's

The Site of the Royal Victoria Gardens, North Woolwich, about 1839.

in Hamme is called 'a manor and 2 hides of land, containing always one caracute of arable, worth in Saxon times
20 shillings, but in Norman sixty shillings, then 3 bordars,
afterwards five, and woodland to find pannage for eight
hogs.' In 1542 this property of the Dean and Chapter of
Westminster was described as 'a farm in the marshes of
East Ham, near Barking.' It was part of this estate that
was sold by the Dean and Chapter, and afterwards converted into the North Woolwich Gardens, alienated, after a
possession of 1,200 years, by a corporation of clergymen to
become a tea-garden.†

* Not the great King of Mercia, but probably a King of the East
Saxons of the same name.
† Catherine Fry, 'History of East and West Ham,' 1888.

To show how this side of the Thames has changed in character, we may quote from a local historian,* who, in deploring the completion of the railway to North Woolwich, adds: ' It is singular to hear the whistle of the locomotive and the clatter of the iron wheels where, twelve months since, the heron, the plover, and the bittern roamed in almost undisturbed solitude.' Some little allowance for imagination must be made in reading this extract, but it will give some idea of the change from stagnation to activity which has taken place.

There is a story told of this former swamp, the place of which has been taken by the gardens, which would give them the sanctity of haunted ground. It is of a handsome young huntsman and his bride who elected to spend their wedding-day in boar-hunting. The lady, who was foremost in the chase, forgetful in her excitement of impending pit-falls, dashed wildly on, till she found herself beyond reclaim sinking slowly but surely in the quagmire from which no escape was possible. Her lover plunged gallantly in to save her, but he was too late, and he also was lost in his efforts to extricate his young bride. On this sad honeymoon is based the superstition that a skeleton horseman on the boniest of steeds is to be seen here at nightfall—in fact, that

> ' A hideous huntsman's seen to rise
> With a lurid glare in his sunken eyes ;
> Whose bony fingers point the track
> Of a phantom prey to a skeleton pack,
> Whose frantic courser's trembling bones
> Play a rattling theme to the hunter's groans ;
> As he comes and goes in the fitful light,
> Of the clouded moon on a summer's night.
> Then a furious blast from his ghostly horn
> Is over the forest of Hainault borne,
> And the wild refrain of the mourner's song
> Is heard by the boatman all night long,
> That demon plaint on the still night air,
> With never an answering echo there.' †

---

* Ruegg, ' Woolwich and its Environs.'

† Irving Montagu, ' Ghosts,' *Strand Magazine*, 1891, vol. xi.

The marshes now occupied by the gardens were once the site of a military encampment. In 1667, just after the Fire of London, war was declared against the Dutch, who, after being defeated in a battle off Lowestoft, took Sheerness, and sailed up the Thames, threatening London itself. At this time Sir Allan Apsley, who had distinguished himself as a staunch Royalist, was quartered with his regiment at the point of land over against Woolwich. His position is still further explained by a letter dated June 17, 1667, from 'the *marsh over against Woolwich*,' in which he complains that his men are deserting him, and 'cannot be persuaded that they are obliged to stay.'*

Traces of the Roman occupation are very evident at this part of the Essex shore. During some excavations in 1863, for ballast to form the embankment which carries the northern high-level sewer to Barking, the remains of what must have been an extensive Roman cemetery were discovered by the workmen. Among the spoils thus exhumed were three leaden coffins, a stone coffin with a coped lid, and skeletons supposed to have been interred in wooden coffins, together with cinerary urns, and broken fragments of Samian pottery. This discovery would point pretty conclusively to the fact that the Romans had a considerably large colony just about this part.

The river-wall which protects the gardens from inundation is under the jurisdiction of the Essex Commissioners of Sewers. Various conjectures have been made as to whom the credit is due for having first embanked the Thames. Some say the Romans, others the many religious bodies who had lands bordering upon the river. In 1707 a serious breach in the river-wall flooded the whole of this district, and the present gardens must then have been entirely under water.

This inundation, commonly known as the Dagenham breach, 'happened 17th of December, 1707, at an extraordinary high tide, accompanied with a violent wind, and

* Vincent, 'Records of the Woolwich District.'

was occasioned by the blowing up of a sluice, made for the drain of the land waters in the wall and banks of the Thames. If proper and immediate help had been applied, it could have been easily stopped . . . but through the neglect thereof, the constant force of the water setting in and out of the levels soon made the gap wider, so that a large channel was torn up, and a passage made for the water, of 100 yards wide, and twenty feet deep in some places. By which unhappy accident about 1,000 acres of rich land in the levels of Dagenham and Havering . . . were overflowed. The expense of repairing this breach was, at first, laid upon the proprietors of the lands, but after many wearied and unsuccessful attempts of theirs for about seven years, until they had expended more than the value of the land, it was given wholly over as impracticable. However, being deemed a public concern, upon application to Parliament, an Act was obtained for the speedy and effectual preserving the navigation of the river Thames by stopping the breach in the levels of Havering and Dagenham. By which Act, for ten years from 10th of July, 1714, the master of every ship or vessel (with some specified exceptions) coming into the port of London was obliged to pay threepence per ton.'* The cost of repairing this breach amounted to over £40,000.

* Morant, ' History of Essex.'

CHAPTER XXIV.

LEICESTER SQUARE.

THE task of writing the history of Leicester Square is a difficult one, owing to the wealth of materials at the chronicler's disposal. Any casual passer through the square must have noticed the plates affixed to two houses by the Society of Arts, marking them as the residences of former celebrities, and we find at once that we are upon historical ground. At the present day Leicester Square is looked upon as the headquarters of the French colony in London. After the revocation of the Edict of Nantes (1685) this neighbourhood became a favourite resort of the more aristocratic French Protestant exiles, and their descendants have remained here ever since.

There was a time when fields covered the site of the square. In Aggas's survey of London in the time of Elizabeth, dated 1592, the land is shown as open pasture. Leaving Charing Cross, with St. James's Park and its deer on the left, a small lane—Hedge Lane (now Whitcomb Street)—leads to the fields, which are occupied by two pedestrians, a woman laying out clothes to dry, and two animals, one of which appears to be deformed, either intentionally or through an error on the part of the engraver. This map was drawn at a time when both St. Martin's and St. Giles' could legitimately claim their distinguishing titles of ' in the fields.' A few years soon made a great difference, as will be seen by comparing Aggas's map with Faithorne's,

General View of Leicester Square.

dated 1658.*   The whole of the neighbourhood to the east of the fields is now seen to be covered with houses, with here and there a square dotted.   In Leicester Fields a mansion has appeared, with its gardens reaching back to what is denominated ' Military yard,' where Prince Henry, the eldest son of James I., exercised his troops.   This mansion is Leicester House, around which clings much of the history of Leicester Square.   The land upon which the mansion was built was called Lammas land, *i.e.*, land open to the poor after Lammas-tide, and the Earl of Leicester had to pay rent for the ground to the overseers of the poor of St. Martin's-in-the-Fields.   An entry in the time of Charles I. records : ' To receewed of the Honble Earle of Leicester, for ye Lamas of the ground that adjoins to the Military Wall—£3. . . . The Rt. Honble the Earl of Leicester, for the Lamas of the ground whereon his Lordship's house and garden are, and the field that is before his house neare to Swan Close.'†   This field is, of course, the present Leicester Square, and Swan Close is identified by some as the ground occupied by the house, corresponding therefore with Leicester Place, Leicester Street, and Lisle Street.

Leicester House was built about 1632-36 by Robert Sidney, Earl of Leicester, the father of Algernon Sidney, and of Lady Dorothy, the Sacharissa of the poet Waller. An item of interest about this time regarding the use to which the present garden was put may be taken from the Stafford Letters, vol. i., p. 377.

' *March* 5, 1635.—There was a difference like to fly high betwixt my Lord Chamberlain and my Lord of Leicester about a Bowling Green that my Lord Chamberlain had given his barber leave to set up, in lieu of that in the Common Garden, in the field under my Lord of Leicester's house ; but the matter after some ado is taken up.'

---

* Both of these maps are reproduced in Mr. Tom Taylor's exhaustive ' History of Leicester Square.'

† Quoted in Wheatley and Cunningham's ' London : Past and Present,' vol. ii., p. 380.

The Earl of Leicester was an absentee landlord. When not engaged on his frequent embassies abroad, he was at his favourite country seat at Penshurst in Kent. Whilst my Lord of Leicester was away from town, his house was rented by several illustrious personages. Among these was Elizabeth, Queen of Bohemia, and eldest daughter of James I. She was living at Craven House when her fatal illness struck her, and moved here only a fortnight before her death, in February 1662. Another occupant was Colbert, the French Ambassador in the reign of Charles II. Pepys was one of a deputation who should have waited upon him, but he had been to a house-warming, and he continues : ' I rose from table before the rest, because under an obligation to go to my Lord Brouncker's, where to meet several gentlemen of the Royal Society, to go and make a visit to the French Ambassador Colbert, at Leicester House, he having endeavoured to make one or two to my Lord Brouncker, as our President, but he was not within, I came too late, they being gone before, so I followed to Leicester House ; but they are gone in and up before me.'*

Evelyn was more successful than his brother diarist. He had gone to Leicester House to take leave of Lady Sunderland, whose husband was Ambassador to Paris, and was there edified by the feats of a fire-eater named Richardson. ' He devour'd brimston on glowing coales before us,' says Evelyn. ' chewing and swallowing them ; he mealted a beere-glasse, and eate it quite up ; . . . then he mealted pitch and wax with sulphur, which he drank down as it flamed ; . . . with diver other prodigious feates.'†

The Earl of Leicester died in 1677, and his successors spent little of their time here. The house and gardens passed from the Sidneys altogether at the end of the last century, when they were sold to the Tulk family for £90,000, to pay off the encumbrances on Penshurst.‡ But we are hurrying on too quickly. The German Ambassador was in

* Pepys' ' Diary,' October 21, 1668.   † ' Diary,' October 8, 1672.
‡ Tom Taylor, ' Leicester Square,' p. 125.

residence here in 1708, and to this house came Prince
Eugene in 1712, on his fruitless mission to prevent the Peace
of Utrecht. He was very popular with the people, although
' the Queen used him civilly, but not with the distinction
that was due to his high merit ; nor did he gain much ground
with the ministers.' * He failed in his mission, but had

Leicester Square in 1700.

some consolation in returning to Holland with a diamond-
hilted sword, presented him by the Queen.

Once more in 1717 the mansion changed tenants, the next
occupant being George Augustus, Prince of Wales, and it
remained as the town-house of the heirs to the throne for

* Bishop Burnet, quoted in *English Illustrated Magazine*, August
1886.

another forty years. Pennant very happily calls Leicester House the 'pouting place of princes.' The first Prince came here in a temper, and when he succeeded to the throne his son followed his splendid example. This Prince ended his days at the mansion in Leicester Fields in 1751, from the bursting of an abscess in his throat, said to have been caused by a blow from a cricket-ball at Cliveden.

His widowed Princess remained here till 1766, when she removed to Carlton House. Leicester House was then in the occupation of the Duke of Gloucester.

In the meantime there were some gay doings in the square. Leicester House witnessed a gorgeous state ceremonial in 1760, when George III. was proclaimed King in the presence of the nobility and high officers of the State. The first stopping-place of the procession was in Leicester Fields; it then moved on to Charing Cross, where the ceremony was repeated, and so on to the Royal Exchange.* One more state function, and the connection of royalty with Leicester House closes. The large drawing-room of the mansion was the scene of the marriage of Princess Augusta to the popular Prince of Brunswick.

Passing now into private hands, Leicester House became a British Museum on a small scale. Mr. (afterwards Sir) Ashton Lever removed his collection of objects of natural history (to which he gave the name of ' Holophusikon ') from Manchester to London, thinking to obtain the same popularity for his curiosities in the capital as at their former home. In this he was sadly disappointed. After the exhibition had been kept open from 1771 to 1784, the collection was offered to the nation at a moderate price, but refused. Sir Ashton then obtained an Act of Parliament to dispose of it by a lottery of 36,000 tickets at a guinea each, the winner to have the collection, and the other subscribers four admission tickets. Only 8,000 of the tickets were taken, and the winner exhibited the contents under the title of the Museum Leverianum, in a building called the Rotunda, on

* Tom Taylor, ' Leicester Square,' p. 263.

the south side of Blackfriars Bridge. But the experiment
of combining instruction with amusement was not a successful bait for the public, and at Blackfriars once more the
museum was a failure. In Sir Ashton's time the price of
admission was 5s. 3d., and the following advertisement
frequently appeared : ' Sir Ashton Lever's Museum, containing many thousand articles, displayed in two galleries,
the whole length of Leicester House, is open every day
from ten o'clock till four. Admission 5s. 3d. each person.'
—*Morning Post*, November 16, 1778.* The charge was subsequently reduced to half a crown, and then to a shilling,
but all to no effect, and the new proprietors had to close
their doors. The contents of the museum were then sold
by auction, the sale lasting sixty-five days. Perhaps the
most interesting things in the collection, which filled sixteen
rooms at Leicester House, were the curiosities brought home
by Captain Cook from his many voyages.

Soon after the dispersal of these treasures Leicester House
itself disappeared. It was pulled down in 1790, and Leicester
Place and Lisle Street now occupy the site.

Another great house in the square was Savile House,
which adjoined the mansion of the Sidneys. It was so
named after its later occupants, the Savile family, although
built by the Earl of Aylesbury. Among the distinguished
visitors to this house we must give foremost place to that
eccentric personage Peter the Great. There is another side
to his character than the one generally known, in which he
is regarded as the industrious zealot anxious to obtain an
insight into everything that concerned the good of his
empire. Bishop Burnet, who had, as he says, ' much free
discourse with him,' summed him up as a ' man of very hot
tempers, soon inflamed, and very brutal·in his passion.' To
all his practical intelligence he added ' the habits of a sot
and the manners of a savage.' † When Peter the Great

---

* Quoted in Wheatley and Cunningham's ' London : Past and Present,'
vol. ii., p. 381.

† Tom Taylor, ' Leicester Square,' p. 165.

came to England in 1698 Savile House was occupied by the
Marquis of Carmarthen, and he was chosen to be the guide
and companion of the Emperor. The chief occupation of
the Russian visitor at Savile House was the drinking of
oceans of sack, varied with brandy spiced with pepper. After
his departure the house quieted down again, and in 1718
it was hired by the Prince of Wales for the use of his
children, a communication being made between this and the
adjoining Leicester House.

In later years the house came into prominence during the
Gordon Riots of 1780. Sir George Savile, who was living
here at the time, had brought in his Bill for the relief of the
Roman Catholics, which in a great measure brought about
the riots, and his house was one of the first attacked by the
mob, 'carried by storm and given up to pillage, but the
building was saved. The railings torn from it were the
chief weapons and instruments of the rioters.'* Sir George
Savile's intimate friend Burke, in writing about this anxious
time, relates how he kept watch for four nights at Lord
Rockingham's or Sir George Savile's, whose houses were
garrisoned by a strong body of soldiers, together with numbers of true friends of the first rank who were willing to
share the danger.

When it had lost its fashionable occupants, Savile House
became the home of one of the numerous exhibitions
associated with the square. After being rebuilt, and used
for a while as a place of entertainment, it was opened for
the splendid collection of pictures in needlework executed
by Miss Linwood. These were copies of the best pictures
of the masters, both ancient and modern, represented by
coloured worsted upon white linen. The exhibition continued
for forty-seven years, and then the various pictures were sold
by auction on the death of Miss Linwood at the ripe age of
ninety.

Savile House was burnt to the ground in 1865, and for a

* Walpole to Rev. W. Cole, June 15, 1780, quoted in 'London : Past
and Present.'

long while the site remained empty, but about 1880 it was
utilized for a panorama, and subsequently it was adapted for
the Empire Theatre, which, from its present flourishing
condition, seems to have come to stay.

Having thus followed the fortunes of Savile House and
Leicester House, we must retrace our steps and continue
the history of Leicester Square or Fields. The square was
built about the same time as Leicester House, although the
south side was not completed till 1671. Towards the end
of the seventeenth century the ground in the centre was
railed in, and the square at this time is thus described by
Strype : ' Leicester Fields, a very handsome, large square,
enclosed with rails, and graced on all sides with good built
houses, well inhabited, and resorted unto by gentry, especially
the side towards the north, where the houses are larger,
amongst which is Leicester House, the seat of the Earl of
Leicester, and the house adjoining to it (Savile House),
inhabited by the Earl of Aylesbury.'*

The enclosure, thus railed in, like other similar squares,
was more than once the scene of a duel in the times when
sudden quarrels were settled by an appeal to the sword.
At night time there were no lamps to shed light upon any
such encounters which might take place, and, apart from
this, the enclosure generally seems to have fallen into a
neglected condition. In 1737 an attempt was made to
remedy this, and the first laying out was accomplished.
The cost of this was defrayed by a voluntary subscription of
the inhabitants, which probably originated in the desire to
encourage the fashionable resort to the square. A contemporary print in the British Museum shows us the stiff,
formal style in which this was done. A path parallel with
the sides of the square runs round the outside, and two
other paths at right angles divide it into four plots, with a
round basin in the centre. This basin was originally intended
for a fountain, as at present, but its place was taken some
eleven years later by a gilded equestrian statue of George I.,

* Strype, book vi., pp. 68, 86.

brought from the Duke of Chandos' seat at Canons. It is
said by Walpole to have been erected here by Frederick,
Prince of Wales, to vex his father George II. This gilded
statue remained as one of the sights of London, till
another exhibition nearly settled the fate of the garden of
the square. In 1851 Mr. Wylde, the celebrated geographer,
entered into an agreement with the Tulk family (who, as we
have seen, acquired the property of the Earls of Leicester),
under which he erected a huge globe, 60 feet in diameter,
with accessory rooms on the site of the garden. For ten
years this exhibition was carried on, and various historical
and similar collections were also on view. Then in 1861,
under the terms of the agreement, 'Wylde's Great Globe'
had to be taken down. The old statue was then re-erected
in a dilapidated condition, and the garden was once more
allowed to fall into sad neglect. The statue began to fall to
pieces, and was kept up by a wooden prop. A practical
joker afterwards fitted it with a broom in one hand and a
saucepan on its head. But nobody interfered, and the
garden and its statue became a disgrace to this part of
London. During the time the statue was here another
more modest entertainment was noticed by Wordsworth in
the square :

'What crowd is this? What have we here? We must not pass it by;
A telescope upon its frame, and pointed to the sky ;
    *   *   *   *   *
The showman chooses well his place—'tis Leicester's busy square.'*

The attention of the late Metropolitan Board of Works
was drawn to the state of the square in 1863, when it was
proposed to build a market on the enclosure, and the Bill
was successfully opposed. Subsequently the Board, under
the provisions of the Gardens in Towns Act, 1863, took
steps with a view to taking charge of the garden. Mr.
Tulk, who claimed a right of property in the garden, denied
the Board's right to interfere, and commenced an action in

* Wordsworth, ' Star-gazers.'

1865. After many delays, the action was tried, and verdict given against the Board in 1867. This was appealed against, but after appeal the judges decided that Leicester Square did not come within the scope of the Gardens in Towns Act, inasmuch as there had been no irrevocable setting apart

The Last of the Old Horse, Leicester Square.

or dedication of the ground to the public use. The result of this was that a Bill was prepared to vest the garden in the Board, and shortly before the Leicester Square Act, 1874, was passed, the following letter was received from Mr. Albert Grant, then M.P. for Kidderminster (afterwards Baron Grant) :

'To the Chairman and Members of the Metropolitan Board of
Works.

'41, QUEEN'S GATE TERRACE,
'SOUTH KENSINGTON, W.,
'January 21, 1874.

' GENTLEMEN,

'The deplorable state of Leicester Square has for
years drawn the attention of the inhabitants of London to
the absolute necessity of something being done to remove a
state of things discreditable to the Metropolis.

' Accordingly various attempts from time to time have
been made to acquire the rights of the freeholders of the
square, but hitherto without success.

' The idea that the square could be converted into building
land has, under this impression, induced the persons holding
the ownership constantly to refuse to sell their rights, except
for such an enormous sum, based on its value per foot as
building land, as to render acquisition impracticable.

' Notwithstanding these discouragements, for some months
past my agents have been in negotiation with the various
owners, having for object the purchase of their interests in
the square, with a view to my handing the same to the
Metropolitan Board of Works after I had laid out the
grounds—as a gift to the Metropolis.

' During the later negotiations the decision of the Master
of the Rolls came, decreeing the land not to be available for
building, but bound to be kept as an open space.

' The owners were entitled to take, and in fact did take,
the necessary preliminary steps to appeal against this
decision, a course which might have involved a delay of two
years before the decision of the ultimate Court of Appeal
could be obtained, or the alternative of the Metropolitan
Board of Works, in the event of their being authorized by
Parliament to acquire the square compulsorily, being obliged
to pay for the land on the basis of a possibility of the decree
not being sustained, in which case a comparatively high
valuation might by a jury have been awarded to the owners.

'As you are aware, a meeting of the various occupiers of houses in Leicester Square was held, at which it was resolved to apply to the Metropolitan Board of Works to ask them to lodge a Bill in the next session of Parliament to obtain power to buy the site, with a view to their placing the square in proper repair.

'Notwithstanding that the period for lodging the Bill, according to the standing orders of the Houses of Parliament, was past, your Board decided to comply with the request, and accordingly the notice of application for an Act has been duly advertised.

'Meantime, the effect of the decision of the Master of the Rolls was to make the owners moderate considerably their views as to the amount they would accept for the surrender of their rights whatever they were, and ultimately I came to terms, and on the 5th instant acquired all the rights—viz., one undivided moiety or seven-fourteenths—owned by the principal proprietor, Mr. J. A. Tulk, and with such rights possession of the square.

'I am also in negotiation for the acquisition of the other seven-fourteenths which are vested in various persons, but who, having now, according to the decision of the Master of the Rolls, only a nominal right, will no doubt come to satisfactory terms with me for a sale and surrender of such rights.

'In anticipation of these arrangements, I had plans prepared by my architect, Mr. James Knowles, for laying out the grounds as a public garden, and these plans are being carried out by Mr. John Gibson, who, as the designer of the Subtropical Gardens at Battersea, and other works, is favourably known; it is also my intention to enclose the square by a handsome railing, and in the centre to place an ornamental fountain, both specially designed for the purpose, and to provide seats for the public capable of accommodating about 200 persons.

'I further intend to erect at the four corners granite pedestals, on which busts in marble, of a suitable size, will be placed of the following celebrated men, all known to have been locally connected with the traditions of Leicester Square.

'These will be : Hogarth and Sir Joshua Reynolds, both of whom lived and died in houses in the square; Dr. Samuel Johnson, the friend and constant visitor of Sir Joshua Reynolds; and Sir Isaac Newton, who lived in Leicester Place, adjoining the square, for many years after he became President of the Royal Society—men who, it will be admitted, are worthy of being illustrated by the sculptor's art, but who have not, that I am aware of, yet received any recognition of their greatness in that form in any public open space in London.

'These busts have been entrusted by me for execution to the following well-known sculptors, viz., that of Sir Joshua Reynolds to Mr. H. Weekes, R.A.; Hogarth to Mr. J. Durham, A.R.A; Sir Isaac Newton to Mr. W. C. Marshall, R.A.; and Dr. Johnson to Mr. T. Woolner, A.R.A.; and I have every reason to hope they will prove to be at once worthy of the men represented and representing them.

'Workmen have already commenced on the ground in the square, and all works are to be finished at the latest by the 15th of June next.

'By that time I trust the Metropolitan Board of Works will have obtained their Act, empowering them to take over the square on behalf of the public.

'I shall then have much pleasure in signing—I hope in the square itself—a deed of transfer to the Metropolitan Board of Works, as a free gift to the Metropolis, of what will then be a public garden, fitted up in a way which will, I trust, illustrate how much may be done towards embellishing London through her many public squares and other open spaces.

'I think it right, in conclusion, to add that should, contrary to my expectation, any of the remaining holders not have arranged with me for the sale of their rights by the time the Act for the compulsory acquisition of the square has been passed, and upon the Metropolitan Board of Works putting their powers into force for acquiring such rights, I will pay the amount which may become payable under such compulsory purchase, so that the square may vest in the Metropolitan Board of Works free of .cost to them.

'I am, etc.,

'(Signed) ALBERT GRANT.'

There could have been no doubt in accepting this munificent offer, and except for changing the bust of Dr. Johnson for that of the eminent surgeon, John Hunter, this scheme was carried out in its entirety.   The chief ornament of the handsome central fountain is a statue of Shakespeare, the whole executed in white marble, and an immense improvement on the old golden horse.   The opening day was fixed for July 2, 1874, and the square was as gay on that day as ever it had been.   Flags were flying everywhere, and the ceremony took place in one of the many pavilions erected for the occasion.   And so the hoardings which used to flaunt with tattered advertisements are gone, the bulged and battered railings have been renewed in graceful modern guise, and there are flowers and grass on what was formerly the refuse-heap of the neighbourhood.

Leicester Square can boast of having been the abode of two illustrious painters, Hogarth and Sir Joshua Reynolds, many of whose masterpieces are now lodged in the National Gallery, within a stone's-throw of their former studios. Hogarth was associated for the greater part of his life with the square.   He was apprenticed to a silversmith in Cranbourne Alley, and no doubt in his apprentice days spent most of his spare time in Leicester Fields, 'with his master's sickly child hanging its head over his shoulder.'*   When he started in business on his own account, he gave up engraving on silver for the higher branch of the art on copper, and obtained much work in the way of book illus-

* Smith, 'Nollekens and his Times,' vol. i., pp. 46, 47 ; quoted in Tom Taylor's 'Leicester Square.'

tration. He found engraving, however, such a miserable profession that he forsook it for portrait-painting, but he first came into prominence through his satirical and moral sketches, on which his fame rests. His house in Leicester Square was distinguished by the sign of the Golden Head (a bust of Vandyck), which he had made himself from carved pieces of cork, glued together and gilded. This was succeeded

Statue of George I. and Hogarth's House, 1790.

by a plaster head, and afterwards by a bust of Newton. The fashionable life of Leicester Square, in the garden enclosure of which he was often seen in his scarlet roquelaure, gave him plenty of opportunities for studying the ways of society, which he so mercilessly satirized. The last scene of his eventful life took place in Leicester Square. He had returned, on October 25, 1764, from his country villa at

Chiswick to his house in the square, and, exhausted from his journey, he found a letter from Dr. Franklin, to which he drafted an answer. On retiring to bed he was seized with a vomiting fit, upon which he summoned his house-keeper, and died in her presence some two hours afterwards. After his death his widow still kept up the house, taking in lodgers, chiefly artists, for a living, and his housekeeper, Mary Lewis, sold prints here.

The site of Hogarth's house, marked with a memorial tablet, is now used for Archbishop Tenison's Grammar School, founded in 1685 by Tenison, Vicar of St. Martin's-in-the-Fields, afterwards Archbishop of Canterbury. It was formerly in Castle Street, immediately behind the National Gallery, but when that street was pulled down the school was transferred to Leicester Square. This house only dates from 1870, when the Sablonière Hotel, which succeeded Hogarth's occupation, was pulled down to make room for it. Tradition says that Hogarth's studio was used as the billiard-room of the hotel.

From Hogarth's house a walk across the square brings us to another painter's home, that of Sir Joshua Reynolds, now used as Puttick and Simpson's auction-rooms. He was at the height of his fame as a portrait-painter when he came to Leicester Square in 1760. Thus, for four years these two painters, Hogarth and Reynolds, both great in the respective branches of their art, lived opposite one another in the same square. But there could not have been much sympathy between them. 'Hogarth, whose own efforts as a portrait-painter were little appreciated in his lifetime, must have chafed at the carriages which blocked up the doorway of his more fortunate brother; and Reynolds, courtly amiable though he was, capable of indulgence even to such a raw caricaturist as Bunbury, could find for his illustrious rival, when he came to deliver his famous Four-teenth Discourse, no warmer praise than that of "successful attention to the ridicule of life."'* Sir Joshua was as

* Austin Dobson in *English Illustrated Magazine*, August, 1886.

famous in his time for his dinners and drawing-room parties as for his skill in painting. These fashionable receptions attracted to Leicester Square the leading men of the day, Dr. Johnson, the poet Goldsmith, Garrick, and Burke, being amongst the painter's most intimate friends. There was something in the man apart from his genius—his patience, his geniality, his imperturbability of temper—which made him the confidant of all these great men. His biographer, North-cote, could write of him, 'If the devil was on his back, no one would learn it from his face,' and we can believe him in

Staircase in Sir J. Reynolds's House, Leicester Square.

this; though when he comes to write that 'to the grandeur, the truth, and simplicity of Titian, and to the daring strength of Rembrandt, he has united the chasteness and delicacy of Vandyck,' we fear that his admiration for the great master has exceeded his prudence. Upon the foundation of the Royal Academy, in 1768, the post of honour as President was unanimously given to Reynolds, who was thereupon knighted by George III. Like Hogarth, the last scenes of his life are connected with the square. One very pathetic incident endears him at once to us. He had lost the sight of one of his eyes, and had been compelled to give up painting. During this enforced idleness he had tamed a canary, which had one morning flown out of the window, and Sir Joshua might have been seen pacing the garden enclosure for hours with a green shade over his eyes, in the hopes of recovering his lost pet.* It is but a few years to the last scene of all. On July 23, 1792, he breathed his last in Leicester Square, and was accorded a splendid state funeral in St. Paul's Cathedral. The bust in the square, though less pretentious than this state ceremony, is a more lasting memorial to one of the most illustrious personages it is ever likely to have within its walks.

From Reynolds we must pass to another genius whose bust also adorns the square. This is the eminent surgeon John Hunter. He had passed the prime of life when, in 1783, finding his anatomical collection increasing so rapidly, he had to look out for larger premises, and so moved from Earl's Court to Leicester Square. The house he occupied was next to Hogarth's, but the caricaturist had been dead twenty years when Hunter and his museum entered upon the scene. Among other things he did in the ten years he lived in the square was to have his portrait painted by Sir Joshua, in which he is shown sitting at a table in a reverie, with sufficient background to determine his profession. In the garden of his house he built a museum in which to place his collection (without an equal in any other country), which was purchased by the Government after his death, and handed over to the Royal College of Surgeons, where Reynolds' portrait of him is now hung. He died, not in the square, but at St. George's Hospital. It appears that he was annoyed at something said at a Board meeting of the hospital, and left the room to control his rage, and immediately, with a sudden groan, fell dead into the arms of a friend standing by. He is buried in the vault beneath the Church of St. Martin's-in-the-Fields, and his statue on the façade of the University of London is one of twenty

* Tom Taylor, 'History of Leicester Square,' pp. 374, 375.

placed there in honour of the most distinguished men in philosophy, science, and letters that the world has ever known.

The remaining bust is that to Sir Isaac Newton, who did not actually reside in the square, but in St. Martin's Street. His house is in a neglected condition, and may easily be distinguished by the memorial tablet affixed to it. Newton was nearly seventy when he came to reside here in 1710. He was chiefly distinguished then by his official titles of Master of the Mint and President of the Royal Society.

If Leicester Square had no other inhabitants to boast of, the names of these four men would have made it famous. But the list of celebrities who at one time or another lived here is a very lengthy one, so that we can do no more than make a passing mention of their names.*

Commencing with noblemen and prelates, we have Dr. Lloyd, Bishop of St. Asaph in 1681; in 1683 the (second) Earl of Strafford writes from Leicester Fields to the (second) Earl of Clarendon. The law is represented by Lord Chancellor Somers (died in 1716), and the poets by Dryden. Painters are very numerous, for, in addition to Hogarth and Reynolds, there are William Aikman, the portrait-painter (died 1731); Sir James Thornhill, whose daughter Hogarth married; and Theodore Gardelle, the enamellist and portrait-painter, who murdered his landlady, Mrs. King, in 1761. After the death of Hunter, another surgeon lived and died in the square (1800), Cruikshank, who attended both Dr. Johnson and Sir Joshua; and with his name we can associate that of Sir Charles Bell, who discovered the 'distinction of the nerves of sensation and motion, a discovery deserving to be classed, in the opinion of Müller, the famous German physiologist, with Harvey's of the circulation of the blood.'† Thomas Dibdin, the song-

---

* For the details of this list we are indebted to the excellent article on Leicester Square in 'London : Past and Present,' Wheatley and Cunningham, vol. ii., p. 382.

† Tom Taylor, 'Leicester Square,' p. 437.

writer, built his Sans-Souci Theatre in Leicester Square, on the ground now occupied by the Hôtel de Paris et de l'Europe. Edmund Kean, when little more than a child, distinguished himself here by readings and recitations. Newton's house was afterwards the home of Dr. Burney, the musician, and father of Fanny Burney.*

In the present day, when royalty have deserted Leicester Square, when there are no fashionable painters or men of scientific genius to shed their brilliance around, perhaps it is chiefly associated with its two great music-halls, the Empire and the Alhambra. The Empire, as we have seen, is on the site of Savile House. The Alhambra was originally erected as a rival to the Polytechnic Institution, and was called the Panopticon of Science and Art. This is not the only place in the square to remind us that the general public is more interested in ballets than in scientific museums. The Panopticon building was sold in 1857, and converted into a circus and music-hall under its present name. This building was burnt down in September, 1883, but was at once rebuilt, to the delight of the admirers of the terpsichorean art.

* Tom Taylor, ' Leicester Square,' p. 225.

CHAPTER XXV.

LINCOLN'S INN FIELDS—RED LION SQUARE.

Lincoln's Inn Fields.

L INCOLN'S INN FIELDS is often described as one of the largest and finest squares in the world, a character which it well deserved in the past, and which it may attain to in the future. At present it may be considered as in a transitional state, for many of the fine houses on the western side, having become too decayed and unsuitable for modern purposes, are being replaced by huge sets of chambers. The contrast between the sombre gravity of these dingy mansions and the bright and staring newness of their successors is painfully apparent; but perhaps it is not too late to hope that this side of the square will at some future date be restored to its earlier dignity.

After having been enclosed for years as a private garden for the use of the few residents in the houses of the square, it has now been converted to its former use as a public spot for recreation. There are many who affirm that the public ought never to have been excluded, and in nearly all the old maps of the district the Fields are shown as apparently common land, intersected with public footpaths. It must be a matter of congratulation to London generally that this fine garden has been added to the number of its recreation-grounds, where it is possible to step out of the whirl of life and dream for a few moments of the charm and repose of

the country.   The glory of Lincoln's Inn Fields is the
number and size of its fine plane-trees, which seem to thrive
on the smoke and fog of London.   In 1843 the garden was
nearly lost to the public, when the late Sir Charles Barry
prepared a magnificent design for the new Law Courts,
which he proposed to erect in the centre of Lincoln's Inn
Fields.

The struggle for the acquisition of Lincoln's Inn Fields
was a difficult and lengthy matter.   The trustees were
approached with this object in view in 1890, but they stated
that the terms of their Act prevented them from allowing

Lincoln's Inn.   (From a drawing by Herbert Railton.)

the gardens to be used by the public, and that no other Act
would enable them to do so except the Metropolitan Open
Spaces Act, 1881; but they were satisfied that it would be
quite impossible for them to obtain the necessary consent to
the proposed arrangement.

The Act referred to by the trustees was the 8 Geo. II.,
1735, by which they were to preserve and maintain the Fields,
and had power to levy a rate for the purpose.   As, therefore,
Parliamentary powers were necessary to enable the trustees
to make arrangements for the admission of the public, the
London County Council decided in November, 1891, to

insert a provision in their next General Powers Bill for the purchase of the land by agreement; but when the Bill was considered by the Committee of the House of Commons, it was decided (in the absence of four of the members) that ' so much of the preamble as refers to Lincoln's Inn Fields is not proved.' The Committee, however, desired the chair-

Gateway, Lincoln's Inn Fields.

man to state that they would see with pleasure the opening of Lincoln's Inn Fields, but that they declined to override the provisions of the Metropolitan Open Spaces Act of 1881.

As these steps to arrive at a voluntary agreement had failed, the only course that remained was to apply for compulsory powers to acquire the gardens in the usual

way, which involved the liability to purchase, unless otherwise agreed, any estate in the land which might be a subject
measure for compensation. A clause was accordingly inserted
in the Open Spaces Bill of 1893 for the acquisition of
Lincoln's Inn Fields. This Bill successfully passed the
Committee of the House of Commons, but the Select
Committee of the House of Lords, after hearing counsel
and witnesses for and against the proposal, decided to
strike out the portion relating to Lincoln's Inn Fields.
Each rejection of this clause involved a delay of a year,
and the hope of acquisition now seemed as far off as at the
commencement; but the London County Council decided
to make a third attempt, and to once more introduce in its
1894 Bill a clause on the lines passed by the House of
Commons.

After negotiations with the trustees, the purchase-money
was fixed at £12,000, which certainly seemed somewhat
high, having regard to the limitations named in the Act of
Parliament; but as no other terms were obtainable, the
desirableness of the acquisition outweighed other considerations, and the clause in the London County Council
(Improvements) Act, 1894, was framed on these lines. It
was provided that this amount should be paid into court, for
payment to such claimants as might legally prove their title
to receive compensation. The Bill received the Royal Assent
on August 17, and possession was obtained on November 7,
1894. The legal costs and stamp duty, amounting to about
£1,000, made the total cost of acquiring this garden, which
is 7¼ acres in extent, £13,000. It was formally opened to
the public by Sir John Hutton, the Chairman of the London
County Council for that year, on Saturday, February 23,
1895.

On approaching this noble square from the neighbourhood
of Clare Market, we come across a small establishment
which couples the information that it is ' The Old Curiosity
Shop immortalized by Charles Dickens ' with the announcement that the highest prices are given for white and coloured

rags, bones, waste paper, etc. This combination of the romantic and the practical is one of the characteristics of the historic houses of Lincoln's Inn Fields. What were formerly the mansions of Prime Ministers, Lord Chancellors, and nobility of every degree, are now split up into in-

'*The Old Curiosity Shop*,' *Lincoln's Inn Fields.*

numerable chambers and offices for the lights of the legal profession. Lincoln's Inn Fields went originally by the name of Ficket's Field, Fikattesfeld, or Ficetsfeld, which name may have been derived from some very remote owner. This was in 1657 divided into two fields; the dividing-line passing through the site of the present square would stretch from about the centre of the Soane Museum to the centre of the College of Surgeons. The land on the east side of this line was called Cupfield, and that on the west side Pursefield. From time immemorial it has been a place devoted to the recreation of the students of Lincoln's Inn and the general public.

In all early deeds it is referred to as a field or fields, and it was probably laid out with walks at a very remote period. An ancient petition presented to Parliament during the Inter-

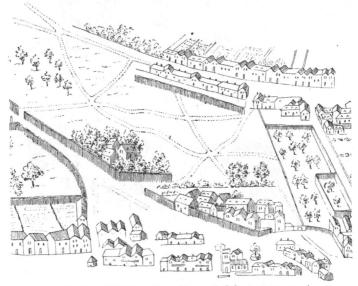

Lincoln's Inn Fields in 1560. (From Ralph Aggas's map.)

regnum gives us some interesting particulars as to the uses to which it was put in the reign of Edward III. This petition states that it appears from record that ' in those times ' (about 1376) ' this field was a common walking and sporting place for the clerks of the Chancery, apprentices, and students of the law, and citizens of London ; and that upon a clamorous complaint made by them unto the King, that one Roger Leget, had privily laid and hid many iron engines called caltrappes, as well in the bottome as the top of a certaine trench in Fiket's Fields' (*i.e.*, Lincoln's Inn Fields), ' neere the Bishop of Chichester's house, where the said clerkes, apprentices, and other men of the said city, had wont to have their common passage, in which place he knew that they daily exercised their common walks and disports, with a malicious and malevolent intent, that all who came upon the said trench should be maimed or else most grievously hurt; which engines were found by the foresaid clerkes, apprentices, and others passing that way, and brought before the King's councell, in the Chapter-house of the Friars, preachers of London, and there openly shewed; that hereupon the said Roger was brought before the said councell to answer the premises; and being there examined by the said councell, confessed his said fault and malice in manner aforesaid, and thereupon submitted himselfe to the King and his councell. Whereupon the said Roger was sent to the King's prison of the Fleete, there to expect the King's grace.' The petition then concludes ' that any device to interrupt or deprive such clerks, and citizens, of their free common walking or disport there, is a nuisance and offence punishable by the King and his councell by fine and long imprisonment; and that the King and councell have ever been very careful of preserving the liberties and interests of the lawyers and citizens in these fields, for their cure and refreshment.'* As in course of time London began to enlarge its bounds, and land thereby became more valuable, owners of property in Lincoln's Inn Fields began to erect buildings here, which were of a mixed character. A proposal to add more led to the Lords of the Privy Council sending a protest to the county justices in September, 1613. Five years later James I. granted a commission to Francis Bacon, Lord Chancellor, and others, ' to reduce Lincoln's Inn Fields into walks,' his idea being to make it like Moorfields. The Commissioners had the aid of the King's architect, Inigo Jones, who only lived to design the west side, which was called the Arch Row. His work can easily

* Quoted in ' History of St. Giles's-in-the-Fields,' J. Parton, 1822, p. 140.

be distinguished by the carved roses and fleurs-de-lys which ornament the houses, and by the stone pilasters and capitals on a brick ground, of which he was very fond.   We shall have occasion to refer later on to his architectural work here, but we can gather from the terms of this commission some description of the Fields at this time.   It showed ' that the grounds called Lincoln' Inn Fields were then much planted round with dwellings and lodgings of noblemen and men of qualitie ; but at the same time it was deformed by cottages and   mean   buildings,   encroachments   on   the   field,   and nuisances to the neighbourhood.'   The Commissioners were therefore directed to reform those grievances, and ' according to their discretion to frame and reduce those fields called '' Cup Field and Purse Field,'' both for sweetness, uniformitie and comeliness, into such walkes, partitions or other plottes, and   in   such   sorte,   manner   and   forme,   both   for   publique health and pleasure, as by the said Inigo Jones, etc., is or accordingly shall be done by the way of map.'*   It is a popular tradition that the square was reduced to the size of the base of the Great Pyramid, but the fallacy of this is seen at once in comparing the respective areas.   The troubles of the succeeding reign prevented the improvement works being completed and laid the way open for more building. This led to the petition before referred to, which resulted in a peremptory proclamation by Oliver Cromwell, dated Whitehall, August 11, 1656 :

' Upon   consideration   of   the   Humble   Petition   of   the Society of Lincoln's Inn, and of divers persons of quality, inhabitants in and about the fields, heretofore called by the several names of Pier's Field,† Cup Field, and Fitchet's Field, and now known by the name of Lincoln's Inn Fields, adjoining to the said Society . . . setting forth among other things that divers persons have prepared very great store of bricks for the erecting of new buildings upon the said Fields; Ordered by his Highness the Lord Protector and the Council

---

* Quoted in ' History of St. Giles's-in-the-Fields,' J. Parton, 1822, p. 141.
† Called ' Purse Field ' in the commission of James I.

that there be a stay of all further buildings . . . and that it be recommended to the Justices of the Peace for the City of Westminster and liberties thereof to take care that there be no such new buildings, nor proceeding in any such buildings already begun.'

In the following year the unfinished state of the square was taken into consideration, and an agreement was accordingly entered into between Sir William Cowper, Bart., Robert Henley, and James Cowper, who had taken a lease of these and other fields for building purposes, and the Society of Lincoln's Inn. One of the clauses states 'that the said Society of Lincoln's Inn were interested in the benefit and advantage of the prospect and air of the said field, but were willing and contented' that the parties to whom the building was to be entrusted 'might proceed in their said design and undertaking . . . with such caution and provision for the beautifying and adorning of the said intended building, and for levelling and plaining the said field, and casting the same into walks, and for prevention of any future building thereupon.' In

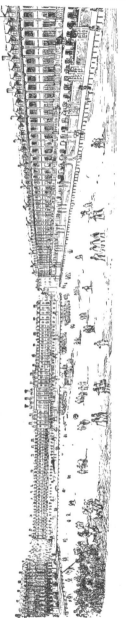

Lincoln's Inn Fields as originally planned by Inigo Jones.

pursuance of these arrangements, a grant was made to Sir Lislebon Long, and other trustees, of 'all the rest, residue or body of the said field therein called Cop Field, alias Cup Field . . . not to be built on' for a term of 900 years.*

The previous history of the ownership is very obscure.

Lincoln's Inn Fields, circa 1780.

At the earliest time of which there is any record, Fikattes-feld was the property of the Knights Templars, and it is often called in old deeds Campus Templariorum. They built the Old Temple on the site of Southampton Buildings in

* Quoted in 'History of St. Giles's-in-the-Fields,' J. Parton, 1822, p. 143.

the reign of Stephen, but in the succeeding reign they commenced a nobler structure opposite the end of what is now Chancery Lane. This they called the New Temple, the church of which was finished and dedicated in 1185.* The Templars were succeeded by the Order of Black Friars, who were granted a piece of ground 'without the wall of the City of Oldborne [Holborn] near unto the old Temple,' in 1221, upon which they built a monastery facing Holborn. When that community removed to the district now called Blackfriars, their house and grounds were granted to Henry Lacy, Earl of Lincoln, from whom we obtain the name of Lincoln's Inn. He built his mansion-house here on the site now occupied by Lincoln's Inn, and some particulars of his grounds may be gathered from a record preserved in the Duchy of Lancaster Office relating to the profits and expenditure of the Earl's garden. We learn from this curious document that apples, pears, large nuts, and cherries, were produced in sufficient quantities not only to supply his table, but also to yield a profit by their sale. The amount realized by one year's sale equalled about £135 of modern currency. The vegetables cultivated in this garden were beans, onions, garlic, leeks, and some others not specially named. Hemp was also grown, and cuttings of the vines were sold, from which it may be gathered that the Earl's trees were held in some estimation. The only flowers named are roses, of which many were sold, and it also appears that there was a pond stocked with pike, for which frogs, eels, and small fish were purchased.† The Earl died without issue in 1310, and in the same year the Society of Lincoln's Inn was founded, but how the property became theirs is not apparent.

The several acts and mandates to which reference has been made did not put an end to the nuisances that had been complained of, because the space was not properly enclosed. In very early times the Fields had been the

* 'History of St. Giles's-in-the-Fields,' J. Parton, 1822, p. 178.
† Spilsbury, 'Lincoln's Inn,' 1850, pp. 32, 35.

scene of several executions. The fourteen conspirators who had plotted to assassinate Queen Elizabeth and set free Mary Queen of Scots were executed here. This attempt was known as the Babington Conspiracy, and in September, 1586, having been found guilty, they were all 'hanged, bowelled, and quartered, in Lincoln's Inn Fields, on a stage or scaffold of timber, strongly made for that purpose, even in the place where they used to meete and to

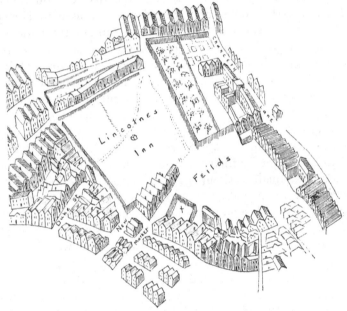

Lincoln's Inn Fields in 1658. (From Newcourt's map.)

conferre of their traitourous practices.'* Nearly a hundred years after, on July 21, 1683, William, Lord Russell, was executed here on the charge of being concerned in the Rye House Plot. It has been said that the Duke of York moved that he might be executed in Southampton Square, before his own house, but the King rejected that as indecent. So Lincoln's Inn Fields was the place appointed for his

* Stow's Annals, p. 1236.

execution.* A brass tablet has been placed in the shelter in the centre of the Fields to mark the exact spot. Through these Fields, in the reign of Charles II., Thomas Sadler, a well-known thief, attended by his confederates, made his mock procession at night, with the mace and purse of the Lord Chancellor Finch, which they had stolen from the Chancellor's closet in Great Queen Street, immediately adjoining, and were carrying to their lodging in Knightrider Street. One of the confederates walked before Sadler with the mace of the Lord Chancellor exposed on his shoulder, and another followed after him, carrying the Chancellor's purse, equally prominent. Sadler was executed at Tyburn for this theft.†

For another fifty years after these events these Fields were the haunt of several worthless characters. Cripples of all kinds made this a regular hunting-ground; not content with extorting money by the display of their apparent misfortunes, they took to intimidating passers-by with their crutches. The literature of the seventeenth century contains frequent allusions to the 'mumpers' and 'rufflers' of Lincoln's Inn Fields, which were the names given to these idle vagrants. An extract from the *London Spy* describes how a party of visitors 'went into the Lame Hospital, where a parcel of wretches were hopping about by the assistance of their crutches, like so many Lincoln's Inn Fields mumpers drawing into a body to attack the coach of some charitable lord.' The 'rufflers' were beggars who assumed the character of maimed soldiers, and imposed upon the credulity of sympathetic passers-by. It was in Lincoln's Inn Fields that 'Lilly, the astrologer, when a servant at Mr. Wright's, at the corner house over against Strand bridge, spent his idle hours in bowling with Wat the cobler, Dick the blacksmith, and such like companions.' Another sport in connection with this place is mentioned by Locke,

* Burnet, 'Own Times,' edition 1823, vol. ii., p. 377.

† Wheatley and Cunningham, 'London : Past and Present,' vol. ii., p. 393.

in his directions for a foreigner visiting England, who could
see ‘wrestling in Lincoln’s Inn Fields all the summer.’*
We have another allusion to the dangers of this spot in
Gay’s ‘ Trivia ’ :

> ‘ Where Lincoln’s Inn wide space is rail’d around,
> Cross not with venturous step ; there oft is found
> The lurking thief, who, while the daylight shone,
> Made the walls echo with his begging tone ;
> That crutch, which late compassion mov’d, shall wound
> Thy bleeding head, and fell thee to the ground.
> Though thou art tempted by the linkman’s† call,
> Yet trust him not along the lonely wall ;
> In the mid-way he’ll quench the flaming brand,
> And share the booty with the pilfering band.
> Still keep the public streets where oily rays,
> Shot from the crystal lamp, o’erspread the ways.’

The rail referred to in these lines was only a wooden post
and rail, not the present iron fencing.

In 1735 the inhabitants obtained an Act for a more rigid
enclosure of the square, which, in spite of the measures
formerly adopted, the Act states ‘ had for some years then
last past lain waste and in great disorder, whereby the same
had become a receptacle for rubbish, dirt, and nastiness of
all sorts, brought thither and laid not only by the inhabitants
of the said Fields, but many others, which had not been
removed or taken away by the several scavengers of the
parishes wherein the said Fields are situate as aforesaid; but
also, for want of proper fences to inclose the same, great
mischiefs had happened to many of His Majesty’s subjects
going about their lawful occasions, several of whom had
been killed, and others maimed and hurt, by horses which
had been from time to time aired and rode in the said Fields ;
and by reason of the said Fields being kept open many wicked
and disorderly persons had frequented and met together
therein, using unlawful sports and games, and drawing in
and enticing young persons into gaming, idleness, and other

* ‘ London : Past and Present,’ vol. ii., p. 394.
† A man carrying a link, or torch, to show the way.

vicious courses, and vagabonds, common beggars, and other disorderly persons, resorted therein, where many robberies, assaults, outrages, and enormities, had been and continually were committed.'

Perhaps an extract from a newspaper, dated June 7, 1733, may throw some light upon this clause. It states that ' Yesterday in the evening His Honour the Master of the Rolls, crossing Lincoln's Inn Fields, was rode over by a boy who was airing an horse there, by which accident he was much bruised.' There is a marked similarity in the wording of these two paragraphs, which makes it more than probable that ' His Honour' played an important part in obtaining the Act. A further clause goes on to state ' that the south, west, and north parts of the said Fields were incompassed with houses, many of which were inhabited by the owners and proprietors thereof, who with the other inhabitants could not go to and from their respective habitations in the night season without danger, and therefore were desirous to prevent any mischiefs for the future, and to enclose, clean, repair, and beautify the said Fields in a graceful manner, and were willing and desirous that an adequate contribution might for that purpose be raised by and amongst themselves.' The Act further empowered the trustees to levy a rate on the inhabitants and owners of the houses in the square, not exceeding 2s. 6d. in the pound, for its maintenance.

The square was laid out under the terms of this Act, completely railed in, planted with trees, and traversed by walks in diagonal directions. In the centre was a pond or reservoir of water. The square was reached by two entrances from Holborn, named Great and Little Turnstile, which sufficiently denote their character, while Gate Street was a way through a gate to admit horses and carriages. These ancient names, it is hardly necessary to add, are still preserved, although the turnstiles and the gate have disappeared. About the year 1820 a fresh laying-out of the garden was rendered necessary, when it assumed its present form.

To come now to consider some of the historical buildings in Lincoln's Inn Fields, we will commence with the west side, formerly called the Arch Row. Lindsay House, No. 59, was built by Inigo Jones, for the Earl of Lindsay, who was the Royalist commander at the outbreak of civil war under Charles I., killed at the Battle of Edgehill. The fourth Earl of Lindsay was created Duke of Ancaster, and the name of the mansion was changed to Ancaster House. It was subsequently purchased by the proud Duke of Somerset.

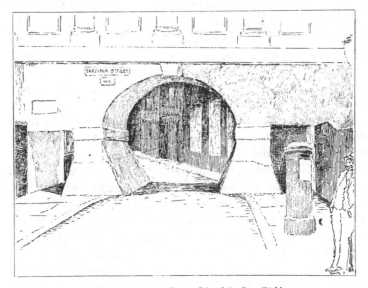

Archway, Sardinia Street, Lincoln's Inn Fields.

'Old Somerset is at last dead. . . . To Lady Frances, the eldest, he has conditionally given the fine house built by Inigo Jones in Lincoln's Inn Fields, which he had bought of the Duke of Ancaster for the Duchess, hoping that his daughter will let her mother live with her.'* The external features of the house are the same, except that the urns which formerly ornamented the balustrade along the front of the roof have disappeared.

* H. Walpole to Mann, December 15, 1748, vol. ii., p. 137.

Powis House, No. 67, at the corner of Great Queen Street, was built in 1686, by William Herbet, Viscount Montgomery and Marquis of Powis, on the site of a former house, which was destroyed by fire. The architect was Captain William Winde. This house also changed its name when it was sold to Holles, Duke of Newcastle, Prime Minister in the reign of George II., when it was called

Lindsay or Ancaster House, Lincoln's Inn Fields.

Newcastle House. A good story is told in connection with this house, which is said to have put an end to the expensive custom of ' vailsgiving,' or the feeing of all the servants, who used to assemble in the hall on the departure of guests. ' Sir Timothy Waldo, on his way from the Duke's dinner-table to his carriage, put a crown into the hand of the cook, who returned it, saying : " Sir, I do not take silver." " Don't you, indeed ?" said Sir Timothy, putting it in his pocket ;

" then I do not give gold.'' '' * In latter years (1827-1879) this house was the head-quarters of the Society for Promoting Christian Knowledge, now in Northumberland Avenue. A

Newcastle House, Lincoln's Inn Fields.

gloomy archway (said to be the work of Inigo Jones) leads to Sardinia Street, formerly Duke Street, on the south side

* Pugh, ' Remarkable occurrences in the Life of Jonas Hanway,' 1787, p. 184.

of which is the Sardinian Roman Catholic Chapel. This building, the oldest of its kind in London, was originally attached to the residence of the Sardinian Ambassador. At one time it was the chief centre of the Roman Catholic worship, but it is now only a church for the immediate neighbourhood. It was severely attacked and partly destroyed in the Gordon riots of 1780.

Reverse of silver medal in the British Museum, struck to commemorate the destruction of the Roman Catholic Chapel in Lincoln's Inn Fields in 1688. The Portuguese Chapel is shown in ruins, whilst the Papal emblems are being burnt in the fields in front.

The principal building on the south side, which was formerly known as Portugal Row, is the Royal College of Surgeons, built on the site of a house belonging to Lord Chancellor Northington. It contains the splendid museum of John Hunter, from whose executors it was purchased by the Government for £15,000. The greater portion of the present building was erected from the designs of Sir Charles

Barry, but the subsequent additions to the museum have necessitated its enlargement on more than one occasion.

One of these extensions led to the demolition of the celebrated Lincoln's Inn Fields Theatre, which was situated at the back of the College of Surgeons. There have been three distinct theatres on this site. The first was originally a tennis-court, and was converted into a theatre by Sir William Davenant, in 1660. Pepys frequently used to go there—in fact, so often that it made Mrs. Pepys 'as mad as

College of Surgeons, Lincoln's Inn Fields.

the devil.' His opinion of it is that 'it is the finest playhouse, I believe, that ever was in England.' After the death of Davenant it reverted to its former use, and became a tennis-court again.

The second theatre on the same site was opened in 1695, and is described by Cibber as 'but small and poorly fitted up within. Within the walls of a tennis-quaree court, which is of the lesser sort.'* The third building was commenced

---

* Cibber, 'Apology,' edition 1740, p. 254.

by Christopher Rich, and opened by his son, John Rich.
This latter actor first introduced the now popular pantomimes here, which were a great success ; but the chief event
connected with this building was the production of the
' Beggars' Opera,' by Gay, which had so great a run 'that
it made Gay rich and Rich gay.' The theatre then had
many changes ; it was used for barracks, as a china deposi-
tory, and, finally, pulled down, as we have seen, for the

Duke's Theatre, Lincoln's Inn Fields.

enlargement of the College of Surgeons. Serle Street,
leading from this side of the square to Carey Street, derives
its name from a former proprietor, Henry Serle, who died
about 1690.

The east side of the square is occupied by the noble
buildings of Lincoln's Inn Hall and Library. This Hall,
commenced in 1843, is one of the finest in London, being
120 feet long, 45 broad, and 64 high. The oak roof, divided
into seven compartments, is a remarkable feature of the

interior. At the northern end is a fresco painted by
G. F. Watts, R.A., entitled 'The Lawgivers,' which is
unfortunately fading. This work was done by the artist
gratuitously, but when it was completed the Inn presented
him with a gold cup containing 800 sovereigns. Among
others in the fine collection of paintings here is Hogarth's
'Paul before Felix,' painted for the Society in 1750, and
removed from the old Hall. The new Hall was opened by
the Queen and Prince Consort, on October 30, 1845. The
total cost was £88,000, the architect being Philip Hardwick, R.A. On this side the Fields have an approach from
Chancery Lane, the gateway of red brick over the entrance
bearing the date 1518. Over this gateway Oliver Cromwell
is said to have lived for some time, and tradition also relates
that Ben Jonson worked as a common bricklayer in the
erection of the adjoining wall about 1617 ; but the truth
of this is very doubtful, as by this time he had written some
of his best plays.

The most notable building on the north side is the Soane
Museum. This was founded by a bequest of Sir John
Soane, the son of a country bricklayer, who rose to great
eminence as an architect. His chief work was the Bank
of England, and he became ultimately Professor of Architecture at the Royal Academy. The museum is crowded
from top to bottom with curiosities of every description.
There are also several masterpieces by Hogarth, Turner, and
Sir Joshua Reynolds. It is strange that this museum, which
is open free to the public, should be as little known as it is.

Parallel with this side of the Fields, and between it and
Holborn, is a narrow roadway known as Whetstone Park,
It derives its name from William Whetstone, a tobacconist.
and also overseer of this parish in the time of Charles I.
and the Commonwealth. The term 'park' certainly seems
out of place as applied to a row of buildings chiefly consisting of stables and workshops. It has borne in times past a
very bad name, owing to the resort here of loose characters.
Several references to these are to be found in the plays of

Dryden, and other allusions in Butler's ' Hudibras ' and the
London Spy.   But Whetstone Park can boast at least one
distinguished inhabitant.   Milton moved, in 1645, from a
house in Barbican, 'to a smaller house in Holborn, which

Sir John Soane's Museum, Lincoln's Inn Fields.

opened backward into Lincoln's Inn Fields.'*   In this case his
garden must have been built over by these houses of ill-fame.
In giving these particulars about the most important
* Philips, ' Life of Milton.'

buildings surrounding the Fields, we have incidentally mentioned some of the eminent inhabitants.   The list is a very
lengthy one, including several Lord Chancellors, Chief
Justices, and Sir William Blackstone, among the legal
world ; the celebrated Duchess of Marlborough, John Locke,
William Pitt, Spencer Perceval, and several other members
of the nobility.   Nell Gwynne was lodging in Lincoln's Inn
Fields when her first son, afterwards Duke of St. Albans,
was born.

Of late years the houses in this square have lost their
residential character, but Tennyson, in his unknown days,
dwelt in lofty chambers up behind the balustraded parapet
of No. 57, and he used to resort to the Cock for his quiet
five o'clock dinner.*

At No. 58 lived John Forster, the friend and biographer
of Dickens.   This house is the original of Mr. Tulkinghorn's
residence described by Dickens in his ' Bleak House.'   He
is another of the celebrities who have helped to make
Lincoln's Inn Fields famous, and though now shorn in
many respects of its former eminence, this historical square
may well be content to live in its past records.

RED LION SQUARE.

On the opposite side of Holborn is Red Lion Square,
half an acre in extent.   Its present appearance is rather
dull, reminding one of a poor relation, as compared to the
more aristocratic squares of the West.   It was built about
1698, and takes its name from the Red Lion Inn, a very
ancient hostelry, and for a long while the largest and best
frequented inn in Holborn.   This was a flourishing institution long before the square was built, when its site was
merely fields.   In the register of St. Andrews, Holborn,
is an entry concerning a foundling ' borne under the Redd
Lyon Elmes in the fields in High Holborn, baptized iij of
August, 1614.'   London at this time only extended to about

* P. Fitzgerald, ' Picturesque London,' p. 186.

this part of Holborn. From Farringdon Street towards Ely House and Gray's Inn Lane the ground was either entirely vacant or occupied in gardens. From Holborn Bridge to Red Lion Street there were houses on both sides, but further up, near Hart Street, the road was entirely open. There was a small colony clustered about St. Giles's Church, which was then worthy of its additional name, 'in-the-Fields,' and after this both to the north and west was open country, the present great thoroughfares being only distinguishable by the avenues of trees. The site of Red Lion Square was then known as Red Lion Fields, as we have indicated above. The first approach towards rendering it habitable seems to have been the laying-out of a bowling-green, and erecting a house of entertainment near it, called the Bowling-Green House. This was built on the site of the present square.*

This has been a public garden since 1885, and its present appearance dates from that time when it was laid out by the Metropolitan Gardens Association. About 150 years before this, when the garden was in a very dirty and neglected condition, the first attempt at any great improvement was made. A newspaper paragraph about that time mentions the subject as being then under consideration: 'Red Lion Square, in Holborn, having for some years lain in a ruinous condition, a proposal is on foot for applying to Parliament for power to beautify it, as the inhabitants of Lincoln's Inn Fields have lately done.' The Act they obtained was 'to enable the present and future proprietors of the houses in Red Lion Square to make a rate on themselves for raising money sufficient to inclose, pave, watch, clean, and adorn the said square.' It must have been in a very bad state according to the preamble of the Act, which runs as follows: 'Whereas the square called Red Lion Square . . . hath for some time past lain in great disorder, and the pales which inclose the area thereof are so ruinous, that the said area is become a receptacle for

* J. Parton, 'History of St. Giles-in-the-Fields,' p. 188.

rubbish, dirt, and nastiness of all kinds, and an encouragement to common beggars, vagabonds, and other disorderly
persons, to resort thither for the exercise of their idle
diversions, and other unwarrantable purposes.' By this
Act fifteen trustees were appointed, who had power to levy
a rate not exceeding 1s. 6d. in the pound on the inhabitants,
seven-tenths of which was to be paid by the tenants or
occupiers, and the remainder by the landlords.

The laying-out consisted of enclosing the area with iron
railings; a stone watch-house was erected at each corner,
and in the centre was a stone obelisk, around which much
mystery hangs. It is said to have covered the remains of
Oliver Cromwell, Ireton, and Bradshaw. The corpses of
these regicides, as they were styled after the Restoration,
were removed in January, 1661, from Westminster Abbey
to the Red Lion, in Holborn, and left for the night. In
the morning they were removed on a sledge to Tyburn,
exhibited on the gallows, and there submitted to other
ignominious treatment. This is not disputed; but a tradition
quoted by Rede in his anecdotes and biography goes on to
say that their mutilated remains were rescued by some of
their followers, and reverently buried in this square, the
stone obelisk marking the exact spot. The leader in this
scheme is said to have been an apothecary, who had considerable local influence, and at the time the square was
built managed to carry out his desires. From researches
which have been made it has been discovered that about
the time of the Restoration an apothecary of the name of
Ebenezer Heathcote was living at the King's Gate, Holborn.
He had married the daughter of one of Ireton's sub-commis-
sionaires, and perhaps this remote connection with that
soldier may have accounted for this enthusiasm on his part.
It has been pointed out that if the body of Cromwell had been
removed from the Abbey and buried in Red Lion Square,
it would not have been possible to have procured another
embalmed body to be sent in its place to Tyburn, as has
been suggested. The ' clumsy obelisk ' is said by Pennant

to have been inscribed with the following inscription : ' Obtusum obtusioris ingenii monumentum. Quid me respicis, viator ? Vade.'

When the first laying-out was completed, the dull effect of the square was not much improved if we may take the opinion of a whimsical author of 1771. He says : ' Red Lion Square . . . has a very different effect on the mind. I never go into it without thinking of my latter end. The rough sod that " heaves with many a mouldering heap," the dreary length of its sides, with the four watch-houses like so many family vaults at the corners, and the naked obelisk that springs from amid the rank grass like the sad monument of a disconsolate widow for the loss of her first husband, all form together a *memento mori* more powerful to me than a death's head and cross marrow-bones ; and were but a parson's bull to be seen bellowing at the gate, the idea of a country churchyard in my mind would be complete.' *

It was a matter of great dispute for some considerable time between the authorities of St. Andrew, Holborn, and St. Giles's-in-the-Fields, as to which parish Red Lion Square was situated in. According to the vestry minutes of the latter body in 1676 and 1777, the inhabitants seem to have been called upon by both parties to pay rates—a luxury which they evidently did not appreciate. Eventually the parish of St. Andrew won the day.†

This garden was originally taken over under an agreement entered into with the owners by the Metropolitan Public Gardens Association, but so greatly did the control of the London County Council improve the garden that the representatives of the inhabitants determined to hand it over to the Council, which they did in 1895 practically as a free gift.

Among the residents at one time or another in Red Lion Square we must mention first of all that eccentric traveller, Mr. Jonas Hanway, who lived at No. 23. He was born at

* ' Critical Observations on the Buildings, etc., of London.'
† ' History of St. Giles's-in-the-Fields,' J. Parton, p. 188.

Red Lion Square in 1800.

Portsmouth in 1712, and made considerable voyages in the course of his mercantile career. He had made sufficient to retire in 1753, when he published a work of some practical interest describing his travels in Russia. His life at home was principally devoted to philanthropic work, the results of which are still seen in the Marine Society and Magdalen Hospital, both of which he founded. So great was the estimation in which he was held that a deputation of merchants waited upon Lord Bute, when Prime Minister, asking him to bestow upon Hanway some mark of the public esteem. As a consequence of this he was appointed a Commissioner of the Navy. Although he is more remembered for his philanthropy than for his authorship, it is worthy of mentioning that he wrote a lengthy attack upon tea, which called forth a sarcastic defence of his favourite drink by Dr. Johnson. The principal rooms in his house were decorated with paintings and various emblematical devices 'in a manner peculiar to himself.' He goes on to say 'to relieve this vacuum in social intercourse' (*i.e.*, the time between the assembling of visitors and the placing of card-tables), 'and prevent cards from engrossing the whole of my visitors' minds, I have presented them with objects the most attractive I could imagine, and when that fails there are the cards.'* After his death in 1786 in Red Lion Square, a monument was erected to him by public subscription. It used to be popularly supposed that Jonas Hanway was the first to introduce the umbrella to public notice. Mr. Peter Cunningham, in his 'Handbook of London,' says: 'Hanway was the first man who ventured to walk the streets of London with an umbrella over his head. After carrying one near thirty years, he saw them come into general use.' But the umbrella must have been common in London in 1712, some years before Hanway was born, for Swift, in 'A City Shower,' published in 1710, says:

'The tucked-up semstress walks with hasty strides,
While streams run down her oiled umbrella's sides;'

* John Pugh, 'Remarkable Occurrences in the Life of Jonas Hanway,' London, 1787.

33—2

and Gay, who is so rich in popular allusions, writes in the
following year :

> ' Or underneath th' umbrella's oily shed,
> Safe thro' the wet on clinking pattens tread.'

Another distinguished resident in Red Lion Square was
Benjamin Robert Haydon, the historical painter.   He was
living here in 1838 in a large house on the west side of the
square.   His life was a peculiar one, but his misfortunes
were chiefly of his own creation.   Through dissatisfaction at
the way his picture of the ' Murder of Dentatus' was hung
in the Royal Academy of 1809, he spent the rest of his life
in open hostility to that body.   His pictures, which were
very numerous, were mostly shown at rival exhibitions of
his own with varying success, as one of the last entries in
his diary tells us : ' Tom Thumb had 12,000 people last
week, B. R. Haydon 133½ (the half a little girl)   Exquisite
taste of the English people.'   Among his pupils he numbered
some of the most distinguished painters of the time : Sir
Edwin Landseer, Sir Charles Eastlake, and George Lance,
the fruit-painter.   His works all suffer from imperfect execution, and his chief lessons must have been in what to avoid
rather than what to imitate.

Another painter, whose speciality was portraits—viz.,
Henry Mayer—lived at No. 3, and it was at this house that
Charles Lamb sat to him in 1826.

The law is represented by Lord Chief Justice Raymond,
born in 1673, died in Red Lion Square 1733.   He was
created Baron Raymond, of Abbots-Langley, Herts, where
he had an estate, but the title is now extinct.   In 1702 he
was counsel for the prosecution of a man named Hathaway,
who was accused of drawing blood from a supposed witch,
and his conduct of the case tended greatly to dispel the
superstitions current with regard to witchcraft.

A medical genius, James Parsons, M.D., born at Barn-
staple in 1705, resided in Red Lion Square, where his house
was for many years the centre of meeting for much of the

literary and scientific society of the period. In 1769, when his health was failing, he moved to Bristol, but returned to his old quarters in the following year, where he died almost at once. He left directions that he was not to be buried till some change appeared in the corpse, and so he was left unburied for seventeen days.

The last inhabitant of Red Lion Square we shall mention is Sharon Turner, the historian. He was intended in early life for the law, but relinquished that profession to follow historical pursuits. The success of his first work, ' History of the Anglo-Saxons,' led him to write many others, but his fame chiefly rests on his first work. He died in Red Lion Square in 1847.

The chief building to notice at the present time in Red Lion Square is the handsome Church of St. John the Evangelist. This was built in 1874 from the designs of the late Mr. J. L. Pearson, R.A., and consecrated in 1878.

Another institution worthy of mention is an ancient Baronial Court at the north-eastern corner of the square. This is held monthly before the Sheriff of Middlesex or his deputy. Its powers are as great as any of the present courts of law, while it is less expensive and more expeditious. This court was instituted by King Alfred on dividing the kingdom into shires, and continued by many statutes, including Magna Charta. It is treated of by several eminent legal authorities, as Judge Hale, Judge Lambert, and many others.*

* *Gentleman's Magazine*, 1829.

CHAPTER XXVI.

RAVENSCOURT PARK—SHEPHERD'S BUSH COMMON.

RAVENSCOURT PARK.

THIS pleasant little park of 32 acres is situated at Hammersmith, and was purchased at the joint expense of the late Metropolitan Board of Works and the Vestry of Hammersmith in 1887 for the sum of £58,000. It comprised a large mansion with well-timbered grounds, meadows, and an orchard, and was therefore well adapted for the purposes of a public park. The London and South-Western Railway had obtained Parliamentary power to carry their Hammersmith and Richmond line through the park on arches, and this fact may have had some influence in the sale of the property. This neighbourhood, like the rest of the suburbs near London, is being rapidly developed for building purposes, and it is extremely fortunate that the estate has escaped the advance of the sea of bricks and mortar, for as far back as 1839 it was proposed to build villas on the line of the present avenue. It is very pleasant now that the ground is safe to read that ' other parts of the park will be let off for detached villas, for which it is particularly adapted from its secluded situation and proximity to London.'*

Since the park was first opened, various additions have been made chiefly for the sake of forming new entrances, so as to give ready access from the many important thorough-

* Faulkner, ' Hammersmith,' 1839, p. 378.

fares adjacent to it. A refreshment-house has been erected and a band-stand, on which performances are given twice a week during the season to densely-packed audiences. Throughout the spring and summer the park is gay with flower-beds and ornamental borders, whilst there is a large area of lawn available for tennis and children's games.

The present mansion in the park has taken the place of the old Manor-house of Pallenswick or Paddenswick, which tradition connects with the name of Alice Perrers, the fair favourite of Edward III. She was Lady of the Bedchamber to Queen Philippa, the worthy consort of the illustrious victor of Cressy, and according to all accounts was a woman of extraordinary wit and beauty. When the Queen died, this woman obtained a great ascendancy over the enfeebled monarch, and his once proud mind was degraded beneath her rule. The ancient Manor of Palingswick or Paddens-wick, which formerly belonged to John Northwyck, gold-smith, of London, was granted in 1373 to certain trustees on her behalf, and the manor-house became her country seat.* The attention which Edward III. paid to her, and the means he took to procure diversions for her, attracted the unfavourable attention of Parliament. The climax was reached when the King held a tournament in her honour at Smithfield, on which occasion Alice appeared in a triumphant chariot as 'Lady of the Sun,' attended by many ladies of quality, each leading a knight by his horse's bridle. When the procession reached West Smithfield, the tournament began, and was continued for seven days.† This led to the Parliament petitioning the King to remove her, which he reluctantly did, but she was soon recalled, and after an eventful career eventually married William Lord Windsor.

A survey of the manor was taken in 1378 upon the banish-ment of Alice Perrers, in which the mansion is described as being well built, in good repair, and consisting of a large hall, a chapel, kitchen, bakehouse, stables, barns and gates.

* Faulkner, 'Hammersmith,' p. 369.
† Barnes, 'Reign of Edward III.,' p. 872.

The manor also comprised two gardens, but these are
described as only worth 1s. 6d. a year, on account of the
apple-trees being blown down by the wind. The remainder
was made up of 40 acres of arable land, valued at £1 6s. 8d.
a year; 60 acres of pasture, at 8d. an acre; and 1½ acres of

The Avenue, Ravenscourt Park.

meadow, valued at 5s. annually, the whole being held by
copy of Court Roll under the Manor of Fulham.*

When Alice Perrers married Lord Windsor, Richard II.
granted this manor to him in the year 1380.

The next mention we have of the manor-house is in 1572,
when it was bequeathed by John Payne to his son William,

* Faulkner, 'Hammersmith,' p. 371.

Lord of the Manor;* and in this house he held the last
Court of the Manor.   He bequeathed one of the small
islands, or eyots, in the Thames for the benefit of the poor
of Hammersmith. A descendant of his, John Payne, sold
the manor or capital messuage of Palingswick, with its
appurtenances, in 1631, to Sir Richard Gurney, citizen, cloth-
worker, and Lord Mayor of London.   He distinguished
himself by his loyalty to Charles I., and of course fell under
the displeasure of Parliament, who preferred several articles
of impeachment against him, for which he was by sentence
of the Peers degraded from the mayoralty, and condemned
to remain a prisoner in the Tower, where he died in 1647.
His widow sold it three years afterwards to Maximilian
Bard, in whose family it continued till 1747, at which time
Hammersmith is described in an old history of Middlesex
as being a small village, near Brentford, containing some
fine seats.† It then passed into the hands of Thomas
Corbett, Secretary to the Admiralty.   The arms of Thomas
Corbett were a raven sable on a white ground. He changed
the name from Paddenswick to Ravenscourt.   At his death
it was sold by public auction, the following being a copy of
the advertisement: 'To be sold by auction, by Mr. Lang-
ford, on the premises (by order of the executor), the beginning
of June next ensuing, the Manor of Paddinswick, at Hammersmith, in the parish of Fulham, and county of Middlesex,
late the estate of Thomas Corbett, Esq., Secretary to the
Admiralty, deceased, consisting of a capital mansion-house,
out-houses, gardens, lands, farms and messuages, thereunto
belonging and adjoining, all copyhold of inheritance, the
situation of which is admirable, the house in the finest
repair, and improved with every conveniency that can be
desired; the lands of a rich and fertile soil, the gardens
elegantly laid out, and the whole calculated to give delight.
At which time will be likewise sold by auction all the genuine
and rich household furniture, linen, china, brewing utensils,

* Faulkner, 'Fulham,' p. 379.
† Faulkner, 'Hammersmith,' pp. 374, 375.

garden-tools, implements of husbandry, and other effects. Further particulars, and timely notice of which, will be given in this and other papers.'*

After being purchased by a Mr. Arthur Weaver, it passed into the hands of Henry Dagge, the author of ' Considerations on the Criminal Laws.' It was leased by him to Lord Chancellor Northington, who had the peculiar distinction of holding the Great Seal nine years, and in two reigns, those of

The Mansion, Ravenscourt Park.

George II. and George III., and during the whole of four administrations, Mr. Pitt's, Lord Bute's, the Duke of Bedford's, and the Marquis of Rockingham's. A romantic story is told of his marriage, which seems quite out of place in the sober dignity of the law. He fell in love with an invalid young lady at Bath, Miss Husband, who fortunately recovered her strength and proved to be an heiress. They were eventually married, and, we presume, lived happily ever afterwards.

The house next passed in 1765 into the hands of John

* *Morning Advertiser*, 1754.

Dorville, who has left his name in the row of houses formerly called Dorville's Row, now part of King Street West, and the Dorville family sold it to the late owner, George Scott. At the time of the purchase of the estate for a public park, it was locally known under the name of Scott's Park. The present mansion was built by the Bard family about 1648 or 1650, and the ancient manor-house which stood a little east of the park was probably pulled down about the same time. It is built in the style of the French architect Mansart; important additions were made to it by the late owner, and the pleasure-grounds and gardens were improved under the direction of Mr. Repton.* A great part of the moat which formerly surrounded the mansion was filled up, and the remainder formed into an ornamental piece of water, which was adapted to form the present lake.

The principal feature of the park is the fine avenue of elms and chestnuts leading from King Street to the mansion, on one side of which is the orchard, a constant source of temptation to the youth of Hammersmith. When one of the ancient elms opposite the mansion was taken down some years ago, a riding-spur was found embedded in a branch nearly of the date when the present house was built. This must have been thrown up and caught in the young tree, and the bark have gradually grown over it, and thus it remained for about 200 years.

It is a great misfortune for the park that it is traversed by the London and South-Western Railway. It need hardly be stated that railway arches do not form a picturesque feature in any park, but as much has been done as is possible to make them both useful and ornamental. Creepers are being trained over the bare bricks, and the arches are used for various purposes, two as a gymnasium, another as an aviary, whilst the remainder serve as shelters in wet weather. A use has been found, too, for the mansion, which is leased to the Hammersmith Library Commissioners at a nominal rent of £10 per annum.

* Faulkner, 'Hammersmith,' p. 375.

The roads adjoining Ravenscourt Park have considerable historical importance. The one to the north, Goldhawke Road (formerly Gould Hawk, after a Mr. Gould, an owner of Gould Farm), is on the site of the old Roman road from Regnum (Chichester) to Pontes (Staines). Its passage through this parish is thus described by Dr. Stukeley: ' It passes now between Staines and London, being the common road at present, till you come to Turnham Green, where the

The Lake, Ravenscourt Park.

present road through Hammersmith and Kensington leaves it, for it passes more northward upon the common, where, to a discerning eye, the trace of it goes over a little brook, called from it Strand Bridge, and comes into the Acton Road at a common at a bridge a little west of Camden House, and so along Hyde Park wall, and crosses the Watling Street at Tyburn.' In confirmation of the correctness of this description, it may be mentioned that the most satisfactory evidence of its existence was discovered in the

year 1834 by the workmen employed in making Goldhawke
Road, for, upon digging down about 10 feet from the surface,
they came to the old Roman causeway, which was very hard
and compact, and consisted of the usual sort of materials
employed in the formation of these roads. Among the
various articles dug up were Roman coins and small square
tiles.*

We have mentioned before how a former owner of Ravenscourt Park was involved in the troubles of the Civil War.
In 1642 the neighbourhood of the park itself was the scene
of one of the encounters between King and Parliament. In
the beginning of November in that year the King marched
with his whole army to Colebrook, and subsequently advanced to Brentford. The historian of the rebellion, Lord
Clarendon, relates how 'the King marched with his whole
army towards Brentford, where were two regiments of their
best foot, for so they were accounted, being those who had
eminently behaved themselves at Edgehill, having barricaded
the narrow avenues of the town and cast up some little
breastworks at the most convenient places. Here a Welsh
regiment of the King's, which had been faulty at Edgehill,
recovered its honour, and assaulted the works, and forced
the barricadoes well defended by the enemy. Then the
King's forces entered the town after a very warm service,
the chief officers and many soldiers of the other side being
killed, and they took there above 500 prisoners, eleven
colours, and fifteen pieces of cannon, and good store of
ammunition. Thus, the Welsh, under Sir Charles Salisbury,
their leader, made true the Greek proverb, " He that flieth
will fight again." Intelligence of the King's progress having
reached London, every possible effort was made by the
Parliamentary party to prevent his entering the capital, and
a large force was drawn together under the Earl of Essex.
This was augmented by the trained bands of London, who
were posted on the heath next Brentford. The Earl of
Essex drew up his forces upon Back Common (Turnham

* Faulkner, ' Hammersmith,' pp. 23, 24.

Green), the whole army consisting of 24,000 men. Both armies continued to face each other the whole day, which was Sunday, yet neither side seemed anxious to attack. King Charles was probably disappointed in the assistance he had expected from London, and the Parliamentary leader was afraid of the desertion of his troops should the battle commence. In the evening the King drew off his forces to Kingston, and on the next day the General gave orders for the citizens to return to London, an order which they did not hesitate to obey. The soldiers had not been forgotten, for their wives and others had sent many cartloads of provisions, wines, and other good things to Turnham Green, with which they were refreshed and made merry.'*

Merry-making under different conditions is one of the objects for which Ravenscourt Park exists. Some of the attractions of the park have already been touched upon—the band performances, lawn tennis, skating, all in their respective seasons—but there is something which is a permanent delight, and that is the natural beauty, which is equally attractive all the year round. The majestic avenue at the King Street entrance is imposing in its grandeur in winter and summer alike. From early spring to late autumn the park boasts of a wealth of flowers, not only in its formal beds, but also in its very extensive borders. Hammersmith may well be proud of this park, which thus affords pleasure and recreation to classes and masses alike.

SHEPHERD'S BUSH COMMON.

Shepherd's Bush Common is a triangular open space of 8 acres, situated at the junction of the Uxbridge and Goldhawke Roads. Some years ago it was simply a village green, around which were clustered the few cottages, shops, and solitary inn which composed the village of Shepherd's Bush. Faulkner, the historian of Fulham and Hammersmith, writing in 1839, laments that a chapel-of-ease is much

* Faulkner, 'Hammersmith,' pp. 86, 87.

wanted here, and just about this time the extensive building operations commenced which have so transformed this part of the Metropolis. Very few traces of the old village remain, and rows of shops and villas have taken the place of the straggling cottages. Even the village green has been changed, and it is now a well-kept suburban common enclosed with post-and-rail fencing. Its very name has been altered to keep pace with the other transformations. In old maps it is marked as Gagglegoose Green, which brings back with it glimpses of flocks of geese strutting across the village green and frightening away timid intruders with their cackling.

The name of the erstwhile village, too, has a rural sound. The meaning of 'shepherd's bush' does not seem to have been discussed by any topographer, it being taken for granted that the bush in question is one similar to that spoken of by Milton:

> 'And every shepherd tells his tale
> Under the hawthorn in the dale.'

But a few years ago the local authorities were anxious to change the name of the district, and this elicited a communication in the *Daily Telegraph* from a clergyman, signing himself 'A Hedgerow Parson,' who was personally acquainted with what is now an extinct archæological curiosity. 'Did you ever see a shepherd's bush?' he asks, and adds, 'I was myself asked the question forty years ago, and replied, "I dare say I have seen one," believing that shepherd's bushes are as other bushes. My questioner at once replied, "Oh, then you don't know what a shepherd's bush is. I'll show you one." He then took me to the top of a hill overlooking extensive sheepwalks, on which stood a solitary and ancient white thorn, its shape that of an inverted mushroom. The upper surface of the bush was worn smooth, forming a shallow cup, by shepherds having lain upon it resting their elbows on its well-defined green edge while watching their flocks. The entrance to this upper surface was by a smoothly-worn hole between the bole and branches. In consequence of this use as a watch-box, the thorny and green growths had been

forced downwards and horizontally outwards, giving the bush
its peculiar shape.' This, then, seems to be the origin of the
name of this little common, and it would be interesting to
know if such a bush as has been described ever existed in
this quarter of the world, and if so, why it became of such
importance as to give its distinctive name to the district.

Shepherd's Bush Common is one of the oldest of the
municipal open spaces. At the time when its acquisition
was first mooted it was nothing better than a swamp surrounded with a ditch. The ditch has been filled in and the
common raised, so that its former objectionable character
has been remedied. It was acquired under a scheme by
which the rights of the Ecclesiastical Commissioners, the
Lords of the Manor of Fulham, in which it is situated, were
purchased for the sum of £505. This scheme was confirmed
by the Metropolitan Commons Second Supplemental Act,
1871.

None of the buildings surrounding the common can lay
any claim to antiquity, but some of the old cottages which
have been pulled down possessed some little historical interest.
Close by the Wellington Tavern in the Uxbridge Road once
stood a house, lately occupied as a butcher's shop. For
many years this was a famous inn for travellers, at a time
when it was the only house standing between Acton and
Kensington Gravel Pit. It was here that the notorious
highwayman called Sixteen String Jack was finally taken
into custody.*

Another old cottage at the corner of Goldhawke Road
occupied the site of a much-frequented inn, which was hired
by Miles Syndercombe for the purpose of assassinating
Oliver Cromwell on his way to Hampton Court in January,
1657. The house was owned by Henry Busby, coachman
to the Earl of Salisbury, and the road at this spot was so
narrow and bad that carriages were forced to go slowly.
Syndercombe, who seems to have been of an inventive turn
of mind, devised an engine or machine-gun, which was to be

* Faulkner, 'Hammersmith,' p. 382.

loaded with twelve bullets, and be discharged at Cromwell's coach as he passed by.* The conspirator had two confederates in Cecil and Toope, two of Cromwell's guard, who were able to keep him well informed of their master's movements. If the plot had been successfully carried out, the coach with Cromwell and the other passengers would have been effectually destroyed, but Syndercombe was betrayed by one of his accomplices. When he was tried he resolutely denied the plot, but he was found guilty, and the sentence of the court was that he 'be put from hence to the prison of the Tower of London, from whence he came, and thence be drawn upon a hurdle through the streets of London to Tyburn, there to be hanged on a gallows untill he be half dead, and then cut down, and his entrails and bowels taken out and burnt in his face or sight, and his body divided into four quarters, and be disposed of as his Highness shall think fit.' As the behaviour of Syndercombe at the trial had given reason to suppose that an attempt would be made to rescue him, the Protector gave particular charge for his being guarded in the Tower, but when his keepers went to call him in the morning he was found dead in his bed. Cromwell was very much disturbed at this, for instead of getting a useful confession out of this man, he found himself under the reproach of causing him to be poisoned, and though he did not make the discovery he expected, he found that he himself was more odious to his army than he believed he had been.† The original inn hired by Syndercombe was pulled down about 1770, and the cottage which took its place was a small thatched building of one story, at one time occupied by Mr. Galloway, the eminent engineer.‡

* *Mercurius Politicus*, January 15 and February 5, 1657.
† Faulkner, ' Hammersmith,' pp. 90, 91.   ‡ *Ibid.*, p. 383.

34

CHAPTER XXVII.

SPA GREEN—WHITFIELD GARDENS.

SPA GREEN.

THE gardens comprised under this name include four separate plots of land situated in Rosebery Avenue, Clerkenwell, which owe their existence to the formation of that thoroughfare. The first of these gardens was acquired in 1891, as the result of an exchange with the New River Company. Some of the surplus land which had been acquired for the purposes of the improvement was offered to that company, which resulted in their proposing to give in exchange the land known as Spa Green proper, which had up to that time been kept by them as an enclosed grass plot, and upon which they intended to build after the completion of the new thoroughfare. Later on in the same year, at the request of the Vestry of Clerkenwell, Upper Gloucester Street was connected with Rosebery Avenue, thus adding a portion of Spa Green to the public way, and dividing it into two. The smaller of these two plots was then paved, planted with trees, and provided with seats, and the remainder was laid out as a garden, with tar-paved paths, flower-beds, and shrubberies.

The remaining two plots were purchased by the Council out of funds which were paid them by the Postmaster-General, under the Post-Office Sites Act, 1889. When the Government decided to purchase the site of Coldbath Fields Prison for Post-Office purposes, the London County Council

endeavoured to obtain a portion for an open space, and
eventually a compromise was effected by which they were
empowered to purchase a portion, or receive a sum of
£10,000 to provide an open space elsewhere. The Post-
Office, after considerable negotiations, decided to pay the
money and retain the whole of the prison site. Although

View in one of the Gardens, Spa Green.

it did not follow of necessity that this money need be
expended in Clerkenwell, it was quite in accordance with
the spirit of the Act to do so, and accordingly two plots of
land in a line with Spa Green were purchased.

After being laid out and enclosed, these two small gardens
were opened to the public on July 31, 1895. The total area
of Spa Green is ¾ acre, and although in itself this seems

34—2

very insignificant, the importance of the gardens in a crowded district like Clerkenwell, where open spaces are few and far between, cannot be over-estimated.

These small plots of land, together with Spa Fields and Wilmington Square, are all that remain of the large open space which once existed here, known under the various names of Spa Fields, Clerkenwell Fields, Pipe Fields, or Ducking Pond Fields.

The appellation Spa Fields was applied more properly to the district round the present recreation ground of that name, whilst the fields extending northwards were called the Ducking Pond Fields, but all four terms were used synonymously. An inquiry into the origin of these many names will tell us a great deal about the history of the place itself.

Taking first of all, Spa Fields: the Spa was a mineral spring of some celebrity in the seventeenth and eighteenth centuries, when the district was sufficiently in the country to attract those who would despise the merits of anything at their doors. The two northernmost plots are part of the site of the grounds of the Spa, which was curiously enough called Islington Spa, or New Tunbridge Wells. The reason why it received this former name is because it was nearer to the town of Islington than that of Clerkenwell, whilst the similarity of the composition of the water to that at Tunbridge gave it its other title. This Islington Spa must not be confounded with another Spa a little to the south, close to Spa Fields, known as the London Spa, nor with the wells on the opposite side of the road, where Sadler's Wells Theatre now stands.

The first mention of the Islington Spa is in various newspaper advertisements from 1685 to 1692, which refer to it incidentally as a place that is open to the public. An advertisement of May, 1690, in the *Gazette*, made the following announcement:

' These are to give notice, That the well near Islington, call'd New Tunbridge, will be open on Monday next, the 25th instant, during the whole season for drinking the medicinal water, where the poor may have the same gratis, bringing a certificate under the hand of any known physician or apothecary. The coffee-house within the garden there is to be lett at a reasonable rate.'*

The price of admission was at first fixed at 3d., which occasioned a burlesque poem by Ned Ward, 'The Islington Wells; or, the Threepenny Academy.' The coffee-house referred to in the advertisement was the humble original of a ball-room for dancing, which became one of the standing attractions of the Spa as it grew in popular estimation. Year after year the opening of the season at the Spa was advertised, together with the announcement that there would be dancing on Mondays and Thursdays; but it obtained no especial hold upon the public till patronized by Lady Mary Wortley Montagu, one of the leaders of society, who professed to have received much benefit from the taking of the waters. It at once became a fashionable resort, and was constantly visited by Royalty and the leading nobility. Among other exalted personages here in 1733 we may mention the Princesses Amelia and Caroline, daughters of George II. The *Gentleman's Magazine* of that year mentions that in the month of June the Princess Amelia ceased her visits to the New Tunbridge Wells, where Her Highness and the Princess Caroline had attended almost every morning in May to drink the water, when she gave the proprietor twenty-five guineas, the water-servers three each, and the other attendants one apiece. This was very liberal on her part, but the proprietor had gone out of his way to give her a proper reception, even to firing a royal salute of twenty-one guns on her arrival. He could well afford to do this, for he is said to have received as much as £30 in a morning.† In addition to the waters, the practice adopted at other spas was followed here, of having entertainments, concerts, and the like. To take an advertisement from the *Daily Post* of May 13, 1740, we find that 'The New Wells

* Quoted in Pink's 'History of Clerkenwell,' p. 399.
† Malcolm, iii., p. 231.

near the London Spaw begin their diversions at five in the afternoon. A new entertainment of singing, dancing, feats of activity.' So the crowds of rich and gouty noblemen continued to resort hither till about 1750, when we find the puffs and advertisements of the Spa getting bigger and bigger as the patients grew steadily fewer, and the receipts proportionately smaller. Aristocracy withdrew its patronage, and as a consequence those who had come more for the sake of society and amusement than to drink the waters had to find other places to gratify their tastes. The proprietor, in order to curtail expenses, had to close the gardens, but in spite of his retrenchments, in 1777 he became bankrupt. A later energetic manager, by dint of providing fresh attractions in the way of a bowling-green and 'astronomical lectures during Lent,' contrived to effect a temporary revival, but the days of its popularity were numbered, and at last the place had to be closed. The greater part of the gardens was then built over, but those who cared to have the water for medicinal purposes were still supplied.

From the derivation of the term of Spa Fields we next pass to that of Pipe Fields, which we have already seen was another name for the locality. The pipes referred to were wooden ones, hollow trunks of elm-trees belonging to the New River Company, which at one time covered a considerable extent of the ground.* Britton, speaking of the fields as he knew them at the close of the last century, says they 'were really fields devoted to the pasturage of cows and to a forest of elm-trees, not standing and adorned with foliage in the summer, but lying on the ground southward of the New River head, destined to convey water in their hollow trunks to the north and western parts of London, in combination with similar pipes laid under the roadway of the streets.'†

The last name of Ducking Pond Fields was the one by which they were known to Pepys. On March 27, 1664, he writes: 'Lord's day. It being church time, walked to

* Pink, 'History of Clerkenwell,' p. 645.
† Britton's 'Autobiography.'

St. James's to try if I could see the belle Butler, but could not. . . . Thence walked through the *ducking-pond fields;* but they are so altered since my father used to carry us to Islington to the old man's, at the " King's Head," to eat cakes and ale . . . that I did not know which was the ducking pond, nor where I was.'* It would have been interesting to have known what these alterations were, but our diarist does not give any further particulars. The ducking-pond was so-called from the barbarous sport of duck-hunting, which consisted of placing a dog upon a duck's back in the water, whilst the spectators watched the struggles of the wretched bird to escape.

Spa Fields in the past, then, seem to have been connected chiefly with the amusements of Londoners. The fields themselves, apart from the places of entertainment, were a specially favourite resort on Sundays. As early as Pepys' time this seems to have been the case, for, curiously enough, his only mention of them is in connection with a Sunday airing. Even as late as 1803 they are incidentally referred to as a favourite place for Sunday promenading, but their appearance must have been much spoilt by the erection of Coldbath Fields Prison in 1794, with its dismal walls frowning down on them.

In addition to their evil reputation for duck-hunting, they were also occasionally the scenes of bull-baiting, pugilism, and other rough sports. As an example of this, we may quote a newspaper extract of 1768: 'On Wednesday last, two women fought for a new shift, valued at half-a-guinea, in the *Spaw Fields*, near Islington. The battle was won by the woman called " Bruising Peg," who beat her antagonist in a terrible manner.'† At Whitsuntide the fields were the scene of a 'gooseberry fair,' where the stalls of gooseberry-fool vied with the tea-booths and the ale of the various public-houses.

The fields at night time had some dangerous characters,

* Pepys' 'Diary' (Cassell's Edition), p. 71.
† *Daily Advertiser*, June 22, 1768.

which were rather a drawback to the visitors to Sadler's Wells Theatre. The proprietors were well aware of this, and in the advertisements of the theatre there are frequent additions like this: ' N.B.—A full moon during the week'; and again, after an announcement of a charity performance: ' A horse patrole will also be sent in the New Road this night by Mr. Fielding (a well-known Bow-Street magistrate), for the protection of the nobility and gentry who go from the squares and that end of the town. The road towards the City will also be properly guarded.'

Thomas Dibdin, son of the celebrated composer of ' Tom Bowling,' who resided in Myddelton Square, in writing the history of his life, makes a passing mention of the locality which will confirm these statements about the lonely character of the fields. He says: ' The site of the square and church, not five years since (1822), was an immense field, where people used to be stopped and robbed on their return in the evening from Sadler's Wells; and the ground floor of the parlour where I sit was as nearly as possible the very spot where my wife and I fell over a recumbent cow on our way home one murky night in a thunderstorm.'* It was a common thing on dark nights for men and boys to wait outside the theatre to light the people home through the fields to the streets of Islington and Clerkenwell.

It is not to be supposed that a place so near London could retain this rural character for very long, especially with the attractions of a fashionable spa and a popular theatre. About 1817 the fields began to be built over, and it was not many years before they were thickly covered with houses, and now they have disappeared altogether except in name.

Facing the green there are two places of interest which deserve a passing mention. These are Sadler's Wells Theatre and the New River Head.

Sadler's Wells Theatre is an outcome of another mineral spring which is of very ancient origin. This spring once belonged to the rich priory of St. John at Jerusalem, and

* ' Autobiography of T. Dibdin,' vol. ii., p. 323.

before the Reformation it was famed for the cures performed here, which were pretended by the monks of Clerkenwell to be due to their prayers. The slanderous stories which are related of the priests, and their supposed pious frauds at this well, are not sufficiently corroborated to be repeated, but at any rate at the Reformation the springs were closed to prevent superstitious persons from visiting them. They then seem to have been quite forgotten till they were accidentally discovered in 1683 by a Mr. Sadler, from whom they take their name. At the time this well was publicly opened a pamphlet was published giving a history of its discovery.* The account runs as follows: ' Mr. Sadler, being made surveyor of the highways, and having good gravel in his garden, employed two men to dig there, and when they had dug pretty deep one of them found his pickaxe strike upon something that was very hard, where-upon he endeavoured to break it, but could not; where-upon, thinking within himself that it might peradventure be some treasure hid there, he uncovered it very carefully and found it to be a broad flat stone, which having loosed and lifted up, he saw it was supported by four oaken posts, and under it a large well of stone arched over, and curiously carved.' After they had told their master, ' Sadler . . . went down to see the well, and observing the curiosity of the stone-work, and fancying within himself that it was a medicinal water formerly held in great esteem, but by some accident or other lost,' he sent some for analysis to an eminent physician. As the water was slightly ferruginous, and was discovered to be beneficial, it was soon recommended, and visitors began to flock here. This spring is entirely distinct from the Islington Spa, or New Tunbridge Wells, on the opposite side of the road. Lysons was misled probably by this pamphlet into supposing that they were one and the same, and as his ' Environs of London ' has formed the foundation for most histories since his time, his error has

---

* ' A True and Exact Account of Sadler's Wells ; or, the New Mineral Waters lately found at Islington.' 1684.

been repeated by many other writers. The old plan here reproduced shows the two wells quite distinct from one another. Sadler's Wells on one side of the New River were reached by a bridge at the extreme end nearest to the reser-

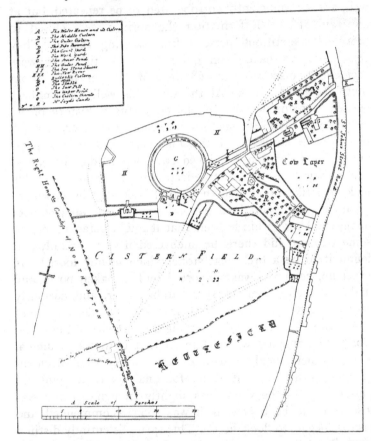

*Plan of the site of Spa Green and surroundings in 1744. The New Tunbridge Wells are shown on the south of the New River (*K K K*) opposite to Sadler's Wells.*

voir, whilst on the other side were the New Tunbridge Wells. Both places had an approach by means of paths across the fields from the London Spaw, which was on the site of the present public-house of that name. Pepys does not record any visit here, but Evelyn mentions under date June 11, 1686: 'I went to see Middleton's receptacle of water (New River Head) and the New Spa Wells, near Islington.' The garden in which the spring was thus brought to light was attached to a small wooden music-house, of which Mr. Sadler was proprietor. As a place of entertainment it was old then, and it lays claim now to being the oldest theatre in London. It was probably frequented long before the Reformation as a place of amusement, and a petition is mentioned as having been presented by the proprietor to the House of Commons, in which it is stated that the site was a place of public enter-tainment in the reign of Queen Elizabeth.* The bill of fare provided here was not of the choicest. Among other items we may mention a gourmand, who, after dining heartily off beef-steak, proceed to eat a fowl, feathers and all, and then offered to bet anyone five guineas that he would do the same again in two hours.

Sadler appears to have stayed here till 1699, when we find the place advertised as Miles's Music House. Subsequent proprietors were Forcer, a barrister, Rosomon (after whom Rosoman Street takes its name), and King, the famous comedian. In 1765 the old wooden theatre was pulled down, and a new one of brick built at a cost of £4,425. The old variety entertainments were kept up with good success till 1804, when the proprietor took advantage of the proximity of the New River Company to turn the stage into an immense tank, and to present aquatic scenes with real water. The first of these was the 'Siege of Gibraltar,' in which the fortress was bombarded by real vessels. The theatre had now estab-lished itself, and the season was extended from six to twelve months. Joe Grimaldi, the well-known clown and actor, whose father had been previously employed here as chief dancer and ballet master, commenced his theatrical career at Sadler's Wells at the early age of one year and four months, and in 1798 he married Miss Hughes, daughter of the proprietor. It was here that he first sung his immortal

* Malcolm, ' Londinium Redivivum.'

song of 'Hot Codlins,' not many years before the total decay of his frame, brought on by his exertions on the stage, rendered his retirement necessary. In 1844 a decided change for the better took place, when Mr. Phelps, aided by Mrs. Warner, rented the theatre. Between this year and 1862 no less than thirty of Shakespeare's plays were given one after another. The greatest success was 'Hamlet,' which claimed 400 performances out of the 4,000 nights.

The theatre was then closed for some time, and afterwards reopened for various short terms with burlesque entertainments and pantomimes. It next appears as a skating-rink, and finally had to be closed because it was not safe. In 1879 it was rebuilt on a larger scale from the designs of

Sadler's Wells, with the New River in front, in 1756.

Mr. C. J. Phipps, but it has had rather a fitful existence ever since, and has never regained the popularity it once had.

Running the whole length of Spa Green are the reservoirs and works of the New River Company, commonly known as the New River head. This gigantic undertaking owes its origin to Sir Hugh Myddelton, who first proposed this scheme about 1608, at a time when London had far outgrown its means of water supply. He persuaded the Corporation to apply for Parliamentary powers to bring the New River from the Chadwell and Amwell Springs, near Ware, in Hertfordshire, to Islington. When they had obtained these powers the difficulties in the way of the undertaking deterred them from taking any more steps in the matter, so Sir Hugh undertook to carry it through on condition that the Corporation transferred their powers to him. This they readily did, and the contract time for finishing the project was four years. The cost of the execution, however, was so great that Sir Hugh, in the course of the third year, found he could not go on without more funds. For these he applied to the City, but they would not risk their money in so hazardous an undertaking; but he was more successful in his application to the King (James I.). The King undertook to pay half the cost, past and future, on condition that he should receive half the profit. The work was now rapidly pushed on, and was completed on September 29, 1613. The property of the company was originally divided into seventy-two shares. Half of these belonged to Sir Hugh, who became in after-life so impoverished that he had to sell his shares, which are known as the adventurer's shares. The remaining thirty-two are called the King's shares, but they were alienated from the crown by Charles I. When the water was first brought to London, the company was not a very paying concern, as the expenses of distribution were very great. It is needless to add how very different this is now, when one of the original shares constitutes a fortune. The New River, as at first executed, was a canal about 10 feet wide and 4 feet deep, with a winding course nearly 40 miles long; but it has subsequently been widened, shortened, and otherwise improved. The present appearance of the works cannot by any stretch of the imagination be termed picturesque. They, too, like the rest of the neighbourhood, have changed considerably during 200 years. When the river was first completed, there was built here 'a house ornamented with vases and quoins, surrounded with a variety of flourishing trees, and fronted by this noble sheet of water, which altogether give it the appearance of a nobleman's villa. This house belongs to the Company, and was originally built in the year 1613, and repaired and newly fronted in 1782, under the direction of Robert Mylne, surveyor to the company, as his place of residence. A large room in this house was fitted up for the meetings of the company about the latter end of the seventeenth century. On the ceiling is a portrait of King William, the arms of Myddelton and Green.'* This last-

The New River at Sadler's Wells. (From an old woodcut.)

named personage was John Green (or Grene), clerk of the company in the time of William III. Green Terrace takes its name from him. It may be mentioned that the New River in its course passes through two municipal parks— viz., Finsbury and Clissold.

* 'London, Westminster and Middlesex,' vol. iii., p. 598.

WHITFIELD GARDENS.

These gardens consist of two small plots of land situated on the western side of Tottenham Court Road, one of the busiest thoroughfares in London. They were acquired, after considerable negotiations, in 1894, at the joint expense of the London County Council and the Vestry of St. Pancras, the total cost being over £5,000. Although this amount appears very large for so small an area, the money has been well spent, considering the crowded neighbourhood in which the gardens are located and the benefit which their acquisition has conferred upon the district. Being a disused burial-ground, the land could not be built over, but a so-called fair was carried on upon it, which became such a nuisance that the Home Secretary had to intervene to put a stop to the disgraceful scenes that occurred here. He wrote to the late Metropolitan Board of Works suggesting that they should (under the powers of the Metropolitan Open Spaces Act, 1881) take such steps as would ensure the ground being kept in order and treated with proper care. The result of this was that in 1889 a clause was inserted in their General Powers Act to enable the acquisition being carried out. For five weary years the negotiations dragged along, the chief difficulty in the way being the pendency of a suit in Chancery. At length all the obstacles were surmounted, and after being laid out, the gardens were opened to the public in February, 1895, by Sir John Hutton, who was then the chairman of the London County Council.

As has been already stated, the gardens were part of the burial-ground attached to Whitefield's Tabernacle. In 1756, when this place of worship was first opened, it stood in the midst of fields. On the opposite side of the road or lane was a farm with market-gardens attached. On the north side of the tabernacle were but two houses, and the next after them was the Adam and Eve public-house, half a mile distant. In place of the busy thoroughfare of Tottenham Court Road there was a country lane passing through fields and meadows, a place to be praised by the poet—

'When the sweet-breathing spring unfolds the buds,
Love flies the dusty town for shady woods,
Then Tottenham Fields with roving beauty swarm.'
                GAY, '*Epistle to Pulteney.*'

These fields extended right down to St. Martin's Lane, which name still preserves the remembrance of its former rural character.

In these fields was held annually at the beginning of August what was known as a 'Gooseberry Fair,' which attracted hither some of the lowest characters.   At this time some of the leading comic actors from the London theatres used to perform here in specially erected booths.   The spectators who preferred to listen to the 'drolls and interludes' given here, to the fare provided for them in the close theatres, were admitted at the modest figure of sixpence each.   In course of time, however, the players or the performances must have degenerated, for it became necessary for the strong hand of the law to intervene and put a stop to them.   An official proclamation issued by the Quarter Sessions of Middlesex, and published in the *Daily Courant* of July 22, 1827, sets forth that in consequence of 'this court having been informed that several common players of interludes having for some years used and accustomed to assemble and meet together at or near a certain placed called Tottenhoe, *alias* Tottenhal, *alias* Tottenham Court, in the parish of St. Pancras in this county, and to erect booths and to exhibit drolls and exercise unlawful games and plays, whereby great number of His Majesty's subjects have been encouraged to assemble and meet together and to commit riots and other misdemeanours in breach of His Majesty's peace,' these interludes were in the future to be prohibited. A quaint old engraving of 1738 gives a representation of a curious race which was usually run at this fair, called 'Running for the smock.'   The competitors—young girls in their teens—had to run 100 yards on the turf with nothing on but a smock, the victor being rewarded with a holland chemise decorated with ribbons. This favourite North-Country pastime was discontinued about the middle of the present century 'in compliance with the proprieties of the age.'

Tottenham Court Road owes its name to the fact that it was the road leading to Tottenham Court, *i.e.*, the court-house or manor-house of the Manor of Tottenham, or, more correctly, Totenhall. This manor was formerly kept by the Prebendary of Totenhall in his own hands. In 1343 John de Carleton held a court baron as lessee. In 1560 the manor was demised to Queen Elizabeth for ninety-nine years, in the name of Sir Robert Dudley, but in the year 1639, twenty years before the expiration of Queen Elizabeth's term, a lease was granted to Charles I. in the name of Sir Harry Vane for three lives. In 1649 this manor was seized as Crown land, and was sold to Ralph Harrison, of London, for the sum of £3,318 3s. 11d. At the Restoration it reverted to the Crown; and in the year 1661, two of the lives in King Charles's lease being surviving, it was granted by Charles II. in payment of a debt to Sir Henry Wood for the term of forty-one years, if the said survivors should live so long. After that the lease became the property of Isabella, Countess of Arlington, from whom it was inherited by her son, Charles, Duke of Grafton. In 1768, the manor then being leased to the Hon. Charles Fitzroy (afterwards Lord Southampton), an Act of Parliament was passed by which the fee simple was also invested in him, subject to the payment of £300 per annum in lieu of the ancient reserved rent of £46, and all fines for renewals. According to the survey of 1649, the demesne land of the manor comprised about 240 acres.*

The site of the manor-house is now occupied by the Adam and Eve tavern in the Hampstead Road; its walls were, in fact, part and parcel of that house. As early as the time of

* Clinch, 'Marylebone and St. Pancras.'

Henry III. the building standing here had some eminence, being owned by William de Tottenhall, and it was in all probability the original manor-house. In course of time the halls and courtyards of the spacious building in which my lord's retainers had shouted their drinking songs degenerated to the uses of a common tavern. Its courtyards were now given up to morality plays or mysteries, one of which may have given its name to the building. Gardens with fruit-trees and shady arbours were laid out to allure visitors from dusty town to the quiet seclusion of Tottenham Court Fields. The Paddington Drag, the only conveyance at the commence-ment of this century between Paddington and the City, would call twice a day for passengers at the Adam and Eve, performing the whole journey in two and a half hours *quick time*, the return occupying three hours, which was fair time, making all allowance for the precaution it was necessary to take against highwaymen and the other evils of night travelling.

Enough has been said to prove the rural character of this sylvan retreat when Whitefield took up his quarters here. He had passed through some wonderful vicissitudes in his eventful life. Born in 1714, at the Bell Inn at Gloucester, where for some time he served as a common drawer or barman, the most violent optimist would not have predicted that this public-house lad would develop into the prince of pulpit orators. His paternal grandfather and great-grand-father were clergymen, it is true, but when his mother was left a widow with a large family, the expense of a University education seemed out of the question. She, however, did the best she could for him by sending him to the Grammar School of St. Mary de Crypt, Gloucester, where he distin-guished himself in elocution, and made fair progress in classical studies. Subsequently, in his nineteenth year, he was admitted as a servitor at Pembroke College, Oxford. Here he became intimately acquainted with the Wesleys and other leading Methodists, and entered so enthusiastically into all their work that he was attacked with a severe illness, which compelled him to return to Gloucester. While home on this visit he received encouragement from Dr. Benson, Bishop of Gloucester, to take Orders, and was ordained by him as deacon in 1736. His first sermon was preached in the Church of St. Mary de Crypt, and gave good promise of his after-career. He then returned to Oxford, and took his degree in due course, and at once commenced an evangelizing tour in Bath, Bristol, and other towns. The same year he received an invitation from the Wesleys to help as a missionary in Georgia, and he occupied the interval before sailing in preaching tours, his eloquence attracting immense throngs. In some of the London churches vast crowds used to assemble long before daybreak in order to hear him. It was not till December, 1737, that he embarked for America, and as our Atlantic liners were not then in existence, the journey took nearly five months. He only stopped three months, returning to England to be ordained as priest, and to raise funds for an orphanage he had founded in America. His popularity had now excited the jealousy of his brother clergy, and the doors of their churches were closed to him. This led him to take up open-air preaching, his first field pulpit being at Bristol, where the colliers flocked to hear him, his audiences being latterly estimated at 20,000. His powerful voice was heard by every one in the crowd, and was particularly adapted for this style of preaching.

It must not be supposed that he went calmly on without opposition. He suffered much from the hostility of brutal mobs. On one occasion when returning from preaching in Ireland, he was attacked by an ignorant rabble. Volleys of stones were thrown at him from all quarters, till he was covered with blood. He only just managed to stagger to the door of a minister's house, or he would certainly have been murdered. Whitefield used to say, when speaking of this event, that in England, Scotland, and America, he had been treated only as a common minister, but that in Ireland he had been elevated to the rank of an Apostle, in having

35—2

had the honour of being stoned.* But perhaps the blow he felt most was caused by a divergence of opinion from his friend Wesley, which led to his withdrawing himself from the Wesleyan communion. They parted company, and their adherents on each side fiercely quarrelled. Whitefield would gladly have abstained from strife. 'Desire, dear brother Wesley,' he used to say, 'to avoid disputing with me. I think I had rather die than see a division between us ; and yet how can we walk together if we oppose each other ?' Whitefield remembered his friends to the end. In his last will and testament, made six months before his death, he says: 'I also leave a mourning ring to my honoured and dear friends, and disinterested fellow-labourers, the Revs. John and Charles Wesley, in token of my indissoluble union with them in heart and Christian affection, notwithstanding our difference in judgment about some particular points of doctrine.'

It was while preaching in Scotland that Whitefield made the acquaintance of the Countess of Huntingdon, who was afterwards so great a benefactress to him. He had to learn that a rolling stone gathers no moss, for during his visits to America, and his frequent preaching tours through Great Britain, his congregation had dispersed, and he had to sell up his furniture to pay the debts of his orphanage. But the Countess appointed him as her chaplain, spent her ample fortune in endowing Calvinist Methodist chapels in various parts of the country, and erected a college for the training of candidates for the ministry. The remainder of his busy life was spent in visiting America, Great Britain, and Ireland. It was hardly to be expected that his life would bear the continual strain he put upon it. It has been stated that in the compass of a single week, and that for years, he spoke in general forty hours, and in very many sixty, and that to thousands. When the demands of his failing health rendered it at the last necessary, he placed himself on what he called 'short allowance,' preaching only once every week-day, and

* Wakeley, 'Anecdotes of Rev. G. Whitefield.'

three times on Sunday. In 1769 he made his last trip to America, and although worn out, he yet went on travelling and preaching. ' I would rather wear out than rust out ' was the answer he gave to those who advised him to rest from his labours. A severe seizure of asthma brought him to his end in the following year, September 30, 1770, at Newbury, in New England. He died in harness, for he had arranged to preach there on the day of his death. In

Whitefield's Tabernacle, Tottenham Court Road, 1756.

accordance with his wishes, he was buried before the pulpit of the Presbyterian church of the town where he died.*

The land upon which the chapel was erected, including the two plots now known as Whitfield Gardens, was first leased to him, in 1755, by the Fitzroy family, and in the following year he commenced collecting funds for his new chapel. The land was formerly the site of an immense pond,

* The details of Whitefield's life are taken in the main from the article in the ' Encyclopædia Britannica.'

called in Pine and Tinney's maps (1742 and 1746), 'The
Little Sea.' The whole of this was covered with a concrete
platform of considerable thickness, which proved a serious
difficulty in the way of treating the ground ornamentally
when it was subsequently laid out as a garden. The foundation-stone was laid in May, 1756, and the building was
completed and opened in November of the same year, the
design being furnished by Whitefield himself. The floor
was of brick, and there were no pews. Two years later
almshouses and a vicarage were built in the burial-ground
adjoining the chapel, and the next year it was enlarged by
the addition of an octagonal front, which gave it the
appearance of two chapels. Whitefield's successors continued here till the lease expired, and in 1831 the freehold
was purchased by trustees, at a cost of about £20,000. The
service, which up to this time had been liturgical, was then
changed to the congregational form. The burial-ground,
after having been used for something like 30,000 interments,
was closed in 1854, and, owing to the falling off of the fees
derived from this source, the prosperity of the church
steadily declined, till it was dissolved in 1862. The property was then put up to auction, and it narrowly escaped
being purchased for a music-hall. It was, however, bought
by the London Chapel-Building Society, and re-opened in
October, 1864, as a Congregational Church. In 1889 the
building showed signs of serious decay, and it was found
necessary, in April, 1890, to pull down the whole of the
structure owing to the insecurity of the foundations. The
trustees of the church have in their possession some interesting
relics, which will doubtless find a place in the new building
which has just been commenced. Among these may be
mentioned Whitefield's arm-chair and several portraits of
former ministers, including a portrait in oils of the founder.
Some of the gravestones from the burial-ground have also
been preserved, among which is one to John Bacon, R.A.,
sculptor, who died in 1799, and also one to Rev. A. M. Top-
lady, who, as the author of the hymn 'Rock of Ages,' has

obtained a celebrity as universal as Whitefield. The inscription is a very simple one:

> 'Within these hallowed walls, and near this spot, are interred the mortal remains of the Rev. Augustus Montague Toplady, Vicar of Broad Hembury, Devon. Born 4th November, 1740 ; died 11th August, 1778 ; aged 38 years. He wrote :
>
> > ' " Rock of Ages, cleft for me,
> > Let me hide myself in Thee." '

We have mentioned before that Whitefield was buried in America, and so his cherished idea of finding a last resting-place here was not realized. On one occasion he told his congregation, ' I have prepared a vault in this chapel where I intend to be buried, and Messrs. John and Charles Wesley shall also be buried there. We will all lie together. You will not let them enter your chapel while they are alive; they can do you no harm when they are dead.'

There is also a tablet to Whitefield's wife, who died in 1768, two years before her husband. She was formerly a Mrs. James, a widow, whom he had met on one of his preaching tours in Wales, and married in 1741. Although he always spoke most highly of her, it is to be feared their married life was not happy. Nor can we wonder at this, for Whitefield's idea of courtship and matrimony was certainly out of the ordinary run. He once wrote a letter to the parents of a girl whom he thought would suit him as a help-meet as follows:

' MY DEAR FRIENDS,

' I find by experience that a mistress is absolutely necessary for the due management of my increasing family (*i.e.*, of orphans), and to take off some of that care which at present lies upon me. . . . It hath been, therefore, much impressed upon my heart that I should marry, in order to have a helpmate for me in the work. . . . This comes (like Abraham's servant to Rebekah's relations) to know whether you think your daughter, Miss E., is a proper person to engage in such an undertaking. If so, whether you will be

pleased to give me leave to propose marriage unto her?
You need not be afraid of sending me a refusal; for, if I
know anything of my own heart, I am free from that foolish
passion which the world calls love. . . .

'GEORGE WHITEFIELD.'

The letter to the daughter was written in the same cold-
blooded way, and after asking her if she could bear the
inclemencies of a foreign climate, separation from her
husband for months at a time, trusting to Providence for
support for herself and children, and other things of a like
nature, it is not to be wondered at that she refused to have
him. It is strange that a man like Whitefield, who spoke
so feelingly as to move the roughest audiences to tears, should
himself have remained so callous to what he calls 'the foolish
passion.'

But this is only a small blemish in a beautiful and self-
sacrificing life, and these gardens, hallowed by their associations with so great a man as Whitefield, will not be the least
important factor in keeping his memory green.

CHAPTER XXVIII.

VICTORIA PARK—MEATH GARDENS.

VICTORIA PARK.

THOUGH Victoria Park has not acquired the prestige of either Hyde or Regent's Park, it is not inferior to either of them in natural beauty or brightness of floral decoration. From end to end it is somewhere about a mile and a quarter long, and it is nearly half a mile wide at its broadest part. Every inch of its large area of 217 acres contributes its quota towards brightening the lives of the teeming thousands who dwell in the densely-populated districts surrounding the park. This splendid playground of the East End is quite as dear to the industrial population who frequent it as the sweeping drives and pleasant walks of the West End parks to their fashionable visitors. Besides, at Victoria Park the hard-working artisan is a bit of a horticultural critic in his way. Somehow, in the small back-gardens and crowded yards he manages to rear many a choice specimen, so that the flowers in the adjoining park have to be kept up to the mark. The ornamental gardening alone is well worth going to see; at almost every season of the year there are bright flowers to be seen. In the spring the beds are gay with tulips, hyacinths, and other showy bulbs imported from Holland to brighten our flower-gardens. These in the summer give way to every possible variety of bedding-out plants. The area is so large, and the beds so numerous, that the skill of the officials is taxed to the utmost to infuse sufficient variety into the whole of the large surface. Something like 200,000 plants are bedded out annually. In the nooks and corners and trim lawns the beds are so arranged as to contrast most favourably with the green verdure of the turf and the dark background of shrubs. Many of them are extremely beautiful, although composed of the simplest and commonest materials. No sooner have the summer flowers faded than the chrysanthemums are ready for exhibition, and here they are to be seen to perfection in every form and variety with which we

The Principal Entrance to Victoria Park, with the Superintendent's Lodge.

have been familiarized in late years. Even in the winter the large decorative house is full of floral life, and when the wings are added which are required to make it complete, it will be a very handsome structure, and form a permanent attraction at a time when flowers are scarce. As the timber of the park matures, it is becoming each year more delightful, and the shrubs, especially the hollies, are certainly some of the finest to be seen in any London park. The laying-out of the park is a standing testimonial to the ability of Sir James Pennethorne, who also designed Battersea Park.

The area of the park is so large that it is possible to pro-

vide for nearly every form of out-door amusement and recreation. Foremost among these must be placed swimming and

Bathing, Victoria Park.

bathing, for which this park affords special facilities. As
many as 25,000 bathers have been counted on a summer

morning before eight o'clock. What an incalculable boon open-air swimming-baths like those provided here must prove to the neighbourhood ! The principal bathing-lake is 300 feet long. It is provided with a concrete bottom, shelters, and diving-boards, and all the accessories to make it a perfect out-door swimming-bath and it has been pronounced the finest in the world. In case of accidents, two boatmen are always on duty during the season, which is a necessary precaution when the number of bathers is taken into account. Apart from the two bathing-lakes, there is another large sheet of ornamental water upon which boating is allowed. This lake has a fine fountain spray playing in it, and the water seems to abound with fish. On one of the islands is a two-storied Chinese pagoda, which produces a pretty effect with the trees and foliage surrounding it. This pagoda was formerly the entrance to the Chinese Exhibition held on the site now occupied by part of St. George's Place, Knightsbridge.

In the summer cricket is amply provided for. There are thirty-two pitches on the match-ground, not to speak of the many games of the youngsters who are allowed to set up their stumps or pile up their jackets on any part of the unappropriated ground. For the followers of lawn-tennis there are some thirty-seven courts, all of them free. In the summer band performances are given, which attract considerable audiences. There are four gymnasia, two of which are specially reserved for children. The children certainly are well looked after, and nothing can be pleasanter than to stroll round from point to point and watch the happy little crowds disporting themselves on swings and see-saws, sailing their boats on the waters of the lake, or digging in the sand-pit, apparently quite as happy as though they were within sight and sound of the sea-waves.

Another feature which is very popular with the children is the introduction of animal and bird life into the park. A recently-erected aviary contains a varied selection of English birds, such as pigeons and doves, chaffinches, linnets, green-

finches, and a pair of golden pheasants. But perhaps the
guinea-pigs afford more amusement to the youngsters. There

The Children's Sandpit, Victoria Park.

are goats in a rockery by themselves, and another enclosure
for deer.

Altogether, Victoria Park forms a splendid playground, and though the cost of maintaining it is considerable, it must be admitted that the money is well spent, seeing that it brings brightness to many lives whose lot is not of the happiest.

Victoria Park was formed in accordance with the provisions of an Act passed in 1840, entitled ' An Act to enable Her Majesty's Commissioners of Woods and Forests to complete a contract for the sale of York House, and to purchase certain lands for a royal park." York House was built for the Duke of York, second son of George III., and by this Act was sold to the Duke of Sutherland for £72,000, and renamed Sutherland House. It stands in St. James's Park, close to the palace, and is considered the finest private mansion in London. The work of laying out the park was commenced in 1842, and it was opened to the public in 1845. At the time of the formation of the park there was much discussion as to the site which should be adopted. The land-owners of Stepney, Limehouse, and its vicinity, urged on the Commissioners that the south side of the Mile End Road would form a more desirable position for a public park, but time has amply proved that the site chosen was the best, for the districts of Bow, Stratford, Hackney, Dalston, Clapton, Kingsland, and Stoke Newington all derive benefit more or less. The Queen visited the park which bears her name on April 2, 1873, and in memory of her reception she presented a clock and peal of bells to St. Mark's Church.

Victoria Park was maintained by Her Majesty's Office of Works till November, 1887, when it was transferred, together with Battersea and Kennington Parks, to the late Metropolitan Board of Works, under the provisions of the London Parks and Works Act passed in that year. These places of recreation had, since their formation, been kept up at the cost of the State, Parliament having annually voted the money required. The vote had often been objected to by the representatives of provincial constituencies in the House of Commons, on the ground that the people of London

The Boating Lake, Victoria Park.

ought to pay for their own parks; but until the year 1886 the objections had never prevailed. In that year (the first of the new Parliament after the wide-spread extension of the suffrage and the redistribution effected by the Act of 1885) the House of Commons, upon being asked to make the usual vote for the parks at first refused it; and it was not till the Government had promised to introduce a Bill to transfer the charge to the ratepayers of London that the money was voted for the year. The London Parks and Works Act, 1887, was the result of this promise of the Government, by which the charge of the places referred to was transferred to the late Board.

Victoria Park as originally laid out contained only 193 acres. The remainder of the lands acquired for the purpose were vested in Her Majesty's Commissioners of Woods and Forests, who were empowered by their Act to "lease any part of the said royal park, not exceeding in the whole one fourth part, for the purposes of the same being used as sites for dwelling-houses, or ornamental buildings and offices and gardens thereto to be annexed." By an Act of Parliament passed in 1852, the quantity of land to be set apart for building was reduced to one-sixth of that actually purchased. In the year 1872, when steps were about to be taken to let this reserved land for building purposes, a number of persons interested in the welfare of the inhabitants of that portion of the Metropolis formed themselves into a society for the purpose of extending the area of the park so as to meet the requirements of the immense and growing population of the East End. Their first step was to ask the Chancellor of the Exchequer to submit to Parliament a proposal that the project of building on the ground not included in the park might be abandoned, and that this area might be added to the park. The Chancellor of the Exchequer expressed his inability to adopt the course suggested to him, and the society thereupon applied to the late Board for its assistance in attaining the end in view. The Board, fully sympathizing with the society's object, resolved to do what it could in the matter, and waited upon the Chancellor by deputation. But their object was not attained, for the right honourable gentleman, whilst admitting the desirability of preserving as far as possible all the open spaces in the Metropolis, was not able to assent to the proposition that this should be done at the cost of the State, which would be the case were the Crown to forego its right of letting for building purposes the land specially reserved to it under the statute. At that very time negotiations were pending for letting portions of the land, and it was promised that these should be suspended for a week, in order to enable the Board to determine whether it would purchase the land. The Board was equal to the occasion, and it was resolved to purchase of the Crown such portions of the ground remaining unlet as could properly and conveniently be included in the park. The quantity was about 23¾ acres, and the price agreed on was £20,450. The agreement was that the land so bought should be annexed to, and form part of, the park, and be maintained by Her Majesty's Office of Works. This scheme was confirmed by the Victoria Park Act, 1872. Some ten years before this, another improvement in connection with Victoria Park had been carried out by the late Board viz., the formation of an approach from Limehouse. This road, called Burdett Road, is 70 feet in width, and nearly a mile long, and extends from the point of junction of the East and West India Docks Road to Mile End Road, and thus affords direct access to the park by way of Grove Road. Burdett Road was opened to the public on May 25, 1862.

Before passing to the historical associations of Victoria Park, there are one or two buildings, in addition to those already mentioned, which ought to be described. The principal lodge, adjoining the Regent's Canal at Bonner Hall Bridge, is a handsome building in the Elizabethan style. It is of red bricks, with stone dressings, and was designed by Sir James Pennethorne. Its principal feature is a lofty square tower and entrance porch, which together make it an

36

imposing building. The chief structure in the park, however, from an architectural point of view is the large and
ornate drinking-fountain presented to the park by Lady
Burdett - Coutts, so well known for her endeavours to
ameliorate the condition of the East - End poor. It is
situated in an open part of the park, where its beauties are
not hidden in any way, and can be seen to great advantage.
The architect of this handsome fountain was Mr. H. A.

The Victoria Fountain, Victoria Park.

Darbishire, who also designed Columbia Market. The fountain, approached by a flight of steps, is octagonal in shape
in the Gothic style of architecture, and is said to have cost
over £5,000. The shafts and bases are of polished granite,
relieved with coloured marbles. Within these are niches
containing marble figures, which pour the water from vases
into basins beneath. The whole structure is surrounded by
flower-beds, and altogether forms one of the principal features
of the park. The fountain was inaugurated June 28, 1862,

on which occasion Lady (then Miss) Burdett-Coutts was present.

Facing the cricket-ground, some of the semi-octagonal recesses which, according to the inscription upon them, came from Old London Bridge have been placed in position, and serve as alcoves. A doubt has been expressed about the accuracy of the statement contained in this inscription, and

Alcove on Old Westminster Bridge, now in Victoria Park.

some authorities assert that the alcoves came from Old Westminster Bridge. H.M. Office of Works was consulted on the point, but without clearing up the doubt. The only other building to be mentioned is an arcade furnished with seats, which faces the ornamental lake.

No account of Victoria Park would be complete without some reference to the position it occupies as the forum of

36—2

the East End.   Victoria Park on Sunday is one of the great revelations and surprises of out-door life.   Strange to say, the attractions on this day are not the beautiful scenery or the fresh air, but those of public discussion and debate.   At the head of the lakes, close to the Victoria Fountain, is the place for public meetings, which is a regular sea of heads on Sunday, where the working men in their thousands crowd round their favourite speakers.   Here may be seen the National Secularist Society, with their banner and portable tribune, always sure of a large audience.   The Roman Catholics, too, are ably represented by the members of the Guild of Our Lady of Ransom.   Politics are introduced by the Social Democratic Federation and the Independent Labour party, but their following is not large when compared with those who are interested in religious discussion. Occupying one of the most picturesque positions, under the group of trees known as 'the Eight Sisters,' will be found the Tower Hamlets Mission, from the Great Assembly Hall in the Mile End Road.   They rely not only on the eloquence of their speakers, for in addition they have a splendid brass band, and here, and here only, many women form part of the audience.   Lastly, in point of order, but by no means in numbers and importance, is the meeting held by the Christian Evidence Society.   This East London branch of the Christian Evidence Society was established by the late Mr. Celestine Edwards, the well-known coloured lecturer of Victoria Park.

It will be noticed that these meetings are not held to provide amusement.   Apart from one group of boys and girls listening to a young man who is reciting burlesque melodrama, all are engaged in strenuous controversy on social questions as seen from the religious, political, or economic point of view.*

The chief historical associations of Victoria Park centre round the portion near the principal lodge and entrance by

* For further details as to the Sunday meetings, see an admirable article on ' Sunday in East London : Victoria Park.' in *Sunday at Home*, October, 1895.

Bonner Hall Bridge across the Regent's Canal. Close to this spot, between the ornamental lake and the Hospital for Diseases of the Chest, stood an ancient and famous building known as Bishop's Hall or Bishop Bonner's Hall. This was in all probability the manor-house of the extensive Manor of Stebonheath or Stepney. The Bishops of London, to whom this manor belonged, formerly resided at the Manor-house of Bishop's Hall, where they had a private chapel. Roger Niger, an early Bishop, is said to have died

Main Walk, Victoria Park.

here in 1241. Bishop Baldock, who dates many of his public acts from Stepney, died here in 1313, and another Bishop, Ralph Stratford, died at Stepney in 1355. Bishop Braybrooke, who was Lord Chancellor, spent much of his time at this mansion.* He died in 1404, and no authentic account can be found of any Bishop residing here after this date, although tradition always connects the name of cruel Bishop Bonner with the manor-house. In 1548 we find that this Bishop granted to Sir Ralph Warren a ninety-nine

* Rev. W. H. Frere, M.A., 'Two Centuries of Stepney History'

years' lease of a messuage and chapel at Bethnal Green, evidently the manor-house in question.*

In 1588 we find it occupied by another layman, for in that year John Fuller, of Bishop's Hall, gave £50 towards the Armada defence fund. He was a Judge, and erected some almshouses in Stepney. He formerly lived in the City at Paul's Wharf, but transferred his residence afterwards to Bishop's Hall.† In his will made March 29, 1592, he bequeaths 'all my capital messuage, house, buildings, lands, tenements, profits, easements, fishing and commodities called Bishopshall unto Jane my wife.'

From Stow we learn that the land round the manor-house was well wooded, and one of the Bishops slightly anticipated the scheme of the Commissioners for forming a park of the site ; but there was this difference, that he wanted it for his private use, and wished to exclude the inhabitants who had enjoyed the right of hunting here since very ancient times. This Bishop was Richard de Gravesend, who in 1292 procured a grant of free-warren from Edward I., and also a license from him to enclose these woods and put wild beasts or deer therein. When the petition of the Bishop to the King was shown to the Aldermen of the city, they reported after consideration ' that from the time whereof no memory is extant, they had used to take and hunt within the said woods, and without, hares, foxes, conies, and other beasts, where and when they would. And they say, that they do not believe that the lord the King granted him anything in prejudice of the city's liberties ; wherefore they say that they desire to use the liberties which hitherto they have used. And they pray that the same Bishop may hold his woods in the form and manner as his ancestors and predecessors have held them. And they will not consent that he may enclose them, nor will they grant him any warren.'‡ This project was therefore abandoned till the present century, when the

---

* Hill and Frere, ' Memorials of Stepney Parish,' p. viii.
† *Ibid.*, p. 24, note.
‡ Quoted in Robinson's ' Hackney,' p. 202.

Government decided to make this popular improvement. The Bishops, thus deprived of their sport, amused themselves with tournaments, which were often held near the Bishop's palace between the years 1305 and 1331.*

The site of these merry-makings cannot now be determined exactly, but from the description given it is as likely as not that they were held on the lands now forming the park.

But we have already seen that the Bishop's palace came into the hands of laymen, and the old place was partially pulled down in 1800. With the materials a farmhouse was erected to the east of the former site, which was removed in connection with the laying out of the park and the formation of approach-roads.† In a notice of sale this property was described as ' the very desirable leasehold farm known by the name of Bishop Bonner's or Bishop's Hall, advantageously situated in the parishes of St. Matthew, Bethnal Green, and St. John, Hackney, containing about 102 acres of rich arable, meadow, and pasture land, in eight enclosures, lying within a ring fence, in the centre of which is a spacious, convenient, new-built brick dwelling-house, with a brew-house, stabling for twelve horses, etc.'‡

The fields around the farmhouse were open to the public till a few years after the opening of the park. These were included in the lands described in the Act for the formation of the park, although they are not now comprised in its area. The chief event for which they will now be remembered is a Chartist demonstration or fiasco, of which we reproduce a contemporary account :

' Whit-Monday, 1848, which was predicated to figure as a " white-stone " day in the annals of Chartism, will, alas ! only be remembered as the date of the most signal but most quiet and noiseless triumph of law and order over the grossest and most presumptuous folly and stupidity that

---

* Hill and Frere, ' Memorials of Stepney Parish,' p. v.

† Robinson, ' Hackney,' pp. 203, 456.

‡ From a book of newspaper extracts in the Tyssen Library.

ever disgraced a political design.   For months past the
most absurd threats, couched in the vilest language, had
been indulged in with regard to the determination of the
Chartists to concentrate their forces on the Metropolis on
this day.   At Clerkenwell, Islington, Finsbury, and all the
other districts announced to be the scenes of early meetings
during the morning of the day in question, the most marked
tranquillity prevailed, and it was soon understood that the

The Boating Lake, Victoria Park.

special honour was reserved for Bishop Bonner's Fields of
receiving the concentrated chivalry of all the Chartist clubs.
Accordingly, a squadron of 1st Life Guards, having ridden
past the anticipated scene of action, took up their quarters
in a farmyard at the south-east side of Victoria Park, adjoining the bridge which crosses Duckett's Canal.   To aid them
in keeping the peace, a detachment of 80 mounted police,
together with 1,100 constables (among whom 350 cutlasses

were distributed) and a battalion of 400 pensioners were drafted to the scene. Up till one o'clock in the day the number of persons assembled was perfectly insignificant, and was evidently composed of persons attracted rather by curiosity than by any sympathy which they entertained in the objects of the Chartist leaders. At one o'clock or a little later, Dr. Macdouall, one of the Chartist leaders, accompanied by several other well-dressed persons, said to be associated with him in the management of the demonstration, arrived on the ground in a cab. He appeared to be considerably agitated, and anxious to ascertain whether or not the authorities were really determined to put a stop to the meeting under any circumstances. Of this fact he received several very strong assurances from persons in authority, and it was made known that, besides the police being considerably out of temper from the great fatigue and annoyance inflicted on them by the freaks of the Chartists, orders had been given to the military that, in the event of their services being called into requisition, they were to act "effectively." When Dr. Macdouall understood this, he expressed his intention of immediately preventing the assemblage, and left the ground with his friends, followed by a considerable crowd of boys. During this period there was a heavy drizzle of rain, which had the effect of chasing the mob beneath the trees for shelter; and at three o'clock, the hour appointed for the Chartist meeting, the rain descended so heavily, and there being no appearance on the part of the Chartists to adhere to their original design, that it was considered advisable to march the unmounted police off the ground to a neighbouring church for shelter. At intervals, when the severity of the weather in some degree moderated, several small knots of persons formed at different parts of the grounds for the purpose of discussion, but they were at once dispersed by the horse patrol. About four o'clock, however, there came on a dreadful thunderstorm, and the rain descended in torrents. Instantly the remaining crowd ran away in all directions, seeking shelter where they could, and choking the already crowded taverns of the neighbourhood.   At six o'clock the fields and the neighbourhood were quite deserted, by which time several of the approaches were flooded to an extraordinary extent, and nothing could be more striking than the contrast afforded by the deserted appearance of Bishop Bonner's domain at that hour, compared with its gay and lively aspect when occupied with the military and spectators in the morning. Thus ended the Chartist fiasco of 1848 in Bishop Bonner's Fields.'*

The remainder of the park does not present any feature of historical interest.   The site was previously market-gardens and brick-fields.   The ornamental lake is made over one of these rough brick-fields.   The few cottages at the north of the park, near Victoria Park Road, were formerly the residences of some of these market-gardeners.   When the lands were purchased for the park, these cottages were retained, and are now occupied by officials of the park.†

Shore Road, which commences at the north - western corner of the park, and runs into Well Street, preserves the memory of a tradition that Jane Shore once lived here. Strype mentions this fact, and states that he was told this was formerly the manor-house, and that the lord's court for the Manor of King's Hold was held in this house.   He thinks, however, that the true name should be Shoreditch Place, named after the owner of the mansion, Sir John Shoreditch, a knight of the fourteenth or fifteenth century, who was buried in Hackney Church, but whose monument and inscription have now disappeared.‡   The name of Shoreditch seems to have been shortened to Shore Place, and was given to the row of houses which have taken the place of the old mansion.   This place is said to have been one of the greatest remains of antiquity in the parish of Hackney, which, with

* Condensed from a contemporary newspaper cutting in the Tyssen Library, Hackney.

† 'Glimpses of Ancient Hackney,' by F.R.C.S., p. 169.

‡ Strype's edition of Stow, vol. ii., p. 796.

the lands formerly belonging to it, is supposed to have been a grant from Sir John Shoreditch in the year 1339 to William de Corstone, chaplain to Edward III.*

MEATH GARDENS.

In close proximity to Victoria Park, to which they form a valuable adjunct, are Meath Gardens, $9\frac{1}{2}$ acres in extent, originally known as Victoria Park Cemetery. Under their former name, the gardens will long be remembered as a disgrace and scandal, and many attempts have been made in the past to compel the owners to properly maintain this disused cemetery; but as it was entirely of a private nature, it was exempt from the legislation which affects such places. Entrances to the ground had been burrowed from neighbouring back-yards, and it became the resort of the loafers and roughs of the East End, who came here to gamble and amuse themselves by the wanton destruction of the decaying property. It appears that the ground was originally purchased in 1840 for building purposes from Mr. W. W. Gretton by the late Charles Salisbury Butler, Esq., M.P. for the Tower Hamlets. Before building operations were commenced, a company offered to purchase the ground for the purposes of a cemetery. The purchase-money was to be paid by annual instalments, and the company was duly incorporated about 1845, and took over the land from Mr. Butler. But as the annual payments were not forthcoming, Mr. Butler was compelled to resume possession in 1853. As interments had taken place, and the land generally arranged for purposes of burial, including the erection of a chapel, he was practically obliged to continue the ground as a cemetery. This he did until the year 1876, when, for want of further accommodation, it was finally closed. When Mr. Butler died, his trustees found the cemetery a white elephant, as they had to pay rent-charges amounting to £43 10s. per annum, and they had no return from the property. For

* Robinson, 'Hackney,' p. 84.

some time the cemetery was maintained in fair order, but it was soon given up to the mercies of the roughs of the East End. The cemetery was never consecrated, and the passing of the Disused Burial-Grounds Act, 1884, prevented the erection of any buildings upon it, otherwise the land would have been worth some £40,000 as a building site. But the provisions of the Act just quoted would not have prevented

Meath Gardens before Laying-out.

this cemetery from being used (as many private graveyards have been) as store and lumber yards, carters' yards, or even as sites for low-class fairs and entertainments.

In order to prevent this, the Metropolitan Public Gardens Association in April, 1885, approached the Rev. J. B. M. Butler, the son of the former proprietor, and asked him to permit the association (if it could raise the funds) to lay out the ground as a public garden. Mr. Butler expressed his cordial sympathy with the project, and in February, 1886, through his solicitors, stated that he would be quite willing to hand over the ground for the purpose indicated, provided some arrangement were made which would relieve him of the maintenance of the disused cemetery, and of the pay-

Meath Gardens after Laying-out.

ment of the rent-charges. The Bethnal Green Vestry, who were asked to undertake this, did not see their way clear to do so, and as the funds of the association did not permit of their meeting annual liabilities of this nature, the scheme had to remain in abeyance for some time. So the cemetery continued in a very neglected and deplorable condition. The appointment of the newly-formed London County Council gave a fresh impetus to the negotiations, and finding that this body were sympathetically inclined towards the scheme, the Association again proceeded in the matter, and set to work to obtain the necessary funds for the laying-out of the ground. Among the principal donations were £500 promised by a former Duke of Bedford shortly before his decease (which promise was loyally redeemed by his successor, the late Duke), and an anonymous gift of £1,000 ' in memoriam Sidney Gilchrist Thomas.' Other smaller sums came in, and in January, 1891, the association was in a position to offer to lay out the ground, provided the London County Council would undertake to maintain it and pay the rent-charges. This offer was accepted in February, 1891, but various legal difficulties involving much delay had to be surmounted, and the works of laying-out could not be commenced till the end of March, 1893. The sum spent was about £3,000 exclusive of the cost of repairing the outer boundary railings, which was borne by the Council, who also redeemed the rent-charge by a payment of £1,005.*

The ground was re-named Meath Gardens out of compliment to the Earl of Meath, the energetic chairman of the Metropolitan Public Gardens Association, of whose zeal and perseverance for obtaining open spaces London has many examples. The transformed cemetery was opened to the public by H.R.H. the Duke of York, K.G., on July 20, 1894. The greater portion of the ground is laid out as a garden, and the remainder is devoted to two large playgrounds for boys and girls, fitted with swings, see-saws, and gymnastic apparatus. All who remember the gruesome state of this disused burial-ground in years past, with its yawning chasms, rank grass, and mutilated tomb-stones, will recognise what a thorough transformation has taken place.

* These particulars are taken from the printed statement prepared by the Metropolitan Public Gardens Association for the opening ceremony.

CHAPTER XXIX.

WATERLOW PARK.

T HE first event in the history of this place as a municipal park dates back to a certain Tuesday in November, 1889, when Lord Rosebery, the first Chairman of the London County Council, read the following letter:

'29, CHESHAM PLACE,
'LONDON, S.W.

'MY DEAR LORD ROSEBERY,

'On the southern slope of Highgate Hill, in the parish of St. Pancras, I own an estate of nearly 29 acres in extent, which was for many years my own home. This property, if judiciously laid out, would, I think, make an excellent public park for the North of London. The grounds are undulating, well timbered with oaks, old cedars of Lebanon, and many other well-grown trees and shrubs. There is also 1½ acres of ornamental water, supplied from natural springs. The land is freehold, with the exception of 2¾ acres, held on a long lease, of which thirty-five and a half years are unexpired. It is bounded almost entirely by public roads and a public footpath. Commencing the work of my life as a London apprentice to a mechanical trade, I was, during the whole seven years of my apprenticeship, constantly associated with men of the weekly-wage class, working shoulder to shoulder by their side. Later on, as a large employer of labour, and in many various other ways, I have seen much of this class and of the poorer people of London, both individually and collectively. The experience thus gained has from year to year led me more clearly to the conviction that one of the best methods for improving and elevating the social and physical condition of the working classes of this great Metropolis is to provide them with decent, well-ventilated homes on self-supporting principles, and to secure for them an increased number of public parks, recreation-grounds, and open spaces. This latter object can, I think, be best accomplished by the kindness of in-dividuals acting through the agency of the London County Council, and with as little burden as possible on the public rates. Therefore, to assist in providing large " gardens for the gardenless," and as an expression of attachment to the great city in which I have worked for fifty-three years, I desire to present to the Council, as a free gift, my entire interest in the estate at Highgate above referred to. On the day when the conveyance is executed (and that may be as soon as your solicitors have prepared the necessary legal documents), I will, in addition, pay over to the Council the sum of £6,000 in cash (the estimated value of the freehold interest in the 2¾ acres of leasehold), this sum of money to be used in purchasing this interest, or in defraying the cost of laying out the estate as a public park in perpetuity as the Council may deem most desirable. If your lordship is of opinion that this proposal is one which the members of the Council are likely to accept, this letter may be communi-cated to them as soon as you may deem expedient.

' I remain,

' Yours faithfully,

' (Signed)   SYDNEY H. WATERLOW.'

' To the Earl of Rosebery,
  ' President of the London County Council.'

This generous offer, it is needless to say, was at once accepted, and the London County Council took early steps to protect the inhabitants of London against the loss of this property through the operation of the Mortmain Act, which provides that if any person makes any gift of land to a public body, and dies within twelve months of doing so, his heirs may recover possession of any such property. Sir Sydney Waterlow is still alive, and it must be the wish of everyone who visits this charming little park that he may long be spared to continue his good work.

Although certain alterations had to be made in the grounds to adapt them for public use, they still retain much of their original character, and were it not for the numbers of visitors, it would be easy to imagine one's self in the garden of some country mansion. Owing to the undulating nature of the park, it is not possible to play any games which require a large level surface, with the exception of lawn tennis, for which several courts have been provided. The park, therefore, rests for its attractions mainly upon its natural features. There are two particular points in the gardening which call for special attention. These are the herbaceous border, and the old flower-garden where all the floral favourites in which our grandfathers delighted may be seen amidst the novelties of the present day. In the autumn there is a chrysanthemum show, to the success of which the climate contributes in no small degree. Many of the fruit-trees have been allowed to remain, and the fruit from these, together with the grapes grown in the vineries, are given to the hospitals and similar institutions of the neighbourhood. Bird and animal life is much encouraged here, and there are several aviaries stocked with British birds, and a guinea-pig house, much to the amusement of the youthful generation. On an elevated position is a rustic bandstand, around which is a gravelled promenade for the convenience of the many visitors attracted by the music.

The principal building in the park is a quaint, picturesque mansion known as Lauderdale House, rich in its associations, as we shall see later. Apart from the cost of restoring this, a sum of nearly £5,000 was spent in laying out the park, which was opened by Sir John Lubbock in the presence of the donor and a brilliant company in October, 1891. Since this time the lake which was originally in the grounds has been supplemented with two other sheets of water, much to the improvement of the park.

We can now pass to the history of Waterlow Park, which is as interesting as the place is charming, making it one of London's permanent attractions.

The district of Highgate, including the site of the park, was in ancient times part of that huge forest which surrounded the northern side of the Metropolis. This forest of Middlesex was the haunt not only of thieves and robbers, but of dangerous wild beasts, such as wolves and boars. Mention is made by Mathew Paris in his 'Life of the Twelfth Abbot of St. Albans' of the dangers experienced by the pilgrims proceeding to that shrine from London 'in consequence of the impenetrable woods which adjoined it (*i.e.*, the road), and which were also full of beasts of prey.'* One of the oldest of London topographers, Fitz-Stephen, writing between 1170 and 1180, tells us this forest of Middlesex 'was full of yew-trees, the growth of which was particularly encouraged in those days, and for many succeeding ages, because the wood of them was esteemed the best for making bows.' According to Maitland, this ancient forest was disafforested in 1218, in the reign of Henry III., but as late as Henry VIII.'s time a considerable portion remained, for we find a proclamation of his, dated July 7, 1546, running as follows: 'Forasmuch as the king's most royall ma^tie is much desirous to have the games of hare, partridge, pheasaunt, and heron, p'served in and about his honor, att his palace of West^m for his owne disport and pastime; that is to saye, from his said palace of West^m to St. Gyles in the Fields, and from thence to Islington, to o^r Lady of the Oke, to Highgate, to Hornsey Parke, to Hamstead Heath, and from thence to his said palace of West^m, to be preserved and kept for his owne disport, pleasure, and recreac'on; his highness therefore straightlie chargeth and commaundeth all and singuler his subjects, of what estate, degree, or condic'on soev' they

* Quoted in Prickett's 'Highgate,' p. 5.

be, that they, ne any of them, doe p'sume or attempt to hunt or to hawke, or in any meanes to take or kill any of the said games within the precinctes aforesaid, as they tender his favor, and will estchue the ymprisonment of their bodies, and further punishment at his ma^{ts} will and pleasure.'* As the wolves and boars and other wild beasts are not mentioned,

The Grove, the second residence of Coleridge at Highgate.

we may suppose they had withdrawn by this time to some safer retreat. This forest has been gradually disappearing from this date, but has not yet entirely gone, Bishop's Wood, opposite Caen Wood, Highgate Wood, and other similar places, being parts of this once extensive tract.

The name of Highgate is said by Norden to be derived

* Quoted in Prickett's ' Highgate,' p. 7.

37—2

from the toll-gate, through which traffic passed on its way to and from the north. 'Highgate, a hill over which is a passage, and at the top of the same hill is a gate through which all maner passengers have their waie; the place taketh its name of this high gate on the hill, which gate was erected at the alteration of the way, which was on the E. of Highgate. When the way was turned over the said hill to leade through the parke of the Bishop of London, as now it doth, there was in regard thereof a toll raised upon such as passed that way with carriage. And for that no passenger should escape without paying toll by reason of the widenes of the way, this gate was raised through which of necessitie all travellers pass. This toll is now farmed of the said Bishop at £40 per annum.'*

The same authority also states: 'Upon this hill is most pleasant dwelling, yet not so pleasant as healthful, for the expert inhabitants there report that divers who have long been visited by sickness not curable by physicke, have in a short time repayred their health by that sweet salutaire aire.' This testimonial of the sixteenth century to the salubrity of Highgate is confirmed at the present day by the number of convalescent homes established here.

The high gate referred to was really a brick archway extending across the road, with rooms over it, which were reached from a staircase in the eastern buttress. This archway was so narrow that waggons with high loads could not pass through it, but had to be taken through the yard at the rear of the Gatehouse Tavern. Although it was afterwards widened for carriages, it became such an obstruction that it had to be taken down in 1769, when an ordinary turnpike gate was substituted.

We must not omit to mention another derivation of the word Highgate. It has been suggested that this is an example of the use of the word 'gate' in the sense of road, so that Highgate would simply mean the high road.† The

* Norden, 'Speculum Britannicæ.'
† Taylor, 'Words and Places,' p. 252.

formation of this road over the hill taking the place of the old one by Crouch End, Muswell Hill, and Friern Barnet, was in main part the origin of the village of Highgate, or at any rate the cause which led it to attain any considerable importance.

Of the buildings now remaining in the park, the only one that lays any claim to antiquity is Lauderdale House. This only narrowly escaped destruction. At the time when the park was opened this house had become quite unsafe, and it

Lauderdale House, Waterlow Park.

was an open question whether it should be pulled down, or whether a large sum should be spent in restoration. It was eventually decided to restore it at a cost of nearly £3,000, which course was adopted mainly through the influence of the architectural profession. The external features remain the same, but the interior, in which is preserved all the old panelling, has been fitted up so as to serve for a refreshment-room, and as model dwellings for some of the workmen employed in the park. The refreshment bars are naturally on the ground-floor, and occupy the whole of it with the

exception of a large room fitted with seats, which is used as
a shelter in case of rain.   Here will be found some interesting
relics of bygone days.   The principal of these is generally
known as ' Nell Gwynne's bath.'   This is placed in a recess
in the hall, the oak pillars and architecture of which are
richly carved.   The bath itself is of marble, and is in a good
state of preservation.   Over the fireplace, which is fitted with
an ancient iron stove, is another specimen of old carving, with
figures in high relief, the subject of which is much disputed.

Lauderdale House was built probably about 1660 for the
Duke of Lauderdale, one of the notorious Cabal ministry
of Charles II.   His qualities were such that he was detested
by Royalists and Roundheads alike.   He was mainly instrumental in selling Charles I. to the English army, at which
time he posed as a Covenanter ; but after the Restoration he
turned completely round, and became one of the most ardent
persecutors these hunted people ever had.   He established
the horrors of an inquisition in Scotland, of which country
he was Lord Deputy.   Sir Walter Scott has made us familiar
in his ' Old Mortality ' with the racks, thumbscrews, and iron
boots used by this tyrant and his comrade, Archbishop
Sharpe, whilst his army was pursuing the Covenanters to the
mountains with fire and sword.   One more master-touch
from Carlyle, who describes ' his big red head,' and we have
this monster complete, who enriched himself at the expense
of the people whom he persecuted.   It is pleasant to turn
from this loathsome wretch to another occupant of Lauderdale House, who will always be popular, and that is pretty
Nell Gwynne. While Lauderdale was away in Scotland carrying out his murderous work, his master used often to borrow
his house for his favourite mistress.   Her beauty and her
ready wit had raised her to the stage from being an oyster
and orange wench, and as an actress she attracted Charles's
attention by a droll incident.   A hit had been made on the
stage by an actor who performed the part of Pistol in a hat
of unusually large size.   A rival manager, determined to
outdo this performance, had Nelly appear in a hat as large

as a coach-wheel. This so tickled the King, that she at once took his fancy, and she afterwards gained complete ascendancy over that weak and dissolute monarch. Although never favoured with the wealth and titles conferred on other mistresses of less amiable qualities, she was remembered by her royal lover on his death-bed, who urged his brother not to let ' poor Nelly starve.'

Everyone knows the well-worn anecdote which connects Nell Gwynne with Lauderdale House. She was anxious to obtain a title for her eldest son, a favour which she had long been unsuccessful in gaining. On one occasion, when Charles was walking in the garden, she held the child out of the window, saying, ' If you do not do something for him, I will drop him.' Whereupon he immediately replied, ' Save the Earl of Burford.' And so this title, and afterwards that of Duke of St. Albans, was given to him.

We must not forget another visitor to Lauderdale House, none other than our old friend Pepys, who seems in his wonderful life to have seen everything. On July 28, 1666, he went ' To the Pope's Head, where my Lord Brouncker and his mistress dined. . . . Thence with my Lord to his coachhouse, and there put six horses into his coach, and he and I alone to Highgate. Being come hither, we went to my Lord Lauderdale's house, to speak with him, and find him and his lady and some Scotch people at supper ; pretty odd company, though my Lord Brouncker tells me my Lord Lauderdale is a man of mighty good reason and judgment. But at supper there played one of their servants upon the viallin some Scotch tunes only ; several, and the best of their country, as they seemed to esteem them, by their praising and admiring them ; but, Lord ! the strangest ayre that ever I heard in my life, and all of one cast.' Pepys ought to have been introduced to the bagpipe !

Coming down now to modern times, Lauderdale House was in 1843 the residence of Lord Westbury before his elevation to the Wool-sack, and still later it was granted rent-free by Sir Sydney Waterlow to the trustees of St. Bartholomew's Hospital (of which institution he was treasurer) as a branch convalescent home.

Next to Lauderdale House, on the site of the new circular aviary, stood the residence of Andrew Marvell, poet and patriot. An unpretentious wood-and-plaster cottage, with central bay and porch, it was quite dwarfed by the surrounding mansions. Of all the eminent characters that have been associated with Waterlow Park, Marvell must be accorded the first place.

Andrew Marvell's Cottage, formerly on the site of Waterlow Park. (From a photograph taken in 1848.)

His father was master of the Grammar School at Hull, and is said to have lost his life in crossing the Humber in a storm to assist in the passage of a young couple about to be married. Andrew, born in 1620, was M.P. for Hull from 1660 till his death in 1678, and it was his custom to send a weekly (some say daily) letter to his constituents, by whom he was paid, giving a precise account of each day's parliamentary proceedings. When he first represented Hull he was in no wise unfriendly to the Court; but the unprincipled proceedings and licentious lives of the King and his ministers alienated the honest patriot, and he sternly opposed their arbitrary policy. Such a course naturally brought down upon him the displeasure of the Court, and as he had not spared the King in his attacks, a royal proclamation was made offering a large reward for his arrest. He thought it prudent to retire to Hull, but died suddenly, almost at once, and he is generally supposed to have been poisoned by his enemies.

In addition to his work in Parliament, he was a great writer, exposing in a particularly sarcastic way the corruptions of Church as well as of State. The chief of his writings, which were very voluminous, were 'The Rehearsal Transposed'—a stinging attack on Bishop Parker for his wordliness and persecution of the Nonconformists — and 'An Account of the Growth of Popery and Arbitrary Government in England.'

It is a great loss to Highgate that this cottage should have been taken down in 1869, owing to its unsafe condition. The little slip of garden behind, with its raised walk, where Andrew Marvell used to write his poetry, is now a part of the park.

Ascending the hill once more, we come to a more modern mansion, Fairseat House, which was formerly the home of Sir Sydney Waterlow. We have already explained that these grounds, although intended ultimately to form part of the park, cannot yet be opened to the public. This house is of too recent erection to boast any historical associations. On the occasion of the visit of the Prince and Princess of Wales to Lauderdale House on July 8, 1872, to inaugurate the convalescent home, Sir Sydney had the honour of receiving them here as his guests. The house is spacious, and naturally commands extensive views.

On the opposite side of the road to Lauderdale House is Cromwell House, the octagonal turret of which forms a pleasant feature in the background of the park, although it is out of place in the architecture of the building. This red-brick mansion is a testimony to the good building of our forefathers.   It is said to have been built by Cromwell, and that he himself dwelt in it, within a stone's-throw of Andrew Marvell, another prominent man in the Commonwealth. As no direct evidence can be obtained to support the tradition that Cromwell ever dwelt here, the theory has to be abandoned for another which says that the house was built by Cromwell for his son-in-law, General Ireton, who had

Cromwell House, Highgate.

married his daughter Bridget in 1646.   If this was so, he could not have lived here very long, for soon after his marriage he was called away on active service ; in 1649 he accompanied the Protector to Ireland, and was left in command there as Lord Deputy, but he died of inflammatory fever at Limerick in 1651.   His widow afterwards became the wife of General Fleetwood.

Ireton was as clever a scholar as a soldier, and was certainly Cromwell's right-hand man.   At the victorious

Battle of Naseby he commanded the left wing, but in spite of his bravery and steadiness he was unable to withstand the onslaught of Prince Rupert. Convinced of the treachery of Charles, he had voted strongly for his death, and signed the warrant for his execution. He was incorruptible, and showed his sense of honour by refusing an allowance of £2,000 out of the confiscated estates of the Duke of Buckingham. Cromwell House was evidently built, and internally ornamented, in accordance with the taste of its military occupant. The rooms are large and of good proportion, and have the ceilings moulded in scroll patterns. The fine old oak staircase is a feature of the house. It is richly ornamented with carved balusters, and on the newels are a series of ten carved figures about a foot high representing various types of the Parliamentary army. It is said that there were once twelve figures, the remaining two being Cromwell and Ireton. The balustrades are filled in with devices emblematical of warfare. There are some ceilings on the first-floor executed in rich plaster work, ornamented with a coat of arms said to be Ireton's, together with mouldings of fruit and flowers. The front of the house is rather low, being only of two stories, and it formerly had a platform on the roof, from which a good panoramic view for a considerable distance could be obtained. This platform was removed in the restoration of the house after a fire which occurred in 1864, when it was occupied as a boarding-school. Fortunately, the grand old staircase was preserved from destruction. Externally the house, which is now used as a convalescent home for children, presents few features of interest.

Next to Cromwell House, facing the entrance from High Street, stood another stately mansion, Winchester Hall, which has now disappeared, and with it the fine trees by which it was surrounded.

Going now in the opposite direction, past Cromwell House, we are on the site of Arundel House, the seat of the Earls of Arundel, pulled down in 1825. This, too, was an

ancient mansion dating back to the seventeenth century, around which clings much romantic history. This is supposed to be the same house described by Norden in his account of Highgate written nearly 300 years ago : ' At this place — Cornwalleys, Esq., hath a very faire house, from which he may with good delight beholde the statelie citie of London, Westminster, Greenwich, the famous river of Thamyse, and the country towards the south verie farre.'*
There is in the Harleian MSS.† a letter of Sir Thomas

The Lake, Waterlow Park.

Cornwallis, dated ' Hygat, July 16, 1587.'   He was knighted in 1548, so that the ' — Cornwalleys, Esq.', mentioned by Norden is in all probability his son William.   The Cornwallises during their stay here were honoured by visits from royalty, including Queen Elizabeth, who is said to have visited them in June, 1589.   The bellringers at St. Margaret's, Westminster, were paid sixpence on June 11, when the Queen's Majesty came from Highgate.‡   Whether or no

* Norden, ' Speculum Britannicæ ': ' Middlesex.'   4to., 1593.
† Lysons, ' Environs of London.'
‡ Nichols, ' Progresses of Queen Elizabeth,' vol. iii., p. 30.

Queen Bess was here, it is certain that James I. and his Queen were entertained right royally in 1604 by Sir William Cornwallis at his house in Highgate. On this occasion Ben Jonson was employed to prepare his dramatic interlude of ' The Penates.' At the end of the same year Sir Thomas Cornwallis died, and, to quote a writer in the *Gentleman's Magazine*,* ' it is most probable that Sir William then removed to reside in the Suffolk mansion (at Brome), as we hear no more of his family in Highgate. This residence, it is clear, . . . had been the principal one in the place, and as we find the Earl of Arundel occupying one of a similar description a few years later, whilst we have no information of his having erected one for himself, there appears reason to presume that it was the same mansion. The first mention I have found of the Earl of Arundel at Highgate is of the date 1617. . . . During the absence of the Court, the lords were entertained by turns at each other's houses ; and in Whitsun week . . . the Countess of Arundel—the Earl being with the King in Scotland—made a great feast at Highgate to the Lord Keeper, the two Lords Justices, the Master of the Rolls, and I know not whom else. It was after the Italian manner, with four courses, and four table-cloths one under another ; and when the first course and table-cloth were taken away, the Master of the Rolls, Sir Julius Cæsar, thinking all had been done, said grace, as his manner was when no divines were present, and was afterwards well laughed at for his labour.'

James I. was evidently so pleased with his former reception here that we find him making another visit in 1624. He arrived at the Earl of Arundel's late on Sunday evening, June 2, and slept the night, in order that he might hunt the stag in St. John's Wood early next morning.†

But probably the most important historical connection of Arundel House is with Lord Bacon, who died here in 1626. This statement is made by Aubrey, on the authority

* *Gentleman's Magazine* for 1828, part i., p. 588.
† Nichols, ' Progresses of King James I.,' vol. iii., p. 978.

of Thomas Hobbes, an intimate friend of Bacon's, and the circumstances are certainly interesting enough to be quoted in full :

' The cause of his Lordship's death was trying an experiment, as he was taking the aire in the coach (April 2, 1626) with Dr. Witherborne, a Scotchman, physician to the King. Towards Highgate snow lay on the ground, and it came into my Lord's thoughts why flesh might not be preserved in snow as in salt. They were resolved they would try the experiment presently; they alighted out of the coach and went into a poor woman's house at the bottom of Highgate Hill, and bought a hen and stuffed the body with snow, and my Lord did help to do it himself. The snow so chilled him that he immediately fell so ill, he could not return to his lodgings (I suppose then at Gray's Inn), but went to the Earl of Arundel's house at Highgate, where they put him into a good bed warmed with a panne, but it was a damp bed that had not been laid in for about a yeare before, which gave him such a cold that he died in 2 or 3 days; as I remember he (Hobbes) told me, he died of suffocation.'

In confirmation of this, one of his biographers, Rawley, writing in 1671, says: ' He died on the 9th of April, 1626 . . . at the Earl of Arundel's house in Highgate, to which place he casually repaired about a week before; God so ordaining that he should die there of a gentle fever, accidentally accompanied by a great cold, whereby the defluction of rheume fell so plentifully upon his breast that he died by suffocation.'

There is also extant a letter from Lord Bacon to the Earl of Arundel, who was evidently not here at the time, explaining that he was taken ill after making the experiment referred to, and was staying at his house at Highgate, but was prevented by his fit of sickness from writing personally.*

One other romantic circumstance is connected tradition-

* ' Letters and Remains of the Lord Chancellor Bacon,' collected by Robert Stephens. 1734.

ally with this house, and that is the escape of Arabella Stuart in male attire. The proprietor of the house when this took place is said to have been Mr. Conyers. As this happened in 1611, and the first mention of the Earl of Arundel at Highgate is in 1617, it is supposed that Mr.

An Old-fashioned Gateway, Waterlow Park.

Conyers was the owner or lessee of the mansion after the Cornwallises. Arabella Stuart led a miserable life on account of her dangerous nearness to the throne, and the jealousy of Elizabeth and James I. Both of these did all in their power to prevent her being married. James had discovered that she was attached to William Seymour, second son of Lord Beauchamp, but had forbade them to marry without his permission. Much to his indignation, however, he found out that they had been secretly married without his knowledge. Seymour was at once committed to the Tower, and Arabella was ordered away to Durham, to be looked after by the Bishop, but she had gone no farther than Barnet when she was attacked with a fever, no doubt brought on by the great nervous strain of her troubles. By the permission of the King, very reluctantly given, she was brought back to Highgate, and was allowed to stay here for some time, during which she was actively engaged in concocting a scheme of escape with her husband. It was arranged for her to make her way to Gravesend, disguised in male attire, where her husband was to join her, and together they were to sail to France. Arabella succeeded in reaching the boat in safety, but her husband, although he, too, had effected his escape, did not arrive before the French captain put to sea, who was impatient owing to the risk he was running. But the fates were against her, for the Government, as soon as they were aware that the birds had flown, sent out a number of war-vessels in pursuit, one of which captured the French ship near Calais, and brought the fugitive back. She was confined in the Tower, and died there of a broken heart some four years afterwards. Seymour managed to escape to Flanders, and, after the death of his wife was allowed to return to England, and lived for many years. This, then, is the story connected with Arundel House as given in most histories of Highgate. But in fairness we must point out that Mr. Thorne, in his ' Environs of London,' says that this is a mistake, and affirms that the house was that of Mr. Conyers at East Barnet.

Next to Arundel House, further up the Bank, formerly stood another mansion,* the residence of Sir John Wollaston, and afterwards of Sir Thomas Abney, whose name is chiefly remembered now in Abney Park Cemetery, which is on the

* Prickett, ' Highgate,' p. 108.

site of some of his property. Sir Thomas was Lord Mayor
of London in 1700, and afterwards represented the City in
the Parliament that secured the throne to the House of
Brunswick. Sir Thomas has now probably been forgotten,

A Quiet Nook in Waterlow Park.

but his friend and chaplain, Dr. Isaac Watts, who lived in his
family for upwards of thirty years, has gained undying fame.

The western side of the park is divided from Highgate
Cemetery by Swain's Lane, an alteration for the better of

38

Swine's Lane, by which it was formerly known. The cemetery has the advantage of attractive situation, and is tastefully laid out. Were it not for the numerous monuments, it might be taken for a park, with its shady groves of trees, its flower - beds, and its numerous plantations. Cemeteries laid out as gardens are comparatively modern luxuries, although Evelyn had suggested them after the Great Fire of London. His idea was to have one huge necropolis just outside the city, divided into portions for the various parishes, ' and with ample walks of trees; the walks adorned with monuments, inscriptions, and titles, apt for contemplation and memory of the defunct.' All these features are combined at Highgate, and they certainly make the cemetery a very popular one. Among others interred here are Michael Faraday, chemist and philosopher; Sir William Ross, the celebrated miniature painter; the father and mother of Charles Dickens and his little daughter Dora, familiar to the readers of ' David Copperfield.' But perhaps the tombs which have attracted the greatest numbers are those of Tom Sayers, the pugilist, bearing the portrait of himself and his dog, Wombwell, of menagerie fame, with his lion standing over him, and Lillywhite, the cricketer, whose marble monument, erected by the members of the Marylebone Club, is carved with a wicket struck by a ball, representing the well-known cricketer as ' bowled out.'

Coming back now to the east side again, the classical buildings overlooking the park form the Roman Catholic colony of Highgate. There was at one time a noted roadside inn here, the Black Dog. Afterwards it became a private residence, and was purchased, with its grounds, by the Passionist Fathers for a monastery, known as St. Joseph's Retreat. There are large schools for boys and girls, built of light-coloured brick, with ornamental string courses, and a porch, surmounted by a turret rising high above the roof. The first Superior of the monastery was the Hon. and Rev. George Spencer, who adopted the name of Father Ignatius. Although a clergyman of the Church of England, with very fair prospects of advancement, seeing that his brother was a member of the Cabinet, he joined the Church of Rome, and adopted the cowl, gown, and sandals of the Passionists. The author of the 'Life of Father Ignatius' writes: 'In 1858 we procured the place in Highgate now known as St. Joseph's Retreat. Providence guided us to a most suitable position. Our rule prescribes that our houses shall be out-

St. Joseph's Retreat Entrance, Waterlow Park.

side the town, and yet near enough for us to be of service in it. Highgate is wonderfully adapted to all the requisitions of our rule and constitution. Situated on the brow of a hill, it is far enough from the din and noise of London to be comparatively free from its turmoil, and yet sufficiently near for its citizens to come to our church. The grounds are enclosed by trees; a hospital at one end and two roads meeting at the other promise a freedom from intrusion and

38—2

a continuance of the solitude which we now enjoy.' The new monastery, erected in 1875-6, was designed by Mr. Francis W. Tasker, and was consecrated by the late Cardinal Manning. The accommodation of the building is quite complete. There are forty cells for the monks, rooms for guests, library, refectory, and the usual rooms and offices. These are grouped round three sides of a square, and the design is in the style of the monastic buildings of Central Italy. The walls are faced with white Suffolk bricks, with stone dressings, and the roofs, which project in a remarkable manner, are covered with large Italian tiles.

Opposite the monastery is a building of a very different character—the Old Crown public-house, with its tea-gardens and arbours, which forms a link with the past history of Highgate.

This small park we have found, then, to be very rich in its historical associations, and if we were to slightly enlarge the circle of its surroundings we should find many more places of equal interest. To the north is the Highgate Grammar School and Chapel, both of which replace buildings of great antiquity and history. Underneath the chapel is the vault containing the remains of the poet Coleridge, who lived for nineteen years in the Grove, facing the church. Descending the hill are many relics of Whittington—almshouses founded through his generosity, a stone pedestal marking the place where he heard the bells, and other spots around which clings many a romantic tale. But for these we must refer the reader to the many excellent histories of Highgate.

CHAPTER XXX.

WESTERN COMMONS.

Eel Brook Common, Parson's Green, Brook Green, Wormwood Scrubs.

THESE open spaces were formerly wastes of the Manor of Fulham. Under the name of Fullonham (which is interpreted as 'the habitation of fowls'), this manor is said to have been granted about the year 691 to Erkenwald, Bishop of London, by Tyrhtilus, Bishop of Hereford, with the consent of the Kings of the East Saxons and of the Mercians. History does not relate how it came into the possession of the Bishops of Hereford. This Erkenwald, to whom the manor was given, was son of Offa, King of the East Saxons, and he appears to have been a man of singular learning and attainments for the time in which he lived. He expended large sums in the purchase of lands to augment his see, and he also obtained for it many privileges. through his influence with the kings of the neighbouring districts. Although the original grant of the manor may be rather obscure, it is well established that it belonged to the Bishops of London long before the Conquest. It has remained in their possession ever since, with the exception of the interregnum of the seventeenth century. At the time of the Norman survey we find: 'In Fuleham the Bishop of London held forty hides. . . . Its whole value was forty pounds ; the like when received. In Edward's time the value was fifty pounds. The manor was,

and is, part of the see.'* The present lords of the manor
are the Ecclesiastical Commissioners, whose rights in the
three first-mentioned commons were purchased by the late
Metropolitan Board of Works in 1881 for the sum of £5,000.
A further sum of £2,000 was paid to the Homage Jury of
the Manor of Fulham for the rights of the commoners,
making £7,000 in all.

Brook Green and Parson's Green have many things in
common. They take us back in thought to the early days
of Hammersmith and Fulham, when they were the village
greens of separate communities, and not part of the gigantic
London of the present day. The commons, or rather their
surroundings, are now in a transitional state. There are a
few dilapidated cottages and houses of good standing dating
back to these rural times, and sandwiched in between these
are new and staring modern villas, which will in a few years
swallow up all the remaining space. It will be a wrench to
lose all the relics of a bygone age, but perhaps it will be
better than to gaze on the present incongruous patchwork
around the greens.

Although Parson's Green is identified with the Manor of
Fulham, there is some doubt as to whether it was not part
of the Manor of Rosamunds, of which mention is made in
ancient records. In all probability this manor was a subsidiary one to the Manor of Fulham. We find in 1451 the
Manor of Rosamunds was alienated by Agnes Haseley to
Henry Weaver, and the widow of Sir Henry Weaver died in
1480 'seised of the Manor of Rosamunds in Fulham.'
Nothing later than this has been discovered about this
manor, but it is supposed to be the estate at Parson's Green
adjoining the Rectory House. A tradition states that the
manor-house was a palace of 'Fair Rosamond.' There was
a house once facing the green known as Rosamunds, which
retained the memory of this vanished manor.†

Parson's Green takes its name from the parsonage-house

* Faulkner, 'History of Fulham,' 1813, p. 165.
† *Ibid.*, p. 307.

or rectory of the parish of Fulham, in which the rectors of Fulham used to reside. This rectory is reported to have been the residence of Adoniram Byfield, the noted Presbyterian Chaplain to Colonel Cholmondeley's regiment in the Earl of Essex's army, who took so prominent a part in Cromwellian politics that he became immortalized in 'Hudibras.'* Bowack, writing in 1705, speaks of an old stone building which adjoined it, and which he conjectured was designed for religious use, in all probability as a chapel for the rectors and their domestics. At the time he wrote it was about three or four hundred years old. He continues: ' Before the said house is a large common, which, within the memory of several ancient inhabitants, now living, was used for a bowling-green.'† This ancient stone building was pulled down about 1742, and the parsonage or rectory, after being divided into two, has since shared the same fate. On this side, adjoining Rectory Road, is the church of St. Dionis, built of red brick with stone dressings, having a square castellated tower of pleasing appearance.

Clustering round the green are, or were, several historic houses. Many of these have been pulled down or altered, and others are in a ruinous state. Peterborough House, on the south-west of the green, was built on the site of a famous mansion once standing here. The first name of the older house was Brightwells, which is described in ancient records as the property of John Tarnworth, Privy Councillor of Queen Elizabeth, who died here in 1569 (according to some 1599). It afterwards successively belonged to Sir Thomas Knolles and Sir Thomas Smith, Clerk of the Council, Latin Secretary, and Master of the Requests to James I. The latter owner died here in 1609, and the estate was conveyed by marriage to Hon. Thomas Carey, who married the only daughter of Sir T. Smith. In all probability he rebuilt the mansion, at any rate, it was now known by the name of Villa Carey. Francis Cleyn, who came over to England in

* Croker, 'Walk from London to Fulham,' p. 165.
† Bowack, 'Antiquities of Middlesex,' p. 58.

the reign of James I., and was in great repute for painting ceilings, was employed in the decorations. In 1660 this estate was in the possession of John, Lord Mordaunt, a younger son of the first Earl of Peterborough, who married the daughter of Carey. He was created Viscount Mordaunt by Charles II., in return for his active loyalty in the Civil War. He afterwards fell into disfavour, and spent the

Peterborough House.

greater portion of his time at Parson's Green, where he died in the forty-eighth year of his age, and lies buried in Fulham Church.

He was followed by his son Charles, the celebrated Earl of Peterborough, who succeeded his uncle in the title.* He made a name for himself by his military exploits in Spain,

* Croker, 'A Walk from London to Fulham,' 1860, pp. 166-169.

but he was distinguished no less as an orator than as a soldier. To this house resorted Locke, Addison, Swift, and Pope, who, together with all the distinguished men of the time, were afforded a hearty welcome. There is a tradition that Voltaire visited the Earl in this mansion, and there met Addison, who was suffering on this occasion from one of his fits of taciturnity.

A good story is told which illustrates the eccentricity of this Earl of Peterborough. When his lordship gave a large dinner it was his practice to assume the apron, and to super-intend in person the preparation and arrangement of the various dishes. When the banquet was ready he threw aside his culinary appendages and entered the drawing-room with the grace of a refined courtier, but more proud of having exercised the talent of a skilful cook, which he acquired during his arduous campaigns in Spain.*

Another thing which marked his eccentricity was his curious marriage life. He was twice married, his second wife being the celebrated Anastasia Robinson, the opera-house singer. His pride prevented him from owning this marriage till shortly before his death. She had a separate house close by, which was taken by the Earl for his wife and her mother. While her husband had his literary friends at his table, she for her part held musical parties, at which the most eminent musicians assisted, including Bonancini, Martini, and others.†

The gardens of this house were famous. Swift says in one of his letters that they were the finest he had ever seen about London. They are mentioned by Stow : ' In Parson's Green are very good houses for gentry, where the Right Hon. the Earl of Peterborough hath a large house with stately gardens.'‡ Bowack wrote at the commencement of the eighteenth century : ' The contrivance of the gardens is

---

* Brewer, ' London and Middlesex,' vol. iv., p. 109.

† Hawkins, ' History of Music,' vol. v., p. 305.

‡ Strype's Stow, 'Survey of the Cities of London and Westminster,' 1720, vol. i., p. 44.

fine, though their beauty is in great measure decayed; and
the large cypress shades, and pleasant wildernesses, with
fountains, statues, etc., have been very entertaining.' He
also speaks of a natural curiosity contained in the garden,
which was then said to be unique in Europe. This was a
tree, 76 feet high, which bore a yellow tulip. Its stem was

Richardson's House at Parson's Green in 1799.

about 5 feet 9 inches in circumference, and it had a smooth
gray bark and a very fine green leaf.* This tree died in 1756
of decay, when it was about one hundred years old. In 1794
Peterborough House was purchased by John Meyrick, who
pulled down the old house and erected the last one on the same
site. Of recent years the mansion was used as a private

* Bowack, 'Antiquities of Middlesex,' p. 45.

lunatic asylum, but it is about to be pulled down, and the historic grounds will be split up into building plots.

Near to Peterborough House stood an ancient mansion which formerly belonged to Sir Edward Saunders, Lord Chief Justice of the King's Bench in 1682. This house acquired fame as being the residence of the celebrated novelist, Samuel Richardson, who moved here from North End in 1755. Some doubt has been thrown upon the statement made both by Faulkner and Lysons that Richardson wrote 'Clarissa Harlowe' and 'Sir Charles Grandison' while residing at this house. He died here in 1761. Thomas Edwards, the author of 'Canons of Criticism,' died whilst on a visit to Richardson at Parson's Green in 1757.*

On the east side of the green was a plain white house, built at the end of the seventeenth century by Sir Francis Child, Lord Mayor of London in 1699, whose tomb is in Fulham Churchyard. Among the notable residents of this house we may mention Admiral Sir Charles Wager, Dr. Ekins, Dean of Carlisle, who died here in 1791, and Mrs. Fitzherbert, who is said to have had the porch erected in front of the house. She attracted here George IV., then Prince of Wales, who became a constant visitor.†

Adjacent to Parson's Green, in the King's Road, was Ivy Cottage, built at the end of the last century by Walsh Porter, and afterwards the residence of Mr. E. T. Smith, of Drury Lane fame, who altered the name to Drury Lodge, after his theatre. A few years back this house belonged to that eccentric lady Mrs. Villens, better known in the world of sport as 'Lucky Jack,' whose appearance on the race-course in a Newmarket coat and 'pot' hat was as familiar as that of her friend the late Duchess of Montrose (Mr. Manton). On the site of this cottage was formerly a house, traditionally stated to have been the residence of Oliver Cromwell.‡

At the commencement of the north side of the green is the

---

* Thorne, 'Environs of London,' vol. i., p. 226.
† Faulkner, 'Fulham,' pp. 302, 303.
‡ Croker, 'Walk from London to Fulham,' p. 169.

Holt Yates Memorial Home and Laundry for the friendless and fallen, whilst the adjoining house is a training home for young girls of good moral character; but the strange combination would not be noticed by a passing stranger, owing to the absence of any notice-boards or conspicuous name-plates. Belfield House, a substantial building on the same side of the green, formerly the residence of Mr. Sheridan, M.P., is now occupied by a French artist, Theodore Roussel.

Another distinguished resident of Parson's Green was Sir Thomas Bodley, who lived here from 1605 to 1609. He is famous for having founded the Bodleian Library of Oxford, opened in 1602. It claims a copy of all works published in this country, and for rare works and manuscripts it is said to be second only to that in the Vatican. When Lord Bacon fell into disgrace, he procured a license (dated September, 1611) to retire for six weeks to the house of his friend Sir John Vaughan, at Parson's Green, but when the time expired the King refused to renew the license. Lord Bacon and Sir Thomas Bodley wrote to one another, and several letters are extant from Sir Thomas Bodley dated from his house at Parson's Green.*

The last name we will mention in connection with the green is Sir Arthur Aston, an officer of note in Charles I.'s army, who was son of Sir Arthur Aston, of Parson's Green. He commanded the Dragoons in the Battle of Edgehill, and was successively made governor of the garrison at Reading and Oxford. He had the misfortune to break his leg by a fall from his horse, and left the army. Under Charles II. he was made Governor of Drogheda in Ireland, and was in command there when Cromwell besieged and took the town in 1649. The inhabitants were put to the sword, and poor Sir Arthur was cut to pieces, and his brains beaten out with his wooden leg.†

Eel Brook Common, of fourteen acres in extent, takes its name from the old Eel Brook, which has now been filled up,

---

* Thorne, 'Environs of London,' vol. i., p. 226.
† Faulkner, 'Fulham,' p. 307.

but which was formerly to be seen at the western boundary of the common. Its present name is certainly an improvement upon its former one—Hell Brook Common, the origin of which is unknown. It is mentioned under this name in a list of orders presented at a court held for the Manor of Fulham in 1603 : ' That no person or persons shall put in any horse or other cattle into Helbrook until the last day of April every year henceforth.'* There are no records existing in the parish relative to this common, which seems to be a place without a history. It formed the subject of an action in 1878, Lammin *v.* Ecclesiastical Commissioners, in which Mr. W. H. Lammin, suing on behalf of the freehold and copyhold tenants of the Manor of Fulham, sought to establish their rights of common and customary rights of recreation, and to prevent the enclosure of the common. By an agreement entered into between the parties concerned, this action was to be put an end to if the late Metropolitan Board of Works could obtain a scheme for the establishment of local management under the terms of the Metropolitan Commons Act, 1866. This scheme was certified by the Inclosure Commissioners, and confirmed by the Metropolitan Commons Supplemental Act, 1881. The strips of land in the King's Road fronting Peterborough House nearly join this common to Parson's Green.

Brook Green is a long straggling common of 4¾ acres, plentifully supplied with trees of comparatively modern growth, in addition to a row of six venerable giants, which have been much shorn of their former grandeur. A few shrubberies have been formed at various points to embellish the appearance of the green, but the remainder is open for recreation for children and adults, although the area is too small to permit of organized games or public meetings. Formerly the green was intersected throughout its length by the highroad, and a much-needed improvement was effected some years ago by the Vestry of Hammersmith,

---

* Faulkner, ' Fulham,' p. 24.

who diverted this road, and thus united the two narrow
strips of which this open space formerly consisted.

Facing the green on both sides at its southern end is a
large and flourishing Roman Catholic colony.   It consists
on one side of the Church of the Holy Trinity, a spacious
building of considerable architectural pretension, with a lofty
tower and spire at the north-east.   The almshouses, which
are built of ragstone in the same substantial style, stand
back some way from the green, and form, together with the
church, a spacious quadrangle.   The foundation stones of

Sion House, Brook Green.

the church and almshouses were both laid in 1851: that of
the former by Cardinal Wiseman, and of the latter by the
Duchess of Norfolk.   On the opposite side of the road are
the other buildings, comprising a training college for Roman
Catholic schoolmasters, a practising and other schools.
One of these schools, Brook Green House, bearing a tablet
with the date 1787, had for one of its pupils no less a
personage than Mrs. Stirling, the actress.   A short distance
away is another religious establishment, the old convent of
Sion House.   This is a very ancient edifice, with a pretty
chapel.   It belongs to the Benedictine Order, and was at

the commencement of the century the only educational establishment for Roman Catholic young ladies of the upper classes in England. This is shortly to be pulled down.

Almost the last house on the north side is a very modern building called the Old House, after an old ruinous structure which stood on or near the site. An illustration of this old house is given in Faulkner's 'Hammersmith.' The back part was of wood, and the front wholly brick, having had originally bow windows. A short time before this place was pulled down the owner burnt all the antique furniture and carved ornaments which the house contained owing to the want of fuel during the winter of 1834.*

Perhaps the other side of the green is the most interesting. After leaving the Roman Catholic church, a picturesque Elizabethan house is seen called The Grange. This was for some time the home of our most popular living actor, Sir Henry Irving; but since he left Brook Green the house has been untenanted, and, like the adjoining Sion House, it is about to be demolished to make room for a girls' school in connection with Dean Colet's bequest. Close by are some more almshouses, really in Rowan Road, the end one of which faces the green. These were rebuilt in 1840, and have taken the place of four rural cottages with picturesque gables, built in 1629. Two stone tablets which were taken from the old cottages, founded by John Isles, have been placed in the new building. They bear the following inscription: 'Quod pauperibus datur in Christum confertur. Lutum pro auro, 1629.'

The largest mansion on this side is Bute House, now the residence of Mr. W. Bird, J.P., D.L., formerly known as Eagle House. It is built in the Queen Anne style, screened from the highroad by an iron fence, flanked by two brick piers, surmounted by eagles, from which the house took its former name. The premises were once used as a school, and comprised extensive grounds, subsequently enlarged by

* Faulkner, 'Hammersmith,' p. 392.

the addition of the gardens of adjoining houses, which have
been pulled down.

A French Protestant church once stood in this neighbourhood. It is mentioned in the Court Rolls, but the site

The Grange, Brook Green, formerly the residence of Sir Henry Irving.

cannot be ascertained. The parish registers contain several
entries at the commencement of the eighteenth century
relating to these persecuted Huguenots.*

WORMWOOD SCRUBS.

Wormwood Scrubs proper is a flat open space of some
193 acres, and for the greater portion is subject to the use
of the military. It is made up of three distinct properties,
the largest portion of which, comprising 135 acres, was
common land attached to the Manor of Fulham. The
manorial rights were purchased by the Secretary of State
for War in pursuance of the Military Forces Localization
Act, 1872, with a view to creating a Metropolitan exercising
ground for the troops. In addition to the common, certain

* Faulkner, ' Hammersmith ' pp. 395, 396.

inclosed lands adjoining it containing 53 acres, and another piece of 5 acres, the property of the Great Western Railway Company, were also purchased, making in all an extensive open space of nearly 200 acres. The total cost to the War Office was £52,615, who proposed that all the lands should be vested in the late Metropolitan Board of Works on trust to enable them to be used for such military purposes as might be directed. Subject to this user, the lands were to be held for the perpetual use by the inhabitants of the Metropolis for exercise and recreation. A fringe of the land on the eastern side, known as the non-military portion, is exempt from the use of the troops, and is open to the public at all times. This scheme was confirmed by the Wormwood Scrubs Act, 1879, which effected the transfer without any payment. The Act provides that the military portion can be used for such purposes as the Secretary of State for War from time to time directs, including camps, reviews, drills, training, exercising, firing, or rifle-ranges. Two other important conditions are worthy of note—viz., that no permanent building or erection, except rifle‑butts and all necessary appurtenances, shall be constructed without the consent of both parties, and also that no portion of the Scrubs shall, without consent, be used for military purposes or as a rifle‑range on any public holiday.

A portion of the original common was severed from the remainder by the West London Railway. This land, comprising 22 acres, was vested in the Ecclesiastical Commissioners as Lords of the Manor of Fulham. The copyhold tenants of the manor claimed certain rights of common of pasture over these lands for their cattle and swine at certain seasons, such as were formerly exercisable over the whole tract called Wormwood Scrubs. These rights had ceased to be exercised for several years, and the land was enclosed and let from time to time. The Ecclesiastical Commissioners generously offered to transfer the Little Scrubs, as they are called, to the late Metropolitan Board of Works without any consideration, on condition that they were laid

39

out and maintained as an open space, and that the rights of
the commoners over the land should be acquired and extinguished.   The sum agreed upon for the purchase of these
rights was fixed at £2,000, and the scheme was confirmed
by Parliament under the provisions of the Metropolitan
Board of Works (Various Powers) Act, 1886.

The principal work carried out on the Scrubs was a
complete system of under-drainage, which was necessary
owing to the wet state of the ground.   Before this was done,
at certain seasons of the year the soldiers who came to fire
at the ranges were frequently knee-deep in water.   The rifle-
butts, which are arranged on the Belgian principle, are
situated on the portion railed off from the public.   Certain
improvements were carried out on the Little Scrubs in
1893-94, when the stream was widened and provided with
weirs, so that it might always have water in its bed.   Paths
were formed, and trees and shrubs planted in order to give
the land a more ornamental appearance.

The name of Wormwood Scrubs (A.S. *Scrob*—a shrub)
is a popular corruption of Wormholt Scrubs.   This word
holt (a wood) points back to the old nature of the ground,
when the site of the Scrubs was a wood, of which Old Oak
Common was an extension.   In Rocques' map, 1744, the land
is shown as Warner Wood, with paths through the thicket
in various directions.   Many variations of the name exist—
Woorine-old-Wood, Wormeall Wood, are found in old
documents as well as those already given.

In 1803 there was an important action with regard to a
right of way across the common land.   Some land adjoining
the Scrubs was let by a Mr. Fillingham to a market-gardener,
and it was promised that a road across the Scrubs should be
made for his horses and carts.   The copyholders of the
manor, having heard of this, erected post-and-rail fencing
in order to prevent any trespass.   This led to an action on
the part of Mr. Fillingham, who wanted the parish to make
him proper roads and ways to his land.   The action was
tried before Lord Ellenborough and a special jury, and some

of the most eminent counsel were engaged, among them Erskine, Garrow, Gibbs, and Marryatt. After a lengthy trial, a verdict was found for the defendants, thus establishing the exclusive right of the tenants of the manor to the use of the Scrubs.*

The earliest mention of the land now used as an open space is in a document dated 1189, when Richard Fitzneal was Bishop of London. It is as follows: 'Seventy-eight acris in Wormholt et Herleston (Harlesden), et de ix acris assartorum in Wormeholt, et de iv acris assartorum juxta Wormeholt.'†

The Scrubs, as we have already mentioned, are waste lands of the Manor of Fulham. A part of the demesne lands of this manor form a sub-manor, that of Wormholt Barns, containing 423 acres. This was leased by Bishop Bonner, together with other lands, in 1547, to the Duke of Somerset. The manor is there described as the 'divers messuages, lands, tenements, woods, closes, meadows, feedings, pastures, groves, and other hereditaments in Fulham; and also all that part and portion of lands and woods, with the appurtenances, in Fulham, called Wormeolt Wood, parcel of the possession of the Bishoprick aforesaid.' When the Duke of Somerset was attainted, the Crown obtained possession of the manor. In 1596 it was granted by Queen Elizabeth to Simon Willis, who afterwards assigned one half of his interest to Thomas Fisher, and the other half to Sir Thomas Penruddock. The latter's son subsequently obtained possession of the whole and the manor has remained in private hands ever since.‡

There was a proposal in 1817 on the part of some of the inhabitants to establish races on the Scrubs, to be called 'Wormholt Races.' These were to continue two days, and were to be under the patronage of His Royal Highness the Duke of Sussex. Bills were printed and circulated giving

* Faulkner, 'Hammersmith,' pp. 383, 384.
† Records, 'Dec. et Cap., S. Pauli,' p. 38. Printed by Miss Hackett, London, 1826.
‡ Faulkner, 'Hammersmith,' p. 390.

the following particulars : ' On the first day the Sussex stakes of ten guineas each, twenty to be added for all ages that never won plate, match, or sweepstakes, and sweepstakes of five guineas each, and twenty-five added for all ages. On the second day, fifty pounds for all ages, and sweepstakes of three guineas, with twenty-five added for all ages.' The terms for admission to the race-course were to be for spectators with a horse, sixpence; for a chaise or cart, one shilling; for a four-wheeled carriage, two shillings and sixpence. It was proposed to give the proceeds towards building and endowing additional almshouses, and a considerable sum was subscribed; but in spite of this laudable object, the Government and the magistrates interfered, and prevented the races being held.*

The western boundary of the Scrubs is formed by Old Oak Common, which takes its name from the fine old oaks with which it used to be covered. One of these was standing till about 1830, but it was then cut down and sold. The name of this common is interesting, because it is akin to that of the district in which it is situated, viz., Acton, or the village of oaks. Here, in the early days of Britain, our Druid forefathers, we can easily imagine, had a seat of worship, with an open-air temple surrounded by the huts of the priests and worshippers. In the Doomsday survey the Manor of Fulham is said to have pannage for 1,000 hogs, which doubtless thrived on the acorns of the oak-trees on this common and the Scrubs.†

Before the Scrubs were acquired by the War Office, they had been leased for military purposes for many years previously. The funds derived from the rental of £100 per annum paid by the Government were given towards the Waste Lands Almshouses. This charitable foundation owes its origin to a resolution of the copyholders of the Manor of Fulham, dated April 23, 1810. It was decided then that no grants of the waste land should be made without adequate

---

* Faulkner, 'Hammersmith,' p. 388.
† Walford, 'Greater London,' p. 8.

compensation, and that the money so received should be vested in trustees chosen from the copyholders of the manor for the purpose of building and endowing almshouses. In furtherance of this object, an acre of land was granted in March, 1812, to the trustees of the Almshouse Charity Fund, for the purpose of erecting cottages thereon for the benefit of poor widows belonging to the parish of Fulham. This was a portion of the waste land situated near Starch Green Lane.

At the same court it was resolved to grant a lease of twenty-one years to the War Office of the waste of the manor called Wormholt Wood, for the purpose of training troops. Half of the rental of £100 was reserved for the Fulham, and the other half for the Hammersmith, side of the parish. This lease was renewed at its expiration till the freehold was purchased. From the funds received from these leases, together with other grants for waste lands and some private donations, nine almhouses were built in 1813.*

To turn from almshouses to another asylum of a different kind, we have only to lift our eyes from the Scrubs to see the grim walls of the Wormwood Scrubs Convict Prison. We live in an age of improvements, and thanks to the efforts of reform commenced by John Howard and Elizabeth. Fry, even convicts share in the bettered condition of all classes. Medallion portraits of these two philanthropists form the only relief to the grim, massive, octagonal towers which flank the lofty entrance gates. The prison is the finest specimen of penal architecture in England, and the whole of it was built by convict labour. A recent article† stated that, 'without doubt, England's model prison is the one located at Wormwood Scrubs. It is the most recently erected, and is supposed to represent all that is best in prison construction, and to contain in its interior arrangements everything that experience has proven best calculated to the convenience, the health, the discipline, and the just

---

* Faulkner, ' Hammersmith,' pp. 197, 198.
† *Pall Mall Magazine*, November, 1895.

and humane treatment of its unfortunate inmates.' From
an interesting pamphlet on the prison* some more details
of its construction may be gathered. The bricks are home-
made, having been manufactured by the convicts on the site
of the prison and the land adjoining. The total number was
about 35,000,000. The granite came from Dartmoor, the
stone from Portland, and the iron-castings from Portland
and Chatham prisons, all prepared by convict labour. The
prison contains separate cells for 1,381 male and female
prisoners, besides hospital and other accommodation. It
was opened in 1874, although the final work of erecting the
boundary-wall was not completed till 1883. This wall,
18 feet high, with its flanking towers at the angles, encloses
a space of about 15½ acres. The cost of the prison proper,
i.e., all within the cells, was £97,155, or at the rate of
£70 7s. per cell.

* By Sir Edmund F. Du Cane, Surveyor-General of H.M. Prisons,
published in 1889.

CHAPTER XXXI.

*WAPPING RECREATION-GROUND—CHURCHYARDS AND
SMALL PLAYGROUNDS—PLACES IN COURSE
OF ACQUISITION.*

WAPPING RECREATION GROUND.

W APPING Recreation Ground came into existence
as the result of the clearance of an unhealthy area
under the Artizans' and Labourers' Dwellings
Improvement Act. This Act, passed in 1875,
provides that, whenever a medical officer shall make an
official representation to the local authority that any houses,
courts, or alleys, within a certain area under its jurisdiction,
are unfit for human habitation, or that diseases indicating a
generally low condition of health, and caused by the sanitary
defects of such area, are prevalent therein, and cannot be
effectually remedied otherwise than by an improvement
scheme, the local authority shall consider such representation, and, if satisfied of the truth thereof, and of the suf-
ficiency of its resources, shall pass a resolution to the effect
that the area in question is unhealthy, and shall forthwith
proceed to make a scheme for the improvement of such area.
So runs the legal phraseology of the Act, which obtains its
distinguishing name from the fact that one of its clauses
makes it imperative for the proposed scheme to provide for
the accommodation of at least as many persons of the working class as may be displaced in suitable dwellings within
the improved area or its vicinity.

One of the earliest applications under the provisions of

this Act related to an area in St. George's-in-the-East, comprising a series of unhealthy courts lying close to Wapping Parish Church and the London Docks, bounded on the north by Tench Street and on the south by Green Bank. After full investigation, it was decided to clear this area, a very costly proceeding, the total expense of which amounted to £52,000, and the extent of ground thus acquired was 2½ acres. No adequate offer, however, could be obtained for this ground, on which possibly only workmen's dwellings could have been erected, and owing to other clearances, there were more than enough of these in the neighbourhood. The local authority in this case, the London County Council, therefore decided to lay out the area as a recreation-ground for this poor district, subject to Parliamentary power being obtained. The necessary authority was granted by the Metropolitan Improvements Act, 1889. The greater portion of the ground has been gravelled so as to form a playground for children; the remainder has been embellished with trees, shrubs, and flower-beds, and there is also a children's gymnasium. The total cost of these works was about £1,000, and Wapping Recreation Ground was opened to the public on June 8, 1891.

Wapping, strange to say, is by no means so old as the Metropolis, and previous to 1657 consisted of only one street, extending about a mile from the Tower along the river, almost as far as Ratcliffe. London proper ended at the Tower, and these outlying lands were subject to periodical flooding—in fact, the whole of this neighbourhood was formerly one great wash, covered by the waters of the Thames, and when it was subsequently reclaimed, it was used like the Isle of Dogs as meadow-ground for grazing. The banks of the Thames were then furnished with walls or dykes, in order to defend the land from the incursions of the water. Between the years 1560 and 1570 the embanking wall was broken in several places, and Wapping was once more laid under water. The Romulus of Wapping was found in the person of one William Page, who took a lease of 110 feet of the wall, and spent a considerable sum in rebuilding houses and protecting them from the river till he was stopped by the proclamation of Queen Elizabeth in 1583 to prevent the increase of new buildings. Page petitioned, and some time after was permitted to carry on his scheme of re-edification. But this new Wapping was not the busy and crowded place of to-day, for in the early part of the reign of Charles I. that monarch hunted a stag from Wanstead, in Essex, and eventually killed him in a garden near Nightingale Lane, in the hamlet of Wapping. This attracted so many people to the place that great damage was done in consequence.*

It is a remarkable change from stag-hunting in an open country place to unhealthy, stuffy courts and alleys, and no doubt the principal cause of this transformation is to be found in the formation of so many important docks in this neighbourhood and their attendant busy life. The London Docks, which almost join on to the recreation ground, were commenced in June, 1802, and opened in January, 1805. The first stone of the entrance basin was laid by Mr. Pitt, who was then Chancellor of the Exchequer. He was also present at the opening ceremony, when the *London Packet*, a vessel laden with wine from Oporto, decorated with flags, entered the dock amidst the cheers of the onlookers.

The eastern boundary of the recreation ground—Anchor and Hope Alley—is the place where Judge Jeffreys, the cruel minister of James II., was captured in 1688, when trying to make his escape disguised as a common sailor. His capture was partly due to his tyrannous conduct as a judge, for a case between a sailor and a usurer, who lent money to seafaring men at a high rate of interest, was tried before him. 'The counsel for the borrower, having little else to say, said that the lender was a trimmer. The Chancellor instantly fired : "A trimmer! Where is he? Let me see him. I have heard of that kind of monster. What is it made like?" The unfortunate creditor was forced to stand forth. The Chancellor glared fiercely on him, stormed at him, and sent him

* Rev. J. Nightingale, 'London and Middlesex,' vol. iii., p. 144.

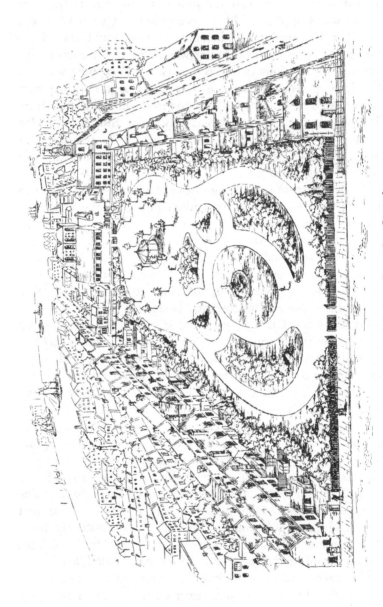

Bird's-eye View of Wapping Recreation Ground.

away half dead with fright. "While I live," the poor man said, as he tottered out of the court, "I shall never forget that terrible countenance." And now the day of retribution had arrived. The trimmer was walking through Wapping, when he saw a well-known face looking out of the window of an alehouse (the Red Cow in Anchor and Hope Alley). He could not be deceived. The eyebrows, indeed, had been shaved away; the dress was that of a common sailor from Newcastle, and was black with coal-dust; but there was no mistaking the savage eye and mouth of Jeffreys. The alarm was given. In a moment the house was surrounded by hundreds of people, shaking bludgeons and bellowing curses. The fugitive's life was saved by a company of the Train-bands, and he was carried before the Lord Mayor,' a simple kind of man, who was bewildered by the greatness his office had thrust upon him. The prisoner frightened him into fits, and the unfortunate Mayor was carried to his bed, from which he never rose. Jeffreys begged to be sent to prison, and with much difficulty he was escorted to the Tower by two regiments of militia, where he ended his days in ignominy and terror.*

Wapping gave rise to a custom which commenced in 1725, and is likely to continue for ever, and that is the annual beanfeast, which has now become an established institution in many places of business. It seems that a Mr. Daniel Day, a block-maker of Wapping, possessed a small estate in Essex, near Fairlop Oak. He used to invite his friends to accompany him to this place on the first Friday in every July, and a feast of beans and bacon was provided for the occasion. The rumour of this annual celebration soon spread, and many other parties imitated his example, the beans and bacon always being to the fore, with the result that these annual summer outings were termed beanfeasts. For several years before the death of the humorous founder it was the custom for the merrymakers to ride in a boat made out of one piece of timber, which was mounted on a

* Macaulay, ' History of England.'

coach carriage, and drawn by six horses. The whole was adorned with ribbons, flags, and streamers, and a band accompanied the coach, so that the modern institution differs very little from the original from which it sprung.*

Wapping Old Stairs, the nearest point on the Thames to the recreation-ground, has been immortalized by Dibdin in his fine old song:

> ' " Your Molly has never been false," she declares,
> "Since last time we parted at Wapping Old Stairs." '

Wapping, too, was a favourite of Dr. Johnson's. In one of his talks with his future biographer on the various modes of life to be found in London, he particularly advised his hearers to ' explore' Wapping, which they resolved to do. ' We accordingly,' says Boswell, ' carried our scheme into execution in October, 1792; but, whether from that uniformity which has in modern times to a great degree spread through every part of the Metropolis, or from our want of sufficient exertion, we were disappointed.'†

CHURCHYARDS AND SMALL PLAYGROUNDS.

The parks, gardens, and other places which have been enumerated so far are either nominally or really the freehold of the London County Council. There are, however, a number of churchyards and small playgrounds which are maintained as a temporary arrangement by them, until some proper decision is arrived at as to their future control by the local authorities or otherwise. The majority of these places have been acquired and laid out by the Metropolitan Public Gardens Association, whose funds are derived from private subscriptions. The maintenance of playgrounds, however, is a work outside their scope, as the large annual cost thus entailed would hamper them in acquiring fresh places, so it is the practice of the association to ask the local authorities to take over the charge of the gardens which they have been

---

* Rev. J. Nightingale, ' London and Middlesex,' vol. iii. p. 146.
† Boswell, ' Life of Dr. Johnson.'

instrumental in obtaining for public use. This has been done in many cases ; and there are a large number of these small playgrounds dotted about London, which, though not of large extent, are nevertheless of considerable advantage, most of them being in crowded districts where the possibility of obtaining any large areas for purposes of recreation is out of the question. With regard to the places under considera- tion, an objection to take over their control has been raised by the local authorities on the ground that the cost of such maintenance should be met out of the. metropolitan rates ; and rather than let London lose the benefit of these play- grounds, the London County Council has decided, for the present, at any rate, to maintain them out of its public funds.

By far the most important of these is the disused church- yard surrounding the parish church of St. Dunstan, Stepney. It is 7 acres in extent, well laid out, and is a valuable addition to the open spaces of the East End: After being laid out as a garden, it was opened by the Duchess of Leeds, July 18, 1887.

In the same year the churchyard of St. Anne, Limehouse, was opened to the public. This is also a large garden, being 3 acres in extent. Limehouse was originally a part of the huge neighbouring parish of Stepney, but was separated from it in 1730. The fine old church, which stands out like a beacon on the banks of the Thames, is one of the fifty authorized to be erected under the reign of Queen Anne, the money being raised from ' several duties upon coals.' The architect was Nicholas Hawksmoor, a well-known pupil of Sir Christopher Wren. The garden is entirely free from tombstones, and is laid out in grass with gravelled walks, and possesses a fountain and seats.

Dealing with the others according to the date of their acquisition, the two earliest to be opened were Carlton Square Garden ($\frac{3}{4}$ of an acre) at Mile End, and St. Bartholomew's Churchyard, Bethnal Green (1 acre). In both cases the ceremony was performed by H.R.H. the Princess Louise in 1885. St. Bartholomew's Churchyard is divided from the church by a public footpath and iron railings. It is open daily for the use of the public, and forms a pleasant garden in the summer time in what is a very crowded district. In the same year St. Paul's Churchyard, Rotherhithe, was opened by H.R.H. Princess Frederica, of Hanover. This garden surrounds the church, which is near the southern bank of the Thames, and adjacent to the timber docks. The whole of the expense of laying out was borne by the late Earl of Leven and Melville. Another disused churchyard at Rotherhithe, viz., that attached to Holy Trinity Church, was transferred to the London County Council in 1896. This is about one-third of an acre in extent, and contains some good trees.

In the following year, 1886, St. Paul's Churchyard, Shadwell (1 acre), and Spa Fields disused burial-ground (1¼ acres) were secured. The former partially surrounds the parish church, and is close to the London Docks. Part of the area is flagged, but is available for recreation, whilst the remainder is laid out as a garden. Spa Fields playground is 1¾ acres in extent, and is entirely gravelled, being used with the consent of the freeholder, the Marquis of Northampton, as a drill-ground for the 21st Middlesex R.V. At other times it is used as a playground for children, for whose benefit gymnastic apparatus has been fitted. It lies at the back of the Countess of Huntingdon's church in Exmouth Street, Clerkenwell, and is approached by an alley at the western side of the church. This little piece of land has an eventful history. The site was formerly part of the Ducking Pond Fields, by the side of which was Ducking Pond House. This was pulled down in 1770, and a large circular assembly room known as the Pantheon was erected in its place. The grounds were then laid out as a sort of minor Vauxhall or Ranelagh, the old ducking-pond being transformed into a lake, upon which boats were let out on hire. The Pantheon acquired an evil reputation, and had to be closed as a place of entertainment in 1776. It was then taken by two Evangelical clergymen, and re-opened as Northampton Chapel.

The lake, being no longer required, was drained, and used with the rest of the grounds for burials. Upon this an action was commenced by the incumbent of the parish to restrain these preachers from holding services in an unconsecrated place. The chapel was then transferred to the Countess of Huntingdon, who took the adjoining house and lived in it, so that clergymen might preach in the chapel under the privilege of peerage, and so render it legal. The old chapel, which had sittings for 2,000, was pulled down in 1879. The burial-ground became notorious in 1845, because the proprietors burnt the bodies in order to make room for fresh interments. It is said that as many as 1,350 bodies were burnt here in one year. It was shortly afterwards closed against burials by an order in council, and was re-opened as a recreation-ground, as we have seen, in 1886. An addition of half an acre was made two years later.

In the Jubilee Year, besides the two fine churchyards of Stepney and Limehouse already enumerated, the disused burial-ground (1¼ acres) attached to the Church of Holy Trinity, Mile End, and a playground at Winthrop Street, Whitechapel, were acquired (½ an acre). The former was opened by Princess Henry of Battenberg on May 9, and the latter two days afterwards by the Countess of Lathom. The purchase money for the latter—£2,300—was given by an anonymous donor. It is situated just at the rear of the Working Boys' Home, in the Whitechapel Road. In 1892 the large churchyard attached to Christ Church, Spitalfields, was dedicated for public use. It is a most useful ground, situated in a particularly bad neighbourhood. It was opened by the Earl of Meath on July 19.

The remaining playground transferred from the Association is the dismal graveyard in Russell Court, Drury Lane, immortalized by Dickens as 'Tom All Alone's' in the 'Poor Joe' episode of 'Bleak House.' By an arrangement entered into between the Rector and the Duke of Bedford, this is to be absorbed in the remodelling of the neighbourhood, and is now closed to the public.

The two other places to be mentioned under this heading are Beaumont Square Garden and Shandy Street Recreation Ground, both the property of Captain Beaumont, to whom a nominal rental of five shillings per annum is paid for each place. Beaumont Square Garden is about 1 acre in extent, and its use was formerly restricted to the inhabitants of the square, but by an agreement with the owner it is now secured for the use of the public until 1928. Shandy Street Recreation Ground (1½ acres) was once known as the East London Cemetery, and some of the tombs at present remain; but a considerable portion has been gravelled, and is available for the recreation of children. In accordance with the terms of the agreement, this place is closed to the public on September 29 each year.

PLACES IN COURSE OF ACQUISITION.

Having now reached the end of the places already opened to the public, it is only necessary to add a word or two about those which are in course of acquisition. Fortunately for London, there is never a time when the list of its parks and open spaces is complete. At the moment of writing there are many places for the acquisition of which negotiations in a more or less advanced state are going on. The two largest of these comprise an estate of 16½ acres at Wells Road, Upper Sydenham, and a river-side park of 16 acres, situate in Putney Bridge Road. There is also a recreation-ground of 7 acres at Bromley Road, Lee, which has been presented by Lords Northbrook and Baring, and a small piece of land in Ivy Street, Hoxton, about to be laid out as a children's gymnasium. Another small ground at Grace Street, Bromley, possesses an old mansion, much in need of repair, the fate of which has not been settled at present.

APPENDIX.

LIST OF ALL THE PLACES IN THE COUNTY OF LONDON
AVAILABLE FOR PUBLIC RECREATION, TOGETHER WITH THOSE
OUTSIDE THE COUNTY MAINTAINED BY THE
LONDON COUNTY COUNCIL.

NAME OF PLACE.	ACREAGE.	BY WHOM MAINTAINED.
Albert Embankment Gardens	$1\frac{1}{2}$ acres	London County Council.
All Saints' Church Ground, Mile End	450 sq. yds.	The Vicar of All Saints.
All Saints' Church Ground, Poplar	4 acres	The Rector of All Saints.*
Avondale Park	$4\frac{1}{4}$,,	Vestry of Kensington.
Baker's Row Recreation-ground	$1\frac{1}{8}$,,	Whitechapel District Board.
Barnsbury Square Garden	1 acre	Vestry of Islington.
Bartholomew Square	$\frac{1}{5}$,,	Vestry of St. Luke.
Battersea Park	198 acres	London County Council.
Beaumont Square, Stepney	1 acre	,, ,, ,,
Benjamin Street, Clerkenwell, Burial-ground (disused)	$\frac{1}{4}$,,	Vestry of Clerkenwell and Holborn District Board of Works.
Bethnal Green Gardens	9 acres	London County Council.
Bishop's Park	14 ,,	Vestry of Fulham.
Blackfriars Bridge Garden	$\frac{1}{20}$ acre	City Corporation.
Blackheath	267 acres	London County Council.
Bostall Heath	$71\frac{1}{4}$,,	,, ,, ,,
Bostall Woods	$62\frac{1}{4}$,,	,, ,, ,,
Boundary Street Central Garden	$\frac{1}{2}$ acre	,, ,, ,,

* Only a portion of this Church ground is open to the public.

NAME OF PLACE.	ACREAGE.	BY WHOM MAINTAINED.
Brockwell Park	84 acres	London County Council.
Brook Green	4¾ ,,	,, ,, ,,
Brunswick Pier, Blackwall	1 acre	Dock Company.
Bunhill Fields Burial-ground (disused)	4 acres	City Corporation.
Camberwell Green	2½ ,,	Vestry of Camberwell.
Camberwell Library Ground	½ acre	,, ,,
Canonbury Square Garden	¾ ,,	Vestry of Islington.
Carlton Square, Mile End	¾ ,,	London County Council.
Chelsea Embankment Gardens	1 ,,	,, ,, ,,
Chelsea Hospital Grounds	9½ acres	H.M. Office of Works.
Christ Church Burial-ground, Southwark	¼ acre	Trustees.
Christ Church Burial-ground, Spitalfields	2 acres	London County Council.
Christ Churchyard, Battersea	½ acre	Vestry of Battersea.
Clapham Common	220 acres	London County Council.
Clapton Common	7½ ,,	,, ,, ,,
Clissold Park	54½ ,,	,, ,, ,,
Covered Mill Pond, Rotherhithe	½ acre	Vestry of Rotherhithe.
Dalston Lane Slips	900 sq. yds.	Hackney District Board.
Dalston Slips, near Police Station	880 ,,	,, ,, ,,
De Beauvoir Square	1¼ acres	,, ,, ,,
Deptford Park	17 ,,	London County Council.
Downs Crescent, Hackney	380 sq. yds.	Hackney District Board.
Drury Lane Garden	¼ acre	Vestry of St. Martin.
Duke Street Garden	½ ,,	Duke of Westminster.
Dulwich Park	72 acres	London County Council.
Duncan Terrace Gardens, Islington	½ acre	Vestry of Islington.
Eastbank, Hackney	½ ,,	Hackney District Board.
Ebury Street Triangle	729 sq. yds.	Vestry of St. George, Hanover Square.
Ebury Square Gardens	¾ acre	Metropolitan Public Gardens Association.
Edward Square, Islington	$\frac{1}{10}$,,	Vestry of Islington.
Eel Brook Common	14 acres	London County Council.
Eltham Common	40½ ,,	War Office.
Eltham Green	7½ ,,	H.M. Office of Works.

NAME OF PLACE.	ACREAGE.	BY WHOM MAINTAINED.
Finsbury Park	115 acres	London County Council.
Fortune Green	2¼ ,,	Vestry of Hampstead.
Fulham Parish Churchyard	2 ,,	Vicar of Fulham.
Garratt Green	9¼ ,,	Lord of the Manor.
Goldsmith Square	¾ acre	Vestry of Shoreditch.*
Goose Green	6¼ acres	London County Council.
Green Park	54 ,,	H.M. Office of Works.
Greenwich Park	185 ,,	,, ,, ,,
Grosvenor Gardens (Lower)	¼ acre	Metropolitan Public Gardens Association.†
Hackney Downs	41¾ acres	London County Council.
Hackney Downs (Enclosure)	¼ acre	Hackney District Board.
Hackney Independent Chapel Ground	⅔ ,,	,, ,, ,,
Hackney Marsh	337 acres	London County Council.
Hackney (West) Churchyard	1⅓ ,,	Hackney District Board.
Hackney Town Hall Garden	½ acre	,, ,, ,,
Hackney Triangle Shrubbery, Mare Street	190 sq. yds.	,, ,, ,,
Hammersmith Recreation-ground, Church Lane	1½ acres	Vestry of Hammersmith.
Hampstead Heath	240 ,,	London County Council.
Haverstock Hill Playground	½ acre	Vestry of Hampstead.
Highbury Fields	27¾ acres	London County Council.
Highgate Road Open Spaces	3 ,,	Owner.
Hilly Fields, Brockley	45½ ,,	London County Council.
Holy Trinity Churchyard, Brompton	1 acre	The Vicar.
Holy Trinity Churchyard, Rotherhithe	⅜ ,,	London County Council.
Holy Trinity Garden, Bow	1¼ acres	,, ,, ,,
Horseferry Road Burial-ground (disused)	1 acre	Vestry of St. Margaret and St. John, Westminster.
Hyde Park	361 acres	H.M. Office of Works.
Ion Square	⅓ acre	Vestry of Bethnal Green.
Island Gardens, Poplar	3 acres	London County Council.

* Freehold of the London County Council.
† Open to the public for six weeks in the autumn.

NAME OF PLACE.	ACREAGE.	BY WHOM MAINTAINED.
Islington Chapel-of-Ease Grounds	4½ acres	Vestry of Islington.
Islington Green	1 acre	,, ,,
Kennington Park	19½ acres	London County Council.
Kensington Gardens	274½ ,,	H.M. Office of Works.
Kenton Road Shrubbery, Hackney	250 sq. yds.	Hackney District Board.
Kidbrooke Green	4¼ acres	Lord of the Manor.
Ladywell Recreation-ground	46¼ ,,	London County Council.
Lambeth Shrubbery (opposite St. Thomas's Hospital)	3/10 acre	Vestry of Lambeth.
Lauriston Road Slips, Hackney	1/5 ,,	Hackney District Board.
Lauriston Road Triangle, Hackney	165 sq. yds.	,, ,, ,,
Lea Bridge Road Waste	½ acre	,, ,, ,,
Lee Old Burial-ground	¾ ,,	Rector of St. Margaret's, Lee.
Leicester Square	½ ,,	London County Council.
Limehouse Churchyard	3 acres	,, ,, ,,
Lincoln's Inn Fields	7 ,,	,, ,, ,,
Lismore Circus, St. Pancras	½ acre	Vestry of St. Pancras.
Lock Burial-ground (disused)	1 r. 12 p.	Rector and Churchwardens of St. George the Martyr.
London Fields	26½ acres	London County Council.
Long Lane Recreation-ground	½ acre	Vestry of Bermondsey.
Maryon Park	11½ acres	London County Council.
Meath Gardens	9½ ,,	,, ,, ,,
Mile End (Brewers' Almshouse) Garden	½ acre	Brewers' Company and London Hospital.
Mile End New Town Parish Churchyard	450 sq. yds.	The Vicar.*
Mill Field (North)	23¼ acres	London County Council.
Mill Field (South)	34¼ ,,	,, ,, ,,
Myatt's Fields	14½ ,,	,, ,, ,,
Natural History Museum Garden	5 ,,	Trustees of the British Museum.

* Open from June 1 to September 1 only.

NAME OF PLACE.	ACREAGE.	BY WHOM MAINTAINED.
Nelson Recreation-ground, Bermondsey	$\frac{5}{8}$ acre	London County Council.
Newington Recreation-ground	$1\frac{1}{2}$ acres	,, ,, ,,
Norfolk Road, Hackney (West side)	560 sq. yds.	Hackney District Board.
Northampton Square Garden	1 acre	Vestry of Clerkenwell.
Nunhead Green	$1\frac{1}{2}$ acres	London County Council.
Paddington Green	$1\frac{1}{2}$,,	Vestry of Paddington.
Paddington Recreation-ground	26 ,,	,, ,,
Palace Road Recreation-ground	9 ,,	Vestry of Fulham.
Parliament Hill	$267\frac{1}{4}$,,	London County Council.
Parliament Square Garden	2 ,,	H.M. Office of Works.
Parson's Green	$2\frac{3}{4}$,,	London County Council.
Peckham Rye	64 ,,	,, ,, ,,
Peckham Rye Park	$48\frac{3}{4}$,,	,, ,, ,,
Penge Recreation-ground	$3\frac{1}{2}$,,	Lewisham District Board.
Penn Road Triangle, Islington	$\frac{1}{4}$ acre	Vestry of Islington.
Pimlico Shrubberies	$\frac{3}{4}$,,	London County Council.
Plumstead Common	100 acres	,, ,, ,,
Pond Square, Highgate	$\frac{3}{4}$ acre	Vestry of St. Pancras.
Poplar Recreation-ground	$3\frac{1}{2}$ acres	Poplar District Board.
Putney Bridge Shrubbery	$\frac{1}{4}$ acre	London County Council.
Putney Old Burial-ground, Upper Richmond Road	1 ,,	Putney Burial Board.
Putney Lower Common	41 acres	Wimbledon Common Conservators.
Putney Upper Common, or Putney Heath	342 ,,	Wimbledon Common Conservators.
Ravensbourne Recreation-ground	$1\frac{1}{4}$,,	Greenwich District Board.
Ravenscourt Park	$31\frac{1}{4}$,,	London County Council.
Red Cross Street Garden, Southwark	1 acre	Trustees.
Red Lion Square	$\frac{1}{2}$,,	London County Council.
Regent's Park and Primrose Hill	$472\frac{1}{2}$ acres	H.M. Office of Works.
Royal Courts of Justice Enclosure	1 acre	,, ,, ,,
Royal Victoria Gardens	10 acres	London County Council.

NAME OF PLACE.	ACREAGE.	BY WHOM MAINTAINED.
St. Alphege Burial-ground, Greenwich (disused)	3½ acres	Greenwich District Board.
St. Andrew's Gardens, St. Pancras	1¼ ,,	Vestry of St. Pancras.
St. Anne's Burial-ground, Soho (disused)	¾ acre	Strand District Board of Works.
St. Bartholomew's, Bethnal Green	1 ,,	London County Council.
St. Botolph's Burial-ground, City (disused)	¾ ,,	Parishes.
St. Clement's Churchyard, Notting Hill	½ ,,	The Vicar.
St. Dunstan-in-the-West, Fetter Lane	¼ ,,	School Board for London.
St. George's Churchyard, Camberwell	1 ,,	Vestry of Camberwell.
St. George's Gardens, St. Pancras	2½ acres	Vestry of St. Pancras.
St. George's Vestry Hall Gardens, Hanover Square	2 ,,	Vestry of St. George, Hanover Square.
St. George's Burial-ground, Hanover Square (disused)	6 ,,	Rector of St. George's, at expense of Duke of Westminster.
St. George's-in-the-East Burial-ground	¾ acre	Vestry of St. George's-in-the-East.
St. George's-in-the-East Parish Churchyard	2 acres	Vestry of St. George's-in-the-East.
St. George's Parish Churchyard, Borough High Street	¾ acre	Rector and Churchwardens of St. George the Martyr.
St. Giles's-in-the-Field Parish Churchyard	1¼ acres	St. Giles's District Board.
St. James's Churchyard, Bermondsey	1½ ,,	Vestry of Bermondsey.
St. James's Churchyard, Clerkenwell	1 acre	Vestry of Clerkenwell.
St. James's Gardens, St. Pancras	3 acres	Vestry of St. Pancras.
St. James's Park	93 ,,	H.M. Office of Works.
St. James's Burial-ground, Pentonville (disused)	1 acre	Vestry of Clerkenwell.
St. James's Churchyard, Ratcliff	¾ ,,	Vicar.
St. John-at-Hackney Old Parish Churchyard	3 acres	Hackney District Board.

NAME OF PLACE.	ACREAGE.	BY WHOM MAINTAINED.
St. John's Churchyard, Hoxton	1 acre	Vestry of Shoreditch.
St. John's, St. Olave's Churchyard, Fair Street	2 acres	St. Olave's District Board.
St. John's Churchyard, Waterloo Road	1¼ ,,	Vestry of Lambeth.
St. John's Wood Old Burial-ground	7 ,,	Vestry of St. Marylebone.
St. Leonard's Parish Churchyard, Shoreditch	1 acre	Vestry of Shoreditch.
St. Luke's Burial-ground, Old Street	1¾ acres	Vestry of St. Luke.*
St. Luke's Burial - ground, Seward Street	1 acre	,, ,,
St. Luke's Parish Churchyard, Chelsea	4 acres	Vestry of Chelsea.
St. Luke's Parish Playground, Whitechapel	⅝ acre	The Vicar.
St. Margaret's Churchyard, Lee	½ ,,	The Rector.
St. Margaret's Churchyard, Westminster	2¼ acres	Vestry of St. Margaret and St. John.†
St. Martin-in-the-Fields	½ acre	Vestry of St. Martin.
St. Martin's Gardens, St. Pancras	1¾ acres	Vestry of St. Pancras.
St. Mary's Churchyard, Bow	½ acre	Metropolitan Public Gardens Association.
St. Mary's Churchyard, Hagerston	1 ,,	Vestry of Shoreditch.
St. Mary's Churchyard, Islington	1¼ acres	Vestry of Islington.
St. Mary's Churchyard, Lewisham	2 ,,	Lewisham District Board.
St. Mary Magdalene's Churchyard, Bermondsey	1¾ ,,	Vestry of Bermondsey.
St. Mary's Churchyard, Newington	2 ,,	Newington Burial Board.
St. Mary's Churchyard, Paddington	1 acre	Vestry of Paddington.
St. Mary's Old Burial-ground, Paddington	3¼ acres	,, ,,
St. Mary's Churchyard, Shoreditch	2 ,,	The Vicar.

* A portion only is open to the public.
† This includes the churchyard of Westminster Abbey.

NAME OF PLACE.	ACREAGE.	BY WHOM MAINTAINED.
St. Mary's Churchyard, White-chapel	¾ acre	The Rector.*
St. Mary's Churchyard, Wool-wich	4 acres	Woolwich Local Board.
St. Mary's Burial - ground, Fulham (disused)	½ acre	The Vicar.
St. Marylebone Church Ground	½ ,,	Vestry of St. Marylebone.
St. Marylebone Burial-ground (disused)	2 acres	,,　　,,　　,,
St. Nicholas' Burial-ground, Greenwich (disused)	2 ,,	Greenwich District Board.
St. Pancras Gardens	7 ,,	Vestry of St. Pancras.
St. Paul's Burial - ground, Deptford	3 ,,	Burial Board of St. Paul, Deptford.
St. Paul's Churchyard, Ham-mersmith	1 acre	Churchwardens.
St. Paul's Churchyard, Rother-hithe	¾ ,,	London County Council.
St. Paul's Churchyard, Shad-well	1 ,,	,,　　,,　　,,
St. Peter's Ground, Fulham	¼ ,,	The Vicar.
St. Peter's Churchyard, Hack-ney Road	¼ ,,	The Vicar.†
St. Peter's Churchyard, New-ington	1 ,,	Newington Burial Board.
St. Philip's Churchyard, Cam-berwell	¼ ,,	Vicar and Church-wardens.‡
St. Philip's Churchyard, Mile End	¾ ,,	The Vicar.
St. Thomas's Square Burial-ground, Hackney	¾ ,,	Hackney District Board.
Savoy Chapel Royal Church-yard	¼ ,,	The Queen.
Sayes Court (portion)	2 acres	W. J. Evelyn, Esq.
Shacklewell Green	¼ acre	Hackney District Board.
Shacklewell Lane Triangle	340 sq. yds.	,,　　,,　　,,
Shaftesbury Avenue Triangle	—	St. Giles's District Board of Works.

* Open during summer months. Admission is gained by ticket or on payment of one farthing.
† Open during summer only.
‡ Open three days a week during certain hours.

NAME OF PLACE.	ACREAGE.	BY WHOM MAINTAINED.
Shandy Street Recreation-ground	1½ acres	London County Council.
Shepherd's Bush Common	8 acres	,, ,, ,,
Shoreditch Old Burial-ground	½ acre	Vestry of Shoreditch.
Shoulder of Mutton Green	5 acres	London County Council.
Silver Street Playground, London Docks	¾ acre	Trustees.
Southwark Park	63 acres	London County Council.
Spa Fields, Clerkenwell	2 ,,	,, ,, ,,
Spa Green, Clerkenwell	¾ acre	,, ,, ,,
Stamford Hill Strips	⅝ ,,	Hackney District Board.
Stepney Churchyard	7 acres	London County Council.
Stepney Green	3¼ ,,	Vestry of Mile End, Old Town.
Stoke Newington Common	5¼ ,,	London County Council.
Stoke Newington Green	1¾ ,,	Vestry of Islington.
Stonebridge Common, Dalston	⅛ acre	Hackney District Board.
Streatham Common	66 acres	London County Council.
Streatham Green	1 acre	,, ,, ,,
Sydenham Recreation-ground	17½ acres	Lewisham District Board.
Telegraph Hill	9½ ,,	London County Council.
Thornhill Gardens, Barns-bury	1 acre	Vestry of Islington.
Tooley Street Garden	¼ ,,	St. Olave's District Board.
Tooting Bec Common	147½ acres	London County Council.
Tooting Graveney Common	63 ,,	,, ,, ,,
Tower Gardens, Tower Hill	2½ ,,	Metropolitan Public Gardens Association.
Tower Wharf	2 ,,	Constable of the Tower.
Trafalgar Square, Mile End	1 acre	Vestry of Mile End, Old Town.
Trinity Churchyard, Poplar	1 ,,	Perpetual trustees.
Vauxhall Old Burial-ground	1½ acres	Vestry of Lambeth.
Vauxhall Park	8 ,,	,, ,,
Vicarage Road Recreation-ground	½ acre	Vestry of Battersea.
Victoria Embankment Gardens	12 acres	London County Council.
Victoria Park	244 acres	London County Council.
Victoria Tower Garden	1½ ,,	H.M. Office of Works.

NAME OF PLACE.	ACREAGE.	BY WHOM MAINTAINED.
Walworth Recreation Ground	$\frac{5}{8}$ acre	London County Council.
Wandsworth Common	183 acres	,, ,, ,,
Wapping Recreation-ground	$2\frac{1}{2}$,,	,, ,, ,,
Waterlow Park	29 ,,	,, ,, ,,
Well Street Common	$21\frac{1}{4}$,,	,, ,, ,,
Well Street, Hackney, Strips	$\frac{5}{8}$ acre	Hackney District Board.
Wendell Park, Shepherd's Bush	4 acres	Vestry of Hammersmith.
Westbank Slips, Hackney	$\frac{1}{4}$ acre	Hackney District Board.
West End Green, Hampstead	$\frac{3}{4}$,,	Vestry of Hampstead.
Whitfield Gardens	$\frac{1}{2}$,,	London County Council.
Wilmington Square Garden, Clerkenwell	$1\frac{1}{2}$ acres	Vestry of Clerkenwell.
Winthrop Street Playground	$\frac{1}{2}$ acre	London County Council.
Woolwich Common	159 acres	War Office.
Wormwood Scrubs	193 ,,	London County Council.
Wormwood Scrubs (Little)	22 ,,	,, ,, ,,

INDEX

THE END.

Elliot Stock, 62, Paternoster Row, London, E.C.

Printed in the United States
By Bookmasters